Lecture Notes of the Unione Matematica Italiana *10*

For further volumes:
http://www.springer.com/series/7172

The Editorial Policy can be found at the back of the volume.

Sergio Albeverio • Ruzong Fan
Frederik Herzberg

Hyperfinite Dirichlet Forms and Stochastic Processes

Springer

UMI

Sergio Albeverio
University of Bonn
Institute for Applied Mathematics and HCM
Endenicher Allee 60
53115 Bonn
Germany
albeverio@uni-bonn.de

Frederik Herzberg
Bielefeld University
Institute of Mathematical Economics
Universitätsstraße 25
33615 Bielefeld
Germany
fherzberg@uni-bielefeld.de

Ruzong Fan
Texas A and M University
Department of Statistics
College Station 77843, TX
USA
rfan@stat.tamu.edu

Current address

Biostatistics and Bioinformatics Branch
Division of Epidemiology, Statistics & Prevention
Eunice Kennedy Shriver National Institute
of Child Health & Human Development
6100 Executive Blvd.
MSC 7510, Bethesda, MD 20892
United States of America

ISSN 1862-9113
ISBN 978-3-642-19658-4 e-ISBN 978-3-642-19659-1
DOI 10.1007/978-3-642-19659-1
Springer Heidelberg Dordrecht London New York

Library of Congress Control Number: 2011928508

Mathematics Subject Classification (2000): 03H05; 60J45

Cover design: deblik, Berlin

Printed on acid-free paper

Springer is part of Springer Science+Business Media (www.springer.com)

In memory of the dear father, Mr. Decai Li (1926–2003), of the second author.

Preface

The theory of stochastic processes has developed rapidly in the past decades. Martingale theory and the study of smooth diffusion processes as solutions of stochastic differential equations have been extended in several directions, such as the study of infinite dimensional diffusion processes, the study of diffusion processes with non-smooth unbounded coefficients, diffusion processes on manifolds and on singular spaces. The interplay between stochastic analysis and mathematical physics has been one of the most important and exciting research areas.

One of the best techniques to deal with the problems of these areas is Dirichlet space theory. In the original framework of this theory, the state space is a locally compact separable metric space, e.g., $\mathbb{R}^d$, or a d-dimensional manifold. This theory has given us a nice understanding about the property of diffusion processes with non-smooth unbounded coefficients. Moreover, it has been fruitfully applied to mathematical physics. This framework has been generalized to state spaces which are more general topological spaces or some infinite dimensional vector spaces or manifolds. Several key problems, such as the closability of quadratic forms and the construction of strong Markov processes associated with quasi-regular Dirichlet forms, have been solved. The study of infinite dimensional stochastic analysis as well as the study of processes on singular structures (like fractals, trees, or general metric spaces) has enriched and extended the Dirichlet space theory. In the meantime, a new framework has been introduced into Dirichlet space theory by the development of nonstandard probabilistic analysis [25, 166]. As is well-known, nonstandard analysis is an alternative setting for analysis (and, indeed, all areas of mathematics), namely by enriching the set of real numbers by infinitesimal and infinite elements. It has its origin in seminal work by Schmieden, Laugwitz [325] and most notably Robinson [310]. By now, several textbooks and surveys exist on this theory and its applications (see, e.g. [25, 63, 125, 217]). Nonstandard analysis gives a novel approach to the theory of stochastic processes. In particular, it has led to hyperfinite symmetric Dirichlet space theory. Besides being interesting by itself, it has also many applications. In the first part of the book, we extend the research to the nonsymmetric case, and remove some restrictive conditions in the previous treatment of the subject (Chap. 5 of [25]). In addition, we shall apply the theory to present a new approach to infinite dimensional stochastic analysis.

In writing this book we have two main aims: (1) to give a presentation of research on nonsymmetric hyperfinite Dirichlet space theory and its applications in (standard) finite and infinite dimensional stochastic analysis, Chaps. 1–4; (2) to find nonstandard representations for a special class of (finite dimensional) Feller processes and their infinitesimal generators, viz. stochastically continuous processes with stationary and independent increments (i.e., *Lévy processes*), Chap. 5. Chapter 6 is a complement to illustrate the usefulness of the hyperfinite probability spaces. The first part (Chaps. 1–4) is based on Chap. 5 of Albeverio et al. [25] and the further in depth research of Sergio and Ruzong; the second part (Chaps. 5–6) is based on results obtained recently by Tom Lindstrøm and their extensions by Sergio and Frederik.

As mentioned earlier, the interplay between stochastic analysis and mathematical physics has been one of the most important and exciting themes of research in the last decades. This is already a sufficient rationale for the research of the first part of the present book. The motivation for including the second part, Chap. 5, into this book is that many of the issues discussed in the more general framework of the first part, such as existence of standard parts of hyperfinite Markov chains, become much less technical to resolve for hyperfinite Lévy processes. Furthermore, the more restrictive setting of the second part also allows one to obtain finer results on the relation between Lévy processes and their hyperfinite analogues, one example being a hyperfinite version of the Lévy–Khintchine formula.

The contents of this book are arranged as follows: In Chap. 1, we introduce the framework of hyperfinite Dirichlet forms. We develop the potential theory of hyperfinite Dirichlet forms in Chap. 2. In Chap. 3, we consider standard representations of hyperfinite Markov chains under certain conditions, and translate the conditions on hyperfinite Markov chains into the language of hyperfinite Dirichlet forms. As an interesting and important application in classical stochastic analysis, we construct tight dual strong Markov processes associated with quasi-regular Dirichlet forms by using the language of hyperfinite Dirichlet forms in Chap. 4. The results show that hyperfinite Dirichlet space theory is a powerful tool to study classical problems. In the first sections of Chap. 5, the notion of a hyperfinite Lévy process is introduced and its relation to hyperfinite random walks as well as to standard Lévy processes is investigated. These results can be used to show that the jump part of any Lévy process is essentially a hyperfinite convolution of Poisson processes. Finally, Chap. 6 is an epilogue, providing a rigorous motivation for the study of hyperfinite Loeb path spaces as generic probability spaces.

The entire book is based on nonstandard analysis. For the reader's convenience, we present some basic notions of nonstandard analysis, such as internal sets and saturation, linear spaces, Loeb measure spaces, structure of $^*\mathbb{R}$ and topology in the appendix. Because of its monographical character centered around the hyperfinite approach, the book has by no means the goal of including all aspects of recent developments in the theory of stochastic processes and its connections with Dirichlet forms theory or the theory of Lévy processes. For this, we rather refer to surveys and proceedings like Albeverio [2], Barndorff-Nielsen et al. [73], and Ma et al. [275], respectively.

The germ of this book goes back to the year 1989 when the second author, Ruzong Fan, worked on the construction of symmetric Markov processes associated with Dirichlet forms at Peking University, Beijing ([165] and Chap. 4). At that time, Ruzong was unaware that Sergio's group was working on the same project using standard methods [41]. The second author, Dr. Zhiming Ma, of [41] did privately inquire Ruzong about the progress of Ruzong's research in 1989 at the Institute of Applied Mathematics, Chinese Academy of Sciences, Beijing. In response to Dr. Ma's request of a private meeting, Ruzong presented his work to Dr. Ma in a classroom with Dr. Ma as the only audience. Dr. Ma, however, did not mention his ongoing work with Sergio in any way. Thus, Ruzong was totally unaware of Sergio's research. In the spring of 1990, Ruzong first realized this when he saw a manuscript of Albeverio and Ma [41] in Beijing with a surprise. These events notwithstanding, Ruzong continued to work on a "symmetric version" of Chaps. 1–4 using non-standard language when he was at Peking University till 1991 and when he visited the Humboldt-University, Berlin, between 1991 and 1992. Under Sergio's supervision and encouragement, Ruzong extended the project to the current "nonsymmetric version" from 1992 to 1994 at Ruhr-University, Bochum. In 2006, Frederik kindly joined the project with a contribution on hyperfinite Lévy processes (Chap. 5) and the Epilogue (Chap. 6). In the summer of 2006, the three authors gathered at the University of Bonn to finalize this monograph. We gratefully acknowledge the manifold support of various institutions in the long process of work on this project.

In the run-up to its completion, Sergio and Frederik were supported partially by the collaborative research center SFB 611 of the German Research Foundation (DFG), Germany; in addition, Ruzong's visit to Bonn was partially funded through a research fellowship from the Alexander von Humboldt Foundation, Germany.

Over the course of his career, Ruzong has received a lot of generous support from Sergio. As a Ph.D candidate in Beijing around 1987–1988, Ruzong was greatly fascinated by Sergio and Raphael Høegh-Krohn's novel work on infinite dimensional stochastic analysis, in which Ruzong finished his Ph.D thesis. Unfortunately, Ruzong got no chance to meet Raphael Høegh-Krohn; right before Ruzong went to Europe, he was shocked to learn that Raphael

Høegh-Krohn died of a heart attack. In a relatively isolated environment, Ruzong mostly worked on himself by reading numerous papers and books of Sergio and Raphael Høegh-Krohn; and many times, Ruzong had to spend a few days on a single equation or lemma to guess and to understand it. Whilst it seemed like a helpless or hopeless situation for Ruzong at that time, Ruzong eventually came to the forefront of research in areas of infinite dimensional stochastic analysis: he studied the hard and central questions regarding Beurling–Deny formulae, representation of martingale additive functionals and absolute continuity of symmetric diffusion processes on Banach spaces, potential theory of symmetric hyperfinite Dirichlet forms, and construction of the symmetric strong Markov processes associated with quasi-regular Dirichlet forms by using the non-standard analysis language. This direction of research was initiated by Sergio, although Ruzong was unaware that Sergio's group already worked on the construction of Markov processes using the language of standard stochastic analysis.

In early 1989, Ruzong applied for a fellowship from the Alexander von Humboldt Foundation from Peking University, Beijing; soon after a rejection from the Foundation in the fall 1989, Ruzong received a warm letter from Sergio with encouragement and a kind offer to nominate, as an academic host, Ruzong for the fellowship and by writing a strong letter of recommendation. This is just one anecdote to illustrate how Ruzong has constantly been able to count on Sergio's help via communications by either mail or face-to-face conversations starting from 1989. Between 1992 and 1994, Sergio generously supported Ruzong at Ruhr-University Bochum to complete the main part of Chaps. 1–4 of this monograph, and helped Ruzong to pass the hard period of time in his career.

The story of Ruzong is an example how Sergio has helped many young mathematicians to grow and to mature. Quite probably, Ruzong would have disappeared from academia a long time ago without the support of Sergio. In a true sense, Sergio has been an academic father figure for Ruzong when he desperately needed one. In recent years, after his departure from Sergio's research group, Ruzong has been mainly working on statistical genetics guided by his beloved American mentor, Dr. Kenneth Lange, at the University of Michigan and UCLA. Nevertheless, Ruzong has fond memories and deep appreciation of numerous communications with his European academic father Sergio; and both Ruzong and Frederik are deeply grateful for Sergio's mentoring.

Thus, especially right after Sergio's 70th birthday in 2009 – which also marks the 50th anniversary of his remarkable scientific career –, Ruzong and Frederik are sure that they will be joined by many other young mathematicians in thanking Sergio for his wonderful role in our professional and personal development and in wishing him all the best for the rest of his life: Not just continued productivity, but most of all good health, happiness, joy, and peace.

We owe a huge debt of gratitude to our families: In the summer of 2006, Dr. Li Zhu (Ruzong's wife) kindly took care of two young children when her husband was visiting Bonn. Their adorable daughter, Olivia Wenlu Fan, was with the second author in Germany for the "hot and interesting" summer of Bonn, where she liked everything except German milk. Frederik thanks his wife, Angélique Herzberg, for her love and manifold support with the words of Proverbs 31,10–12: "A wife of noble character [...] is worth far more than rubies. Her husband [...] lacks nothing of value. She brings him good [...] all the days of her life." We are all very grateful to our families for their love and understanding during the entire process of writing this book.

Finally, we would like to thank Dr. Catriona Byrne as well as Susanne Denskus and Ute McCrory of Springer Verlag for their kind, unfainting editorial assistance in the long process of publishing this work.

Contents

Chapter 1
Hyperfinite Dirichlet Forms

The interplay between methods from functional analysis and the theory of stochastic processes is one of the most important and exciting aspects of mathematical physics today. It is a highly technical and sophisticated theory based on decades of research in both areas. Numerous papers have been written on the standard theory of Dirichlet forms. Apart from the articles and monographs cited below, other notable contributions to the area include: Albeverio and Bernabei [5], Albeverio, Kondratiev, and Röckner [32], Albeverio and Kondratiev [33], Albeverio and Ma [39], Albeverio, Rüdiger, and Wu [54], Bliedtner [94], Bouleau [98], Bouleau and Hirsch [99], Chen et al. [112], Chen, Ma, and Röckner [116], Eberle [149], Exner [154], Fabes, Fukushima, Gross, Kenig, Röckner, and Stroock [155], Fitzsimmons and Kuwae [172], Fukushima [177,179,180], Fukushima and Tanaka [185], Fukushima and Ying [188,189], Gesztesy et al. [191,192], Grothaus et al. [198], Hesse et al. [208], Jacob [218–220], Jacob and Moroz [221], Jacob and Schilling [222], Jost et al. [225], Kassmann [232], Kim et al. [240], Kumagai and Sturm [248], Le Jan [258], Liskevich and Röckner [265], Ma and Röckner [272, 273], Ma et al. [274], Mosco [283], Okura [292], Oshima [294, 295], D.W. Robinson [312], Röckner and Wang [317], Röckner and Zhang [319], Schmuland and Sun [329], Shiozawa and Takeda [331], da Silva et al. [332], Stannat [336, 338], Stroock [340], Sturm [343], Takeda [346, 347], Wu [363], and Yosida [364].

In this monograph, we present the theory of Dirichlet forms from a unified vantage point, using nonstandard analysis, thus viewing the continuum of the time line as a discrete lattice of infinitesimal spacing. This approach is close in spirit to the discrete classical formulation of Dirichlet space theory in A. Beurling and J. Deny's seminal article [87].

The discrete setup in this monograph permits to study the diffusion and the jump part by essentially the same methods. This setting being independent of special topological properties of the state space, it is also considerably less technical than other approaches. Thus, the theory has found its natural setting and no longer depends on choosing particular topological spaces; in particular, it is valid for both finite and infinite dimensional spaces.

S. Albeverio et al., *Hyperfinite Dirichlet Forms and Stochastic Processes*,
Lecture Notes of the Unione Matematica Italiana 10,
DOI 10.1007/978-3-642-19659-1_1, © Springer-Verlag Berlin Heidelberg 2011

Whilst Albeverio et al. [25], Chap. 5, only discussed symmetric hyperfinite Dirichlet forms and related Markov chains (refer to [165, 166] also), we shall extend the theory to the nonsymmetric case. We shall try to follow as much as possible the path suggested by the work on the symmetric case.

An important sub-class of Markov process are Feller processes with stationary and independent increments (*Lévy processes*), and in recent years, these processes have attracted a lot of interest, including from nonstandard analysts. Initiated by T. Lindstrøm [263], a number of articles have been devoted to the investigation of *hyperfinite Lévy processes.* Chapter 5 of this monograph is a detailed exposition of Lindstrøm's theory [263] and its subsequent continuation by Albeverio and Herzberg [14]. The book ends with an expository summary (without proofs) of the model theory of stochastic processes as developed by H.J. Keisler and his coauthors, who formulated and proved the "universality" of hyperfinite adapted probability spaces in a rigorous manner, and a short description of recent fundamental results about the definability of nonstandard universes.

Meanwhile, our purpose in the first chapter is to develop a general theory of hyperfinite quadratic forms. We shall set the scene in Sect. 1.1. Sections 1.2 and 1.3 will study the domains of symmetric parts, the standard parts and resolvents. We shall discuss the property of weak coercive quadratic forms in Sects. 1.4 and 1.7. In Sect. 1.5, we shall study Markov forms and begin the analysis of associated Markov chains and get the basic Beurling–Deny formula. We discuss the hyperfinite lifting theory of standard Dirichlet forms in Sect. 1.6.

1.1 Hyperfinite Quadratic Forms

We shall develop a hyperfinite theory of nonnegative quadratic forms on infinite dimensional spaces. It is well-known that in the Hilbert space case the theory of closed forms of this kind is equivalent to the theory of nonnegative operators. In fact, there is a natural correspondence between forms $E(\cdot,\cdot)$ and operators A given by $E(u,u) = \langle Au, u\rangle$, where $\langle\cdot,\cdot\rangle$ is the scalar product in the Hilbert space. We have chosen to present the theory in terms of forms and not operators for two reasons: partly because forms are real-valued, and this makes it simpler to take standard parts, but also because in most of our applications, the form is what is naturally given.

Let H be an internal, hyperfinite dimensional linear space[1] equipped with an inner product $\langle\cdot,\cdot\rangle$ generating a norm $||\cdot||$. Let ${}^*\mathbb{R}$ be the nonstandard

---

[1] The notions of hyperfinite dimensional linear space are given in Albeverio et al. [25].

real line[2]. We call a map $\mathcal{E} : H \times H \longrightarrow {}^*\mathbb{R}$ *nonnegative quadratic form* if and only if for all $\alpha \in {}^*\mathbb{R}, u, v, w \in H$,

$$\begin{aligned}
\mathcal{E}(u,u) &\geq 0, \\
\mathcal{E}(\alpha u, v) &= \alpha\mathcal{E}(u,v), \\
\mathcal{E}(u, \alpha v) &= \alpha\mathcal{E}(u,v), \\
\mathcal{E}(u+v, w) &= \mathcal{E}(u,w) + \mathcal{E}(v,w), \\
\mathcal{E}(w, u+v) &= \mathcal{E}(w,u) + \mathcal{E}(w,v).
\end{aligned}$$

Since $\mathcal{E}(\cdot,\cdot)$ is a nonnegative quadratic form on the hyperfinite dimensional space H, elementary linear algebra tells us that there is a unique nonnegative definite operator $A : H \longrightarrow H$ such that

$$\mathcal{E}(u,v) = \langle Au, v\rangle \quad \text{for all} \quad u, v \in H. \tag{1.1.1}$$

To see this, let ${}^*\mathbb{N}_0$ be the nonstandard integers[3]. Let $\{e_i \mid 1 \leq i \leq N\}$ be an orthonormal basis of $(H, \langle\cdot,\cdot\rangle)$ for an $N \in {}^*\mathbb{N}$. We put $Ae_i = \sum_{j=1}^{N} \mathcal{E}(e_i, e_j)e_j$. Then (1.1.1) follows immediately. Hence, A is given by the matrix $A = (\mathcal{E}(e_i, e_j))_{1\leq i,j\leq N}$, i.e.,

$$A = \begin{pmatrix} \mathcal{E}(e_1,e_1) & \mathcal{E}(e_1,e_2) & \dots & \mathcal{E}(e_1,e_N) \\ \mathcal{E}(e_2,e_1) & \mathcal{E}(e_2,e_2) & \dots & \mathcal{E}(e_2,e_N) \\ \vdots & \vdots & \ddots & \vdots \\ \mathcal{E}(e_N,e_1) & \mathcal{E}(e_N,e_2) & \dots & \mathcal{E}(e_N,e_N) \end{pmatrix}. \tag{1.1.2}$$

Moreover, $\langle Au, u\rangle \geq 0$ for all $u \in H$. This means that A is a hyperfinite dimensional matrix (not necessarily symmetric). Let $\hat{A}$ be the *adjoint operator* of A, that is,

$$\mathcal{E}(u,v) = \langle u, \hat{A}v\rangle \quad \text{for all} \quad u, v \in H.$$

By (1.1.2), we have that $\hat{A}$ is the transpose of A. If $||A||$ and $||\hat{A}||$ are the operator norms of A and $\hat{A}$, respectively, we have $||A|| = ||\hat{A}||$. We fix an infinitesimal[4] Δt such that

---

[2] ${}^*\mathbb{R}$ is the standard notation for the nonstandard real line, refer to Appendix, Albeverio et al. [25], Cutland [125], Davis [135], Hurd [216], Hurd and Loeb [217], Lindstrøm [262], Stroyan and Bayod [341], and Stroyan and Luxemburg [342].

[3] ${}^*\mathbb{N}_0$ is the standard notation for the nonstandard integers, refer to Appendix, Albeverio et al. [25], Cutland [125], Davis [135], Hurd [216], Hurd and Loeb [217], Lindstrøm [262], Stroyan and Bayod [341], and Stroyan and Luxemburg [342].

[4] In the sense of nonstandard analysis, refer to Appendix, Albeverio et al. [25], Keisler [237,238], Stroyan and Bayod [341], and Stroyan and Luxemburg [342].

$$0 < \Delta t \leq \frac{1}{||A||} = \frac{1}{||\hat{A}||}. \tag{1.1.3}$$

Let us define new operators $Q^{\Delta t}$ and $\hat{Q}^{\Delta t}$ by

$$Q^{\Delta t} = I - \Delta t A,$$
$$\hat{Q}^{\Delta t} = I - \Delta t \hat{A}.$$

The relation (1.1.3) implies that the operators $Q^{\Delta t}$ and $\hat{Q}^{\Delta t}$ are nonnegative. Because A is nonnegative, the operator norms of $Q^{\Delta t}$ and $\hat{Q}^{\Delta t}$ are less than or equal to one. Similarly, we define the *nonnegative quadratic co-form* $\hat{\mathcal{E}}(\cdot,\cdot)$ of $\mathcal{E}(\cdot,\cdot)$ by

$$\hat{\mathcal{E}}(u,v) = \mathcal{E}(v,u) \text{ for all } u,v \in H.$$

Introduce a nonstandard time line T by

$$T = \{k\Delta t \mid k \in {}^*\mathbb{N}_0\}.$$

For each element $t = k\Delta t$ in T, define Q^t and $\hat{Q}^t$ to be the operators

$$Q^t = (Q^{\Delta t})^k,$$
$$\hat{Q}^t = (\hat{Q}^{\Delta t})^k.$$

The families $\{Q^t\}_{t\in T}$ and $\{\hat{Q}^t\}_{t\in T}$ are obviously semigroups. We shall call $\{Q^t\}_{t\in T}$ the *semigroup* and $\{\hat{Q}^t\}_{t\in T}$ the *co-semigroup* associated with $\mathcal{E}(\cdot,\cdot)$ and Δt, respectively. Whenever we refer to $\mathcal{E}(\cdot,\cdot)$, $\hat{\mathcal{E}}(\cdot,\cdot)$, $A, \hat{A}, T, Q^t$ and $\hat{Q}^t$ in the rest of this book, we shall assume that they are linked by above relations.

In applications, the primary objects will often be the semigroup $\{Q^t\}_{t\in T}$ and co-semigroup $\{\hat{Q}^t\}_{t\in T}$. We can then define A and $\hat{A}$ (and hence $\mathcal{E}(\cdot,\cdot)$) by

$$A = \frac{1}{\Delta t}\left(I - Q^{\Delta t}\right),$$
$$\hat{A} = \frac{1}{\Delta t}\left(I - \hat{Q}^{\Delta t}\right).$$

The operator A is called the *infinitesimal generator* of $\mathcal{E}(\cdot,\cdot)$, and $\hat{A}$ is called the *infinitesimal co-generator* of $\mathcal{E}(\cdot,\cdot)$. For each $t \in T$, we may define approximations $A^{(t)}$ of A and $\hat{A}^{(t)}$ of $\hat{A}$ by

$$A^{(t)} = \frac{1}{t}\left(I - Q^t\right),$$
$$\hat{A}^{(t)} = \frac{1}{t}\left(I - \hat{Q}^t\right). \tag{1.1.4}$$

From $A^{(t)}$ and $\hat{A}^{(t)}$, we get the forms

$$\begin{aligned}\mathcal{E}^{(t)}(u,v) &= \langle A^{(t)}u, v\rangle \\ &= \langle u, \hat{A}^{(t)}v\rangle,\end{aligned} \tag{1.1.5}$$

and

$$\begin{aligned}\hat{\mathcal{E}}^{(t)}(u,v) &= \mathcal{E}^{(t)}(v,u) \\ &= \langle \hat{A}^{(t)}u, v\rangle \\ &= \langle A^{(t)}v, u\rangle.\end{aligned}$$

We define the *symmetric part* $\overline{\mathcal{E}}(\cdot,\cdot)$ and *anti-symmetric part* $\mathring{\mathcal{E}}(\cdot,\cdot)$ of $\mathcal{E}(\cdot,\cdot)$ by

$$\begin{aligned}\overline{\mathcal{E}}(u,v) &= \frac{1}{2}\Big(\mathcal{E}(u,v) + \mathcal{E}(v,u)\Big), \\ \mathring{\mathcal{E}}(u,v) &= \frac{1}{2}\Big(\mathcal{E}(u,v) - \mathcal{E}(v,u)\Big).\end{aligned}$$

For $\alpha \in {}^*\mathbb{R}, \alpha \geq 0$, we set

$$\overline{\mathcal{E}}_\alpha(u,v) = \overline{\mathcal{E}}(u,v) + \alpha\langle u,v\rangle.$$

Each of these forms generates a norm (possibly a semi-norm in the case $\alpha = 0$):

$$\begin{aligned}|u|_\alpha &= \sqrt{\overline{\mathcal{E}}_\alpha(u,u)} \\ &= \sqrt{\mathcal{E}_\alpha(u,u)}.\end{aligned}$$

We recall that the original Hilbert space norm on H is denoted by $||\cdot||$. Similarly, we set for $\alpha \in {}^*\mathbb{R}, \alpha \geq 0$,

$$\begin{aligned}\mathcal{E}_\alpha(u,v) &= \mathcal{E}(u,v) + \alpha\langle u,v\rangle, \\ \hat{\mathcal{E}}_\alpha(u,v) &= \hat{\mathcal{E}}(u,v) + \alpha\langle u,v\rangle.\end{aligned}$$

Let $\overline{A}$ and $\{\overline{Q}^t\}$ be the generator and semigroup of $\overline{\mathcal{E}}(\cdot,\cdot)$, respectively. Then

$$\overline{A} = \frac{1}{2}\left(A + \hat{A}\right), \quad \overline{Q}^{\Delta t} = \frac{1}{2}\left(Q^{\Delta t} + \hat{Q}^{\Delta t}\right) \text{ and } \overline{Q}^{k\Delta t} = (\overline{Q}^{\Delta t})^k, \forall k \in {}^*\mathbb{N}.$$

Since $\overline{A}$ and $\overline{Q}^t$ are nonnegative, self-adjoint operators, they have unique nonnegative square roots, which we denote by $\overline{A}^{\frac{1}{2}}$ and $\overline{Q}^{\frac{t}{2}}$, respectively.

In the same manner as (1.1.4) and (1.1.5), we can define approximations $\overline{A}^{(t)}$ of A and $\overline{\mathcal{E}}^{(t)}(\cdot,\cdot)$ of $\overline{\mathcal{E}}(\cdot,\cdot)$ by

$$\overline{A}^{(t)} = \frac{1}{t}\left(I - \overline{Q}^t\right), \quad \overline{\mathcal{E}}^{(t)}(u,v) = \langle \overline{A}^{(t)} u, v\rangle, \quad t \in T.$$

If a nonnegative quadratic form $\mathcal{E}(\cdot,\cdot) : H \times H \longrightarrow {}^*\mathbb{R}$ satisfies

$$\mathcal{E}(u,v) = \mathcal{E}(v,u) \text{ for all } u, v \in H,$$

i.e., $\overline{\mathcal{E}}(u,v) = \mathcal{E}(u,v)$, we shall call it a *nonnegative symmetric quadratic form.* It is easy to see that a nonnegative quadratic form $\mathcal{E}(u,v)$ is symmetric if and only if $A = \hat{A}$ or $Q^t = \hat{Q}^t, \forall t \in T$.

In this book, we shall deal with nonnegative quadratic forms $\mathcal{E}(\cdot,\cdot)$ and the related theory. For the framework, potential theory and applications of nonnegative symmetric quadratic form, we refer the reader to Albeverio et al. [25], Chap. 5, Sect. 5.1 and Fan [165, 166]. We shall utilize the known results of symmetric forms in our study, and extend them to the nonsymmetric case. In particular, we need the notion of the symmetric part $\overline{\mathcal{E}}(\cdot,\cdot)$ of $\mathcal{E}(\cdot,\cdot)$, and the related notations. In Sect. 1.2, we shall define the domain $\mathcal{D}(\overline{\mathcal{E}})$ of the symmetric part $\overline{\mathcal{E}}(\cdot,\cdot)$ by using the semigroup $\{\overline{Q}^t \mid t \in T\}$. We shall introduce the resolvent $\{\overline{G}_\alpha \mid \alpha \in {}^*(-\infty, 0)\}$ of $\overline{\mathcal{E}}(\cdot,\cdot)$ in Sect. 1.3, and characterize the domain $\mathcal{D}(\overline{\mathcal{E}})$ by this resolvent. In Sect. 1.4, we shall define the domain $\mathcal{D}(\mathcal{E})$ of $\mathcal{E}(\cdot,\cdot)$ by its resolvent $\{G_\alpha \mid \alpha \in {}^*(-\infty, 0)\}$; under the hyperfinite weak sector condition, we shall show that $\mathcal{D}(\mathcal{E}) = \mathcal{D}(\overline{\mathcal{E}})$. In Sect. 1.5, we shall introduce hyperfinite Dirichlet forms and related Markov chains. For standard coercive forms, we shall construct their nonstandard representation in Sect. 1.6.

1.2 Domain of the Symmetric Part

In this section, we shall define the domain $\mathcal{D}(\overline{\mathcal{E}})$ of the symmetric part $\overline{\mathcal{E}}(\cdot,\cdot)$ for a hyperfinite nonnegative quadratic form $\mathcal{E}(\cdot,\cdot)$. Before giving a strict definition (Definition 1.2.1), we shall mention an intuitive description. At first, let $\mathrm{Fin}(H)$ be the set of all elements in H with finite norm. By defining $x \approx y$ if $||x - y|| \approx 0$, we know from Proposition A.5.2 in the Appendix that the space[5]

$$^\circ H = \mathrm{Fin}(H)/\approx$$

[5] $\approx$ stands for differing by an infinitesimal, in the sense of nonstandard analysis, refer to Albeverio et al. [25], Cutland [125], Davis [135], Hurd [216], Hurd and Loeb [217], and Lindstrøm [262].

is a Hilbert space with respect to the inner product $({}^{\circ}x, {}^{\circ}y) = \mathrm{st}(\langle x, y\rangle)$, where ${}^{\circ}x$ denotes the equivalence class of x and $\mathrm{st} : {}^*\mathbb{R} \longrightarrow \mathbb{R}$ is the mapping of standard part[6]. We call $({}^{\circ}H, (\cdot,\cdot))$ the *hull* of $(H, \langle\cdot,\cdot\rangle)$.

Consider the standard part $\overline{E}(\cdot,\cdot)$ of the nonnegative symmetric quadratic form $\overline{\mathcal{E}}(\cdot,\cdot)$. If $\overline{\mathcal{E}}(\cdot,\cdot)$ is *S-bounded*, i.e., there exists a constant $K \in \mathbb{R}_+$ such that

$$|\overline{\mathcal{E}}(u,v)| \le K||u||||v|| \qquad \text{for all } u, v \in H,$$

we can simply define $\overline{E}(\cdot,\cdot)$ by

$$\overline{E}({}^{\circ}u, {}^{\circ}v) = {}^{\circ}\overline{\mathcal{E}}(u,v).$$

If $\overline{\mathcal{E}}(\cdot,\cdot)$ is not S-bounded, we shall meet two difficulties. We no longer have that $\overline{\mathcal{E}}(u,v) \approx \overline{\mathcal{E}}(\tilde{u},\tilde{v})$ whenever $u \approx \tilde{u}$ and $v \approx \tilde{v}$, and there may be elements $v \in \mathrm{Fin}(H)$ such that $\overline{\mathcal{E}}(\tilde{v},\tilde{v})$ is infinite for all $\tilde{v} \approx v$. The latter problem should not surprise us. It is an immediate consequence of the fact that unbounded forms on Hilbert spaces cannot be defined everywhere. We shall solve it by simply letting $\overline{E}({}^{\circ}u, {}^{\circ}v)$ be undefined when $\overline{\mathcal{E}}(\tilde{v},\tilde{v})$ is infinite for all $\tilde{v} \in {}^{\circ}v$. The most natural solution to the first problem may be to define

$$\overline{E}({}^{\circ}u, {}^{\circ}u) = \inf\{{}^{\circ}\overline{\mathcal{E}}(v,v) \mid v \in {}^{\circ}u\}, \tag{1.2.1}$$

and then extend $\overline{E}(\cdot,\cdot)$ to be a bilinear form by the usual trick

$$\overline{E}({}^{\circ}u, {}^{\circ}v) = \frac{1}{2}\left\{\overline{E}({}^{\circ}u + {}^{\circ}v, {}^{\circ}u + {}^{\circ}v) - \overline{E}({}^{\circ}u, {}^{\circ}u) - \overline{E}({}^{\circ}v, {}^{\circ}v)\right\}.$$

The disadvantage of this approach is that it gives us little understanding of how the infimum in (1.2.1) is obtained. For an easier access to the regularity properties of $\overline{\mathcal{E}}(\cdot,\cdot)$ and $\overline{E}(\cdot,\cdot)$, we prefer a more indirect way of attack. Our plan is to define a subset $\mathcal{D}(\overline{\mathcal{E}})$ of Fin(H) – we call it the *domain of* $\overline{\mathcal{E}}(\cdot,\cdot)$ – satisfying

$$\text{if } \quad {}^{\circ}\overline{\mathcal{E}}(u,u) < \infty, \text{there is a } \quad v \in \mathcal{D}(\overline{\mathcal{E}}) \text{ such that } v \approx u, \tag{1.2.2}$$

$$\text{if } u, v \in \mathcal{D}(\overline{\mathcal{E}}) \text{ and } u \approx v, \text{ then } {}^{\circ}\overline{\mathcal{E}}(u,u) = {}^{\circ}\overline{\mathcal{E}}(v,v) < \infty. \tag{1.2.3}$$

We then define $\overline{E}(\cdot,\cdot)$ by

$$\overline{E}({}^{\circ}u, {}^{\circ}u) = {}^{\circ}\overline{\mathcal{E}}(v,v), \tag{1.2.4}$$

[6] Refer to Albeverio et al. [25].

when $v \in \mathcal{D}(\overline{\mathcal{E}}) \cap {}^\circ u$. It turns out that the two definitions (1.2.1) and (1.2.4) agree (see Proposition 1.2.4).

If we look at the standard nonsymmetric Dirichlet theory, see Albeverio et al. [9], Kim [241] and Ma and Röckner [270], the domain of a quadratic form is given from the very beginning. After that, the authors such as those of Ma and Röckner [270] introduced the symmetric and anti-symmetric parts (see page 15, [270]). This method makes the domains of the quadratic form and its symmetric part coincide. On the other hand, Albeverio et al. [25] has given us a very nice definition of domain for the symmetric hyperfinite quadratic forms by their semigroups. Therefore, we may define the domain $\mathcal{D}(\overline{\mathcal{E}})$ of $\overline{\mathcal{E}}(\cdot,\cdot)$ via the semigroup of $\{\overline{Q}^t \mid t \in T\}$. In the next section, we shall discuss the property of the resolvent $\{\overline{G}_\alpha \mid \alpha \in {}^*(-\infty, 0)\}$ of $\overline{\mathcal{E}}(\cdot,\cdot)$. We can define the domain of $\mathcal{D}(\overline{\mathcal{E}})$ through $\{\overline{G}_\alpha \mid \alpha \in {}^*(-\infty, 0)\}$.

Now it is very natural to ask: can we as well define the domain $\mathcal{D}(\mathcal{E})$ of $\mathcal{E}(\cdot,\cdot)$ directly from $\{Q^t \mid t \in T\}$? Here we would mention that it seems not easy to do the job. In Sect. 1.4, we shall define $\mathcal{D}(\mathcal{E})$ by means of the resolvent $\{G_\alpha \mid \alpha < 0\}$ of $\mathcal{E}(\cdot,\cdot)$. Under the hypothesis of weak sector condition, we shall prove $\mathcal{D}(\overline{\mathcal{E}}) = \mathcal{D}(\mathcal{E})$ by showing that the two definitions satisfy (1.2.1). This is similar to the procedure in the standard nonsymmetric Dirichlet space theory, see, e.g., Albeverio et al. [9], Albeverio et al. [47], Albeverio and Ugolini [57], Kim [241], and Ma and Röckner [270].

Notice that even when $\overline{\mathcal{E}}(\cdot,\cdot)$ is not S-bounded, $\overline{\mathcal{E}}^{(t)}(\cdot,\cdot)$ is S-bounded for all non-infinitesimal t. One of the motivations behind our definition of the domain $\mathcal{D}(\overline{\mathcal{E}})$ is that we want to single out the elements where $\overline{\mathcal{E}}(\cdot,\cdot)$ is really approximated by the bounded forms $\overline{\mathcal{E}}^{(t)}(\cdot,\cdot), t \not\approx 0$, i.e., those $u \in H$ such that

$$^\circ\overline{\mathcal{E}}(u,u) = \lim_{\substack{t\downarrow 0\\ t\not\approx 0}} {}^\circ\overline{\mathcal{E}}^{(t)}(u,u). \tag{1.2.5}$$

We could have taken this to be our definition of $\mathcal{D}(\overline{\mathcal{E}})$, but for technical and expository reasons we have chosen another one which we shall soon show to be equivalent to (1.2.5) (see Proposition 1.2.2).

Definition 1.2.1. Let $\mathcal{E}(\cdot,\cdot)$ be a nonnegative quadratic form on a hyperfinite dimensional linear space H. The domain $\mathcal{D}(\overline{\mathcal{E}})$ of the symmetric part of $\mathcal{E}(\cdot,\cdot)$ is the set of all $u \in H$ satisfying

(i) $^\circ\mathcal{E}_1(u,u) = {}^\circ\overline{\mathcal{E}}_1(u,u) < \infty$.
(ii) For all $t \approx 0$, $\overline{\mathcal{E}}(\overline{Q}^t u, \overline{Q}^t u) \approx \overline{\mathcal{E}}(u,u)$.

Let us try to convey the intuition behind this definition. Thinking of $\overline{A}$ as a differential operator, the elements of $\mathcal{D}(\overline{\mathcal{E}})$ are "smooth" functions and

$\overline{Q}^t$ is a "smoothing" operator often given by an integral kernel. If an element u is already smooth, then an infinitesimal amount of smoothing $\overline{Q}^t, t \approx 0$, should not change it noticeably, and hence $\overline{\mathcal{E}}(\overline{Q}^t u, \overline{Q}^t u) \approx \overline{\mathcal{E}}(u,u)$. We shall give a partial justification of this rather crude image later, when we show that if ${}^\circ\overline{\mathcal{E}}_1(u,u) < \infty$, then the "smoothed" elements $\overline{Q}^t u, t \not\approx 0$, are all in $\mathcal{D}(\overline{\mathcal{E}})$ (Lemma 1.2.3, see also Corollary 1.2.3).

Our first task will be to establish a list of alternative definitions of $\mathcal{D}(\overline{\mathcal{E}})$, among them (1.2.5). We begin with the following simple identity giving the relationship between $\mathcal{E}(\cdot,\cdot)$ and $\mathcal{E}^{(t)}(\cdot,\cdot)$, and also the relationship between $\overline{\mathcal{E}}(\cdot,\cdot)$ and $\overline{\mathcal{E}}^{(t)}(\cdot,\cdot)$:

Lemma 1.2.1. *For all $u \in H$ and $t \in T$, we have*

(i) $\mathcal{E}^{(t)}(u,u) \geq 0$ *and* $\overline{\mathcal{E}}^{(t)}(u,u) \geq 0$,

(ii) $\mathcal{E}^{(t)}(u,u) = \dfrac{\Delta t}{t} \sum\limits_{0 \leq s < t} \mathcal{E}(Q^s u, u) = \dfrac{\Delta t}{t} \sum\limits_{0 \leq s < t} \mathcal{E}(u, \hat{Q}^s u)$,

(iii) $\overline{\mathcal{E}}^{(t)}(u,u) = \dfrac{\Delta t}{t} \sum\limits_{0 \leq s < t} \overline{\mathcal{E}}(\overline{Q}^s u, u) = \dfrac{\Delta t}{t} \sum\limits_{0 \leq s < t} \overline{\mathcal{E}}(\overline{Q}^{s/2} u, \overline{Q}^{s/2} u)$.

Proof. (i) We have

$$\begin{aligned}
\mathcal{E}^{(t)}(u,u) &= \frac{1}{t}\langle (I - Q^t)u, u\rangle \\
&= \frac{1}{t}\left(\langle u,u\rangle - \langle Q^t u, u\rangle\right) \\
&\geq \frac{1}{t}\left(\langle u,u\rangle - \|Q^t\|\langle u,u\rangle\right) \\
&\geq 0,
\end{aligned}$$

since $\|Q^t\| \leq \|Q^{\Delta t}\|^{\frac{t}{\Delta t}} \leq 1$. It is then easy to see that $\overline{\mathcal{E}}^{(t)}(u,u) \geq 0$.

(ii) By an easy calculation, we have

$$\begin{aligned}
\mathcal{E}^{(t)}(u,u) &= \frac{1}{t}\langle (I - Q^t)u, u\rangle \\
&= \frac{1}{t} \sum_{0 \leq s < t} \langle (Q^s - Q^{s+\Delta t})u, u\rangle \\
&= \frac{\Delta t}{t} \sum_{0 \leq s < t} \mathcal{E}(Q^s u, u) \\
&= \frac{\Delta t}{t} \sum_{0 \leq s < t} \mathcal{E}(u, \hat{Q}^s u). \qquad (1.2.6)
\end{aligned}$$

(iii) In the same way as (1.2.6), we can prove that the first equation holds. The second one is due to the symmetry and the semigroup property of $\overline{Q}^s$. □

Among other things, Lemma 1.2.1 tells us that $\mathcal{E}^{(t)}(\cdot,\cdot)$ and $\overline{\mathcal{E}}^{(t)}(\cdot,\cdot)$ are nonnegative.

Lemma 1.2.2. *Let $B, C : H \longrightarrow H$ be nonnegative, symmetric operators commuting with $\overline{A}$ and each other. Then the functions*

$$t \mapsto \langle \overline{Q}^t Bu, Cu\rangle \quad \textit{and} \quad t \mapsto \overline{\mathcal{E}}^{(s)}(\overline{Q}^t Bu, Cu)$$

are nonnegative and decreasing for all $u \in H$ and $s \in T$.

Proof. We first notice that the $\overline{\mathcal{E}}^{(s)}(\cdot,\cdot)$ part follows from the other one since

$$\overline{\mathcal{E}}^{(s)}(\overline{Q}^t Bu, Cu) = \frac{1}{s}\langle \overline{Q}^t (I - \overline{Q}^s)Bu, Cu\rangle,$$

and the operator $B' = (I - \overline{Q}^s)B$ is nonnegative and commutes with $\overline{A}$ and C. If $t > r$, then

$$\begin{aligned}
\langle \overline{Q}^r Bu, Cu\rangle - \langle \overline{Q}^t Bu, Cu\rangle &= \langle (I - \overline{Q}^{t-r})\overline{Q}^r Bu, Cu\rangle \\
&= (t-r)\overline{\mathcal{E}}^{(t-r)}(\overline{Q}^{r/2} B^{1/2}C^{1/2}u, \overline{Q}^{r/2} B^{1/2}C^{1/2}u) \\
&\geq 0,
\end{aligned}$$

where we used that $\overline{\mathcal{E}}^{(t-r)}(\cdot,\cdot)$ is nonnegative. Hence, $t \longrightarrow \langle \overline{Q}^t Bu, Cu\rangle$ decreases. For the positivity, we observe that

$$\begin{aligned}
\langle \overline{Q}^t Bu, Cu\rangle &= \langle \overline{Q}^{t/2} B^{1/2}C^{1/2}u, \overline{Q}^{t/2} B^{1/2}C^{1/2}u\rangle \\
&\geq 0.
\end{aligned}$$

□

From Lemma 1.2.2 we may now obtain our main inequalities.

Proposition 1.2.1. *For all $u \in H, t \in T$:*

(i) $0 \leq \overline{\mathcal{E}}(u, u - \overline{Q}^t u) \leq \overline{\mathcal{E}}(u,u) - \overline{\mathcal{E}}(\overline{Q}^t u, \overline{Q}^t u) \leq 2\overline{\mathcal{E}}(u, u - \overline{Q}^t u)$.

(ii) $0 \leq \overline{\mathcal{E}}(\overline{Q}^{\Delta t} u, \overline{Q}^{\Delta t} u) - \overline{\mathcal{E}}(\overline{Q}^{2\Delta t} u, \overline{Q}^{2\Delta t} u) \leq \overline{\mathcal{E}}(u,u) - \overline{\mathcal{E}}(\overline{Q}^{\Delta t} u, \overline{Q}^{\Delta t} u)$.

Proof. By trivial algebra, we have

$$\overline{\mathcal{E}}(u,u) - \overline{\mathcal{E}}(\overline{Q}^t u, \overline{Q}^t u) = \overline{\mathcal{E}}(u, u - \overline{Q}^t u) + \overline{\mathcal{E}}(\overline{Q}^t u, u - \overline{Q}^t u).$$

Applying Lemma 1.2.2 with $B = I, C = I - \overline{Q}^t$, we see that

$$0 \leq \overline{\mathcal{E}}(\overline{Q}^t u, u - \overline{Q}^t u) \leq \overline{\mathcal{E}}(u, u - \overline{Q}^t u),$$

and part (i) follows.

(ii) The non-negativity is immediate from (i), and as above we have

$$\overline{\mathcal{E}}(u,u) - \overline{\mathcal{E}}(\overline{Q}^{\Delta t} u, \overline{Q}^{\Delta t} u) = \overline{\mathcal{E}}(u, u - \overline{Q}^{\Delta t} u) + \overline{\mathcal{E}}(\overline{Q}^{\Delta t} u, u - \overline{Q}^{\Delta t} u).$$

Applying Lemma 1.2.2 to each of the latter two terms, using $B = I, C = I - \overline{Q}^{\Delta t}$ in the first case, and $B = \overline{Q}^{\Delta t}, C = I - \overline{Q}^{\Delta t}$ in the second, we get

$$\begin{aligned}
\overline{\mathcal{E}}(u,u) - \overline{\mathcal{E}}(\overline{Q}^{\Delta t} u, \overline{Q}^{\Delta t} u) &\geq \overline{\mathcal{E}}(\overline{Q}^{2\Delta t} u, u - \overline{Q}^{\Delta t} u) + \overline{\mathcal{E}}(\overline{Q}^{3\Delta t} u, u - \overline{Q}^{\Delta t} u) \\
&= \overline{\mathcal{E}}(\overline{Q}^{2\Delta t} u, u) - \overline{\mathcal{E}}(\overline{Q}^{2\Delta t} u, \overline{Q}^{\Delta t} u) \\
&\quad + \overline{\mathcal{E}}(\overline{Q}^{3\Delta t} u, u) - \overline{\mathcal{E}}(\overline{Q}^{3\Delta t} u, \overline{Q}^{\Delta t} u) \\
&= \overline{\mathcal{E}}(\overline{Q}^{\Delta t} u, \overline{Q}^{\Delta t} u) - \overline{\mathcal{E}}(\overline{Q}^{2\Delta t} u, \overline{Q}^{2\Delta t} u).
\end{aligned}$$

The proposition is proved. □

The inequalities above are what we need to establish a reasonable characterization of $\mathcal{D}(\overline{\mathcal{E}})$. We first give our promised list of alternative definitions of the domain of $\overline{\mathcal{E}}(\cdot,\cdot)$.

Proposition 1.2.2. *The following statements are equivalent:*

(i) u is in the domain $\mathcal{D}(\overline{\mathcal{E}})$ of $\mathcal{E}(\cdot,\cdot)$.
(ii) ${}^\circ\mathcal{E}_1(u,u) = {}^\circ\overline{\mathcal{E}}_1(u,u) < \infty$, and for all $t \approx 0$, we have $\overline{\mathcal{E}}(u, u - \overline{Q}^t u) \approx 0$.
(iii) ${}^\circ\mathcal{E}_1(u,u) < \infty$, and for all $t \approx 0$, we have $\overline{\mathcal{E}}(u - \overline{Q}^t u, u - \overline{Q}^t u) \approx 0$.
(iv) ${}^\circ\mathcal{E}_1(u,u) < \infty$, and for all $t \approx 0$, we have $\overline{\mathcal{E}}^{(t)}(u,u) \approx \overline{\mathcal{E}}(u,u)$.

Proof. *(i)* $\iff$ *(ii)*. Follows immediately from Proposition 1.2.1 (i).

(ii) $\implies$ *(iii)*. We have

$$0 \leq \overline{\mathcal{E}}(u - \overline{Q}^t u, u - \overline{Q}^t u) = \overline{\mathcal{E}}(u, u - \overline{Q}^t u) - \overline{\mathcal{E}}(\overline{Q}^t u, u - \overline{Q}^t u),$$

and by Lemma 1.2.2 the term $\overline{\mathcal{E}}(\overline{Q}^t u, u - \overline{Q}^t u)$ is positive.

(iii) $\implies$ *(i)*. We recall that $|u|_0 = \sqrt{\overline{\mathcal{E}}(u,u)}$ is a semi-norm. By Lemma 1.2.2 and the triangle inequality, we have

$$0 \leq |u|_0 - |\overline{Q}^t u|_0 \leq |u - \overline{Q}^t u|_0.$$

Multiplying both sides by $|u|_0 + |\overline{Q}^t u|_0$, we get

$$0 \leq |u|_0^2 - |\overline{Q}^t u|_0^2 \leq |u - \overline{Q}^t u|_0(|u|_0 + |\overline{Q}^t u|_0) \leq 2|u|_0|u - \overline{Q}^t u|_0.$$

Hence if ${}^\circ\mathcal{E}(u,u) < \infty$ and $\overline{\mathcal{E}}(u - \overline{Q}^t u, u - \overline{Q}^t u) \approx 0$, we have that

$$\overline{\mathcal{E}}(u,u) - \overline{\mathcal{E}}(\overline{Q}^t u, \overline{Q}^t u) \approx 0.$$

(ii) $\Longrightarrow$ *(iv)*. Follows at once from Lemma 1.2.1.

(iv) $\Longrightarrow$ *(ii)*. Follows from Lemma 1.2.1 and the fact that $s \mapsto \overline{\mathcal{E}}(\overline{Q}^s u, u)$ is decreasing. □

The characterizations of $\mathcal{D}(\overline{\mathcal{E}})$ given in the Proposition 1.2.2 are useful for different purposes. As an illustration, we use Proposition 1.2.2 (iii) to prove that the domain has the right linear structure.

Corollary 1.2.1. *Let $u, v \in \mathcal{D}(\overline{\mathcal{E}})$, and assume that $\alpha \in {}^*\mathbb{R}$ is a nearstandard*[7] *number. Then αu and $u + v$ are elements of $\mathcal{D}(\overline{\mathcal{E}})$.*

Proof. The αu part is trivial. For $u + v$ we use Proposition 1.2.2 (iii) and the triangle inequality.

$$\begin{aligned} ||(u+v) - \overline{Q}^t(u+v)||_0 &= ||u - \overline{Q}^t u + v - \overline{Q}^t v||_0 \\ &\leq ||u - \overline{Q}^t u||_0 + ||v - \overline{Q}^t v||_0. \end{aligned}$$

The latter two terms above are infinitesimals when $t \approx 0$. □

Corollary 1.2.2. *For any infinitesimal $\delta \in T$, we have $\mathcal{D}(\overline{\mathcal{E}}) \subset \mathcal{D}(\overline{\mathcal{E}}^{(\delta)})$.*

Proof. Let $u \in \mathcal{D}(\overline{\mathcal{E}})$. By Proposition 1.2.2 (iv), we know $\overline{\mathcal{E}}^{(k\delta)}(u,u) \approx \overline{\mathcal{E}}(u,u) \approx \overline{\mathcal{E}}^{(\delta)}(u,u)$ for all k such that $k\delta \approx 0$. By Proposition 1.2.2 (iv) again, we get $u \in \mathcal{D}(\overline{\mathcal{E}}^{(\delta)})$. □

The second part of Proposition 1.2.1 informs us that $\overline{Q}^t u$ is more likely to be in $\mathcal{D}(\overline{\mathcal{E}})$ than u is. The next lemma pins this down more precisely.

Lemma 1.2.3. *Assume ${}^\circ\mathcal{E}_1(u,u) < \infty$. Then for all non-infinitesimals t, we have $\overline{Q}^t u \in \mathcal{D}(\overline{\mathcal{E}})$.*

Proof. By Proposition 1.2.1, we have

$${}^\circ\overline{\mathcal{E}}_1(\overline{Q}^t u, \overline{Q}^t u) \leq {}^\circ\overline{\mathcal{E}}_1(u,u) < \infty.$$

[7] See Appendix and Albeverio et al. [25] for the concept of nearstandard.

To prove that Definition 1.2.1 (ii) is satisfied, we notice that according to Proposition 1.2.1 (ii), the function

$$t \mapsto \overline{\mathcal{E}}(\overline{Q}^t u, \overline{Q}^t u)$$

is decreasing and convex, and hence

$$\frac{1}{s}\Big(\overline{\mathcal{E}}(\overline{Q}^t u, \overline{Q}^t u) - \overline{\mathcal{E}}(\overline{Q}^{t+s} u, \overline{Q}^{t+s} u)\Big) \leq \frac{1}{t}\Big(\overline{\mathcal{E}}(u,u) - \overline{\mathcal{E}}(\overline{Q}^t u, \overline{Q}^t u)\Big)$$

for all $s > 0$. Multiplying through by s, we get

$$0 \leq \overline{\mathcal{E}}(\overline{Q}^t u, \overline{Q}^t u) - \overline{\mathcal{E}}(\overline{Q}^{t+s} u, \overline{Q}^{t+s} u) \leq \frac{s}{t}\Big(\overline{\mathcal{E}}(u,u) - \overline{\mathcal{E}}(\overline{Q}^s u, \overline{Q}^s u)\Big).$$

For $s \approx 0$ and $t \not\approx 0$, the expression on the right is infinitesimal, and the lemma follows. □

We shall now strengthen the lemma above and show that if ${}^\circ\mathcal{E}_1(u,u) < \infty$, then there is an infinitesimal t such that $\overline{Q}^t u \in \mathcal{D}(\overline{\mathcal{E}})$. This is a special case of our next result. First we need to introduce a new definition. A subset F of H is called *$\overline{\mathcal{E}}$-closed* if and only if for all sequences $\{u_n\}_{n\in\mathbb{N}}$ of elements from F such that ${}^\circ|u_n - u_m|_1 \longrightarrow 0$ as $n, m \longrightarrow \infty$, there exists an element u in F such that ${}^\circ|u_n - u|_1 \longrightarrow 0$ as $n \longrightarrow \infty$.

Proposition 1.2.3. *$\mathcal{D}(\overline{\mathcal{E}})$ is $\overline{\mathcal{E}}$-closed. Moreover, if $\{u_n\}_{n\in\mathbb{N}}$ is a $|\cdot|_1$ Cauchy sequence from $\mathcal{D}(\overline{\mathcal{E}})$, and $\{u_n \mid n \in {}^*\mathbb{N}\}$ is an internal extension, then there is a $\gamma \in {}^*\mathbb{N} - \mathbb{N}$ such that $u_\eta \in \mathcal{D}(\overline{\mathcal{E}})$ for all $\eta \leq \gamma$.*

Proof. Let $\{u_n | n \in \mathbb{N}\}$ be a $|\cdot|_1$ Cauchy sequence from $\mathcal{D}(\overline{\mathcal{E}})$, and let $\{u_n | n \in {}^*\mathbb{N}\}$ be an internal extension of it. There is an element $\gamma \in {}^*\mathbb{N} - \mathbb{N}$ such that $|u_n - u_m|_1 \approx 0$ whenever n and m are infinite and less than γ. Let $\eta \in {}^*\mathbb{N} - \mathbb{N}, \eta \leq \gamma$. By the choice of γ, ${}^\circ\overline{\mathcal{E}}_1(u_\eta, u_\eta) < \infty$ and ${}^\circ|u_n - u_\eta|_1 \longrightarrow 0$ as n approaches infinity in $\mathbb{N}$. All that remains is to prove that $u_\eta \in \mathcal{D}(\overline{\mathcal{E}})$.

Assume not, then by Proposition 1.2.2 (iii) there is an $\varepsilon \in \mathbb{R}_+$ and $t \approx 0$ such that

$$|u_\eta - \overline{Q}^t u_\eta|_0 > \varepsilon.$$

Choose $m \in \mathbb{N}$ so large that

$$|u_\eta - u_m|_0 < \frac{\varepsilon}{4}.$$

Then by Proposition 1.2.1 (i), we have

$$|\overline{Q}^t u_\eta - \overline{Q}^t u_m|_0 < \frac{\varepsilon}{4}.$$

Combining the inequalities above, we have

$$\begin{aligned}\varepsilon < |u_\eta - \overline{Q}^t u_\eta|_0 &\le |u_\eta - u_m|_0 + |u_m - \overline{Q}^t u_m|_0 + |\overline{Q}^t u_m - \overline{Q}^t u_\eta|_0 \\ &\le \varepsilon/2 + |u_m - \overline{Q}^t u_m|_0,\end{aligned}$$

but since $u_m \in \mathcal{D}(\overline{\mathcal{E}})$, the last term is infinitesimal by Proposition 1.2.2 (iii). We have the contradiction we wanted. □

Corollary 1.2.3. *If* ${}^\circ\mathcal{E}_1(u,u) < \infty$, *there is a* $t_0 \approx 0$ *such that* $\overline{Q}^t u \in \mathcal{D}(\overline{\mathcal{E}})$ *for all* $t \ge t_0$.

Proof. First we notice that if $\overline{Q}^{t_0} u \in \mathcal{D}(\overline{\mathcal{E}})$, so is $\overline{Q}^t u$ for all $t > t_0$. Put $u_n = \overline{Q}^{\frac{1}{n}} u$. Then the sequence $\{|u_n|_1\}$ is increasing and bounded by $|u|_1$, and we can apply Proposition 1.2.3 to it. The corollary follows. □

Corollary 1.2.4. *If* ${}^\circ\mathcal{E}_1(u,u) < \infty$, *there is a* $\delta_u \approx 0$ *such that* $u \in \mathcal{D}(\overline{\mathcal{E}}^{(\delta)})$ *for all infinitesimal* $\delta \ge \delta_u$.

Proof. By Corollary 1.2.3, there is a $t_0 \approx 0$ such that $\overline{Q}^t u \in \mathcal{D}(\overline{\mathcal{E}})$ for all $t \ge t_0$. Let δ_u be an infinitesimal such that $\delta_u > t_0$ and $t_0/\delta_u \approx 0$. For all infinitesimal $\delta \ge \delta_u$, we have $v = \overline{Q}^\delta u \in \mathcal{D}(\overline{\mathcal{E}}) \subset \mathcal{D}(\overline{\mathcal{E}}^{(\delta)})$ by Corollaries 1.2.2 and 1.2.3. For all $k \in {}^*\mathbb{N}$ such that $k\delta \approx 0$, we have the following

$$\begin{aligned}\overline{\mathcal{E}}^{(\delta)}(\overline{Q}^{k\delta} u, \overline{Q}^{k\delta} u) &= \overline{\mathcal{E}}^{(\delta)}(\overline{Q}^{(k-1)\delta} v, \overline{Q}^{(k-1)\delta} v) \\ &\approx \overline{\mathcal{E}}^{(\delta)}(v,v) \\ &= \overline{\mathcal{E}}^{(\delta)}(\overline{Q}^\delta u, \overline{Q}^\delta u).\end{aligned} \tag{1.2.7}$$

By Lemma 1.2.1 (iii), we have

$$\begin{aligned}\overline{\mathcal{E}}^{(\delta)}(\overline{Q}^\delta u, \overline{Q}^\delta u) &= \frac{\Delta t}{\delta} \sum_{0 \le s < \delta} \overline{\mathcal{E}}(\overline{Q}^{s/2}\overline{Q}^\delta u, \overline{Q}^{s/2}\overline{Q}^\delta u) \\ &= \frac{\Delta t}{\delta} \sum_{0 \le s < \delta} \overline{\mathcal{E}}(\overline{Q}^{s/2+\delta-t_0}\overline{Q}^{t_0} u, \overline{Q}^{s/2+\delta-t_0}\overline{Q}^{t_0} u) \\ &\approx \overline{\mathcal{E}}(\overline{Q}^{t_0} u, \overline{Q}^{t_0} u),\end{aligned} \tag{1.2.8}$$

because $\overline{Q}^{t_0} u \in \mathcal{D}(\overline{\mathcal{E}})$. By Lemma 1.2.1 (iii) again, we have

$$\begin{aligned}\overline{\mathcal{E}}^{(\delta)}(u,u) &= \frac{\Delta t}{\delta} \sum_{0 \le s < \delta} \overline{\mathcal{E}}(\overline{Q}^{s/2} u, \overline{Q}^{s/2} u) \\ &= \frac{\Delta t}{\delta} \left[\sum_{0 \le s < 2t_0} \overline{\mathcal{E}}(\overline{Q}^{s/2} u, \overline{Q}^{s/2} u) + \sum_{2t_0 \le s < \delta} \overline{\mathcal{E}}(\overline{Q}^{s/2} u, \overline{Q}^{s/2} u) \right]\end{aligned}$$

$$\approx \frac{\Delta t}{\delta} \sum_{2t_0 \leq s < \delta} \overline{\mathcal{E}}(\overline{Q}^{s/2-t_0}\overline{Q}^{t_0}u, \overline{Q}^{s/2-t_0}\overline{Q}^{t_0}u)$$

$$\approx \overline{\mathcal{E}}(\overline{Q}^{t_0}u, \overline{Q}^{t_0}u). \tag{1.2.9}$$

By relations (1.2.7), (1.2.8), and (1.2.9), we know $u \in \mathcal{D}(\overline{\mathcal{E}}^{(\delta)})$. □

Remark 1.2.1. Proposition 1.2.3 is rather surprising since there exist standard forms which are neither closed nor closable. In fact, there are numerous applications where the main difficulty is to show that the form constructed is closed, or at least can be extended to a closed form (see, e.g., [11, 16–24, 26, 27, 36, 48, 49, 94, 98, 99, 103, 151, 175, 176, 178, 225, 232, 236, 247, 251, 259, 278, 301, 318, 345, 359]). If we know that a form comes from a hyperfinite form, this follows immediately from Proposition 1.2.3. In Albeverio et al. [25], Chap. 6, we have got various examples of how useful this observation is. For the time being, we only remark that since we shall soon show that all standard, coercive closed forms can be obtained from hyperfinite forms, the method is quite general (we refer to Sect. 1.6 of this chapter).

Notice that if we can show that whenever ${}^\circ\overline{\mathcal{E}}_1(u,u) < \infty$, then for all $t \approx 0$, $||u - \overline{Q}^t u|| \approx 0$, Corollary 1.2.3 will imply the first part of our program, i.e., (1.2.2) above.

Lemma 1.2.4. *Assume* ${}^\circ\overline{\mathcal{E}}(u,u) < \infty$. *Then for all* $t \approx 0$, *we have*

$$||u - \overline{Q}^t u|| \approx 0.$$

Proof. For $t \approx 0$, we have

$$\begin{aligned} ||u - \overline{Q}^t u||^2 &= \langle u - \overline{Q}^t u, u - \overline{Q}^t u\rangle \\ &= t\overline{\mathcal{E}}^{(t)}(u, u - \overline{Q}^t u) \\ &= t\Big[\overline{\mathcal{E}}^{(t)}(u,u) - \overline{\mathcal{E}}^{(t)}(u, \overline{Q}^t u)\Big] \\ &\leq t\overline{\mathcal{E}}(u,u) \approx 0. \end{aligned}$$

□

Let us turn our attention to our second main goal (1.2.3).

Lemma 1.2.5. *If* $u, v \in \mathcal{D}(\overline{\mathcal{E}})$ *and* $u \approx v$, *then*

$$\mathcal{E}(u,u) \approx \mathcal{E}(v,v).$$

Proof. It is obviously enough to show that if $u \in \mathcal{D}(\overline{\mathcal{E}})$ and $u \approx 0$, then $\mathcal{E}(u,u) \approx 0$. But if $u \in \mathcal{D}(\overline{\mathcal{E}})$, we know from Proposition 1.2.2 (iv):

$$^\circ\overline{\mathcal{E}}(u,u) = \lim_{\substack{t \downarrow 0 \\ t \not\approx 0}} {}^\circ\overline{\mathcal{E}}^{(t)}(u,u). \tag{1.2.10}$$

Also

$$\begin{aligned}\overline{\mathcal{E}}^{(t)}(u,u) &= \frac{1}{t}\langle (I-\overline{Q}^t)u,u\rangle \\ &= \frac{1}{t}\Big(\langle u,u\rangle - \langle \overline{Q}^t u,u\rangle\Big) \\ &\le \frac{1}{t}||u||^2,\end{aligned}$$

which is infinitesimal for $t \not\approx 0$. Combining this with (1.2.10), the lemma follows. □

We may now sum up our results on $\mathcal{D}(\overline{\mathcal{E}})$ in one statement.

Theorem 1.2.1. *Let $\mathcal{E}(\cdot,\cdot)$ be a nonnegative quadratic form on a hyperfinite dimensional space H. Then*

(i) If $u, v \in \mathcal{D}(\overline{\mathcal{E}})$ and α is a finite element of ${}^*\mathbb{R}$, then $\alpha u, u+v \in \mathcal{D}(\overline{\mathcal{E}})$.
(ii) $\mathcal{D}(\overline{\mathcal{E}})$ is $\overline{\mathcal{E}}$-closed.
(iii) If ${}^\circ\overline{\mathcal{E}}_1(u,u) < \infty$, then there exists a $v \in \mathcal{D}(\overline{\mathcal{E}})$ with $||u-v|| \approx 0$. Moreover, we have

$$\begin{aligned}{}^\circ\overline{\mathcal{E}}(v,v) &= \lim_{\substack{t\downarrow 0\\ t\not\approx 0}} {}^\circ\overline{\mathcal{E}}(\overline{Q}^t u, \overline{Q}^t u) \\ &= \lim_{\substack{t\downarrow 0\\ t\not\approx 0}} {}^\circ\overline{\mathcal{E}}^{(t)}(u,u).\end{aligned}$$

(iv) If $u, v \in \mathcal{D}(\overline{\mathcal{E}})$ and $u \approx v$, then $\overline{\mathcal{E}}(u,u) \approx \overline{\mathcal{E}}(v,v)$.
(v) If $u \in \mathcal{D}(\overline{\mathcal{E}})$, then ${}^\circ\overline{\mathcal{E}}(u,u) = \inf\{{}^\circ\overline{\mathcal{E}}(v,v) \mid v \approx u\}$.
(vi) If ${}^\circ\mathcal{E}(u,u) < \infty$ and ${}^\circ\overline{\mathcal{E}}(u,u) = \inf\{{}^\circ\overline{\mathcal{E}}(v,v) \mid v \approx u\}$, then $u \in \mathcal{D}(\overline{\mathcal{E}})$.

Proof. We only need to show (v) and (vi), since we have proved the other results.

(v) Noticing (iv), we have $\mathcal{E}(u,u) \approx 0$ if $u \in \mathcal{D}(\overline{\mathcal{E}})$ and $u \approx 0$. This implies the following for general $u \in \mathcal{D}(\overline{\mathcal{E}})$:

$$\begin{aligned}\inf\{{}^\circ\overline{\mathcal{E}}(v,v) \mid v \approx u\} &\le {}^\circ\overline{\mathcal{E}}(u,u) \\ &\le \inf\left\{\left(\sqrt{{}^\circ\overline{\mathcal{E}}(v,v)} + \sqrt{{}^\circ\overline{\mathcal{E}}(u-v,u-v)}\right)^2 \middle| v \approx u\right\} \\ &= \inf\{{}^\circ\overline{\mathcal{E}}(v,v) \mid v \approx u\}.\end{aligned}$$

This is (v).

(vi) From Proposition 1.2.1, we know that $t \mapsto \overline{\mathcal{E}}(\overline{Q}^t u, \overline{Q}^t u)$ is decreasing. This implies (vi). □

The following definition now makes sense.

Definition 1.2.2. The *symmetric standard part of* $\overline{\mathcal{E}}(\cdot,\cdot)$ is the quadratic form $\overline{E}(\cdot,\cdot)$ on ${}^{\circ}H$ defined by:

(i) The domain $D(\overline{E})$ of $\overline{E}(\cdot,\cdot)$ is the set of all equivalence classes ${}^{\circ}u \in {}^{\circ}H$ such that $\inf\{{}^{\circ}\overline{\mathcal{E}}_1(v,v) \mid v \in {}^{\circ}u\} < \infty$.
(ii) If $x, y \in {}^{\circ}H$ are in the domain of $\overline{E}(\cdot,\cdot)$, let $\overline{E}(x,y) = {}^{\circ}\overline{\mathcal{E}}(u,v)$, where $u \in x, v \in y$ are in $\mathcal{D}(\overline{\mathcal{E}})$.

For $\alpha \in [0,\infty)$, let us set

$$\overline{E}_\alpha(\cdot,\cdot) = \overline{E}(\cdot,\cdot) + \alpha(\cdot,\cdot).$$

We recall that $(\cdot,\cdot)$ is the inner product of ${}^{\circ}H$.

An $\overline{E}_1$-*Cauchy sequence* is a sequence $\{x_n\}$ of elements from $D(\overline{E})$ such that $\overline{E}_1(u_n - u_m, u_n - u_m) \longrightarrow 0$ as $n, m \longrightarrow \infty$. We say that $\overline{E}(\cdot,\cdot)$ is *closed* if all $\overline{E}_1$-Cauchy sequences converge in $\overline{E}_1$-norm to an element in $D(\overline{E})$. The next proposition follows immediately from Theorem 1.2.1 and the definition of $\overline{E}(\cdot,\cdot)$.

Proposition 1.2.4. *Let* $\overline{E}(\cdot,\cdot)$ *be the standard part of* $\overline{\mathcal{E}}(\cdot,\cdot)$. *Then* $\overline{E}(\cdot,\cdot)$ *is closed, and for all* $x \in {}^{\circ}H$

$$\overline{E}(x,x) = \inf\{{}^{\circ}\overline{\mathcal{E}}(u,u) \mid u \in x\}, \tag{1.2.11}$$

where we take the value ∞ *on the right to mean that the expression on the left is undefined.*

We point out that (1.2.11) is just our original suggestion (1.2.1) for the standard part of $\overline{\mathcal{E}}(\cdot,\cdot)$. In Sect. 1.4, we shall study the standard part $E(\cdot,\cdot)$ of $\mathcal{E}(\cdot,\cdot)$ under the hyperfinite weak sector condition. We shall show that $D(E) = D(\overline{E})$ and that $\overline{E}(\cdot,\cdot)$ is exactly the *symmetric part of* $E(\cdot,\cdot)$, i.e., $\overline{E}(x,y) = \frac{1}{2}(E(x,y) + E(y,x))$.

1.3 Resolvent of the Symmetric Part

In Sects. 1.1 and 1.2, we have only been interested in the relationship among the form $\mathcal{E}(\cdot,\cdot)$ and the associated semigroup $\{Q^t\}$ and infinitesimal generator A, also $\overline{\mathcal{E}}(\cdot,\cdot)$ and its semigroup $\{\overline{Q}^t\}$ and generator $\overline{A}$, and so on. In this section, we turn our attention to the resolvent $\{\overline{G}_\alpha\}$ of $\overline{\mathcal{E}}(\cdot,\cdot)$. The goal is to give a description of $\overline{\mathcal{E}}(\cdot,\cdot)$ and $\mathcal{D}(\overline{\mathcal{E}})$ in terms of $\{\overline{G}_\alpha\}$, similar to the one we have given using the semigroup. The main result (Theorem 1.3.1) will allow us to reconstruct a form from its resolvent. One may want to notice that the result of Theorem 1.3.1 has played an essential part in the study of singular perturbations of operators in Albeverio et al. [25], Chap. 6.

The operator $\overline{G}_\alpha$ is defined to be $(\overline{A}-\alpha)^{-1}$ whenever this exists. A formal calculation

$$\begin{aligned}\left(\overline{A}-\alpha\right)^{-1} &= \left(I-\left(I-\Delta t(\overline{A}-\alpha)\right)\right)^{-1}\Delta t\\ &= \sum_{k=0}^{\infty}\left(I-\Delta t(\overline{A}-\alpha)\right)^{k}\Delta t\\ &= \sum_{k=0}^{\infty}\left(\overline{Q}^{\Delta t}+\alpha\Delta t\right)^{k}\Delta t \qquad (1.3.1)\end{aligned}$$

tells us that $\overline{G}_\alpha$ will exist if the series on the right hand side converges. Since $\overline{Q}^{\Delta t}$ is a nonnegative and symmetric operator with norm at most one, all its eigenvalues must be between zero and one. We get that the absolute value of all eigenvalues of $\overline{Q}^{\Delta t}+\alpha\Delta t$ must be less than $1+\alpha\Delta t$. Hence if $\alpha<0$ and $|\alpha|\Delta t<2$, the series in (1.3.1) converges, and we get the following proposition.

Proposition 1.3.1. *Let $\mathcal{E}(\cdot,\cdot)$ be a nonnegative quadratic form. Then*

(i) $(\overline{A}-\alpha)^{-1}=\overline{G}_\alpha$ *exists for all* $\alpha\in{}^*(-\infty,0)$. *Moreover, we have* $||\overline{G}_\alpha||\le\frac{1}{|\alpha|}$ *in operator norm.*
(ii) For $\alpha\in{}^*\mathbb{R}$, $-\frac{1}{\Delta t}<\alpha<0$, *we have*

$$\overline{G}_\alpha=\sum_{k=0}^{\infty}\left(\overline{Q}^{\Delta t}+\alpha\Delta t\right)^{k}\Delta t. \qquad (1.3.2)$$

Proof. (i) We notice that for all $\alpha\in{}^*(-\infty,0)$, $u\in H$

$$\langle(\overline{A}-\alpha)u,u\rangle\ge-\alpha\langle u,u\rangle.$$

This implies $(\overline{A}-\alpha)^{-1}=\overline{G}_\alpha$ exists for all $\alpha\in{}^*(-\infty,0)$ by elementary linear algebra. Furthermore, we have

$$\langle\alpha\overline{G}_\alpha u,\alpha\overline{G}_\alpha u\rangle\le-\alpha\overline{\mathcal{E}}_{-\alpha}(\overline{G}_\alpha u,\overline{G}_\alpha u)=-\alpha\langle u,\overline{G}_\alpha u\rangle\le||u||||\alpha\overline{G}_\alpha u||.$$

This implies that $||\overline{G}_\alpha||\le\frac{1}{|\alpha|}$.

(ii) We have already proved this point before stating our proposition. □

We call $\{\overline{G}_\alpha\mid\alpha\in{}^*(-\infty,0)\}$ the *resolvent of* $\overline{\mathcal{E}}(\cdot,\cdot)$.

In the standard Dirichlet space theory, the formula corresponding to (1.3.2) is

$$\overline{G}_\alpha = \int_0^\infty e^{-t(\overline{A}-\alpha)}\, dt = \int_0^\infty e^{-t\overline{A}} e^{\alpha t}\, dt,$$

giving $\overline{G}_\alpha$ as a weighted sum of the elements $e^{-t\overline{A}}$ in the semigroup. Since $-\int_0^\infty \alpha e^{\alpha t} dt = 1$, it is convenient to multiply this equation by $-\alpha$ to obtain

$$-\alpha\overline{G}_\alpha = -\int_0^\infty \alpha e^{-t\overline{A}} e^{\alpha t}\, dt. \tag{1.3.3}$$

It is not quite obvious that this result carries over to the hyperfinite setting, since the equation $(\overline{Q}^{\Delta t} + \alpha\Delta t)^k = \overline{Q}^{k\Delta t}(1+\alpha\Delta t)^k$ (corresponding to $e^{-t(\overline{A}-\alpha)} = e^{-t\overline{A}} \cdot e^{\alpha t}$) is false. But the next result shows that the two operators are close enough for our purposes.

Lemma 1.3.1. *For* $\alpha \in {}^*\mathbb{R}$, $-\frac{1}{\sqrt{\Delta t}} \le \alpha < 0$, *and all* $u \in H$ *with* ${}^\circ\mathcal{E}_1(u,u) < \infty$, *we have*

$$\left| \alpha\overline{G}_\alpha u - \left(\alpha \sum_{k=0}^{\infty} \overline{Q}^{k\Delta t}\left(1+\alpha\Delta t\right)^k \Delta t\right) u \right|_1 \approx 0. \tag{1.3.4}$$

Proof. Let $\{e_i \mid 1 \le i \le N\}$ be an orthonormal basis of eigenvectors for $\overline{A}$, and let a_i be the i-th eigenvalue. Defining $b_i = a_i + 1$, we notice that if $u = \sum_{i=1}^N u_i e_i$, then

$$\overline{\mathcal{E}}_1(u,u) = \sum_{i=1}^{N} b_i u_i^2. \tag{1.3.5}$$

Summing geometric series, we see that

$$\begin{aligned} \alpha\overline{G}_\alpha(e_i) &= \left(\alpha \sum_{k=0}^{\infty} \left(1 - \Delta t a_i + \Delta t\alpha\right)^k \Delta t\right) e_i \\ &= \frac{\alpha}{a_i - \alpha} e_i \end{aligned}$$

and similarly

$$\left(\alpha \sum_{k=0}^{\infty} \overline{Q}^{k\Delta t}\left(1+\alpha\Delta t\right)^k \Delta t\right)(e_i) = \frac{\alpha}{a_i - \alpha + a_i\alpha\Delta t} e_i.$$

This yields

$$\alpha\overline{G}_\alpha(u) - \left(\alpha\sum_{k=0}^{\infty}\overline{Q}^{k\Delta t}\left(1+\alpha\Delta t\right)^k\Delta t\right)(u)$$
$$= \sum_{i=1}^{N}\frac{a_i\Delta t u_i e_i}{\left(1-(a_i/\alpha)\right)\left(1-(a_i/\alpha)-a_i\Delta t\right)}.$$

Taking the $|\cdot|_1$-norm of this, we get from (1.3.5)

$$\left|\alpha\overline{G}_\alpha u - \left(\alpha\sum_{k=0}^{\infty}\overline{Q}^{k\Delta t}\left(1+\alpha\Delta t\right)^k\Delta t\right)u\right|_1^2$$
$$= \sum_{i=1}^{N} b_i u_i^2 \frac{a_i^2\Delta t^2}{\left(1-(a_i/\alpha)\right)^2\left(1-(a_i/\alpha)-a_i\Delta t\right)^2}$$
$$\le \eta\mathcal{E}_1(u,u),$$

where

$$\eta = \max_{1\le i\le N}\left(\frac{a_i^2\Delta t^2}{\left(1-(a_i/\alpha)\right)^2\left(1-(a_i/\alpha)-a_i\Delta t\right)^2}\right).$$

All that remains is to show that η is infinitesimal. First we observe that since $||\overline{A}||\Delta t \le 1$ (recalling (1.1.3) in Sect. 1.1), we have $a_i\Delta t \le 1$ for all i. Hence

$$\eta \le \max_{1\le i\le N}\left\{\frac{a_i^2\Delta t^2}{\left(1-(a_i/\alpha)\right)^2(a_i/\alpha)^2}\right\}$$
$$= \max_{1\le i\le N}\left\{\frac{\alpha^4\Delta t^2}{\left(\alpha-a_i\right)^2}\right\}.$$

Since $-1/\sqrt{\Delta t} \le \alpha < 0$, the latter term is infinitesimal, and the proof is finished. □

Equation (1.3.4) is the nonstandard counterpart of (1.3.3). Notice that

$$\sum_{k=0}^{\infty}(-\alpha\Delta t)\left(1+\alpha\Delta t\right)^k = 1,$$

and that if α is infinite, then there is $t_\alpha \approx 0$ such that

$$\sum_{0\le k\Delta t\le t_\alpha} (-\alpha\Delta t)\left(1+\alpha\Delta t\right)^k \approx 1. \tag{1.3.6}$$

On the other hand, if α is finite, then

$$\sum_{\substack{t\le k\Delta t\\ 0\le k<\infty}} (-\alpha\Delta t)\left(1+\alpha\Delta t\right)^k \approx 1$$

for all infinitesimal t. We can now begin our description of $\overline{\mathcal{E}}(\cdot,\cdot)$ and $\mathcal{D}(\overline{\mathcal{E}})$ in terms of $\overline{G}_\alpha$.

Lemma 1.3.2. *If* $-\frac{1}{\sqrt{\Delta t}} < \alpha < 0$ *and* ${}^\circ\mathcal{E}(u,u) < \infty$, *then*

(i) If α *is infinite, we have* $\|-\alpha\overline{G}_\alpha u - u\| \approx 0$.
(ii) If α *is finite, we have* $-\alpha\overline{G}_\alpha u \in \mathcal{D}(\overline{\mathcal{E}})$.
(iii) There is an infinite α *such that* $-\alpha\overline{G}_\alpha u \in \mathcal{D}(\overline{\mathcal{E}})$.

Proof. According to Lemma 1.3.1, it suffices to prove the statements we get after replacing $\overline{G}_\alpha$ by

$$\overline{R}_\alpha = \sum_{k=0}^{\infty} \overline{Q}^{k\Delta t}\left(1+\alpha\Delta t\right)^k \Delta t.$$

(i) Let $t_\alpha \approx 0$ be as in (1.3.6). Then, we have

$$\begin{aligned}
\| -\alpha\overline{R}_\alpha u - u \| &= \left\| \sum_{k=0}^{\infty} \left(\overline{Q}^{k\Delta t} u - u\right)(-\alpha)\left(1+\alpha\Delta t\right)^k \Delta t \right\| \\
&\approx \left\| \sum_{0<k\Delta t\le t_\alpha} \left(\overline{Q}^{k\Delta t} u - u\right)(-\alpha)\left(1+\alpha\Delta t\right)^k \Delta t \right\| \\
&\approx 0,
\end{aligned}$$

where the last step uses Lemma 1.2.4.

(ii) Choose $t \approx 0$ such that $\overline{Q}^t u \in \mathcal{D}(\overline{\mathcal{E}})$. If $\{e_i\}_{i\le N}$ is an orthonormal basis of eigenvectors for $\overline{A}$, and a_i is the i-th eigenvalue, we have

$$\begin{aligned}
\sum_{0\le k\Delta t<t} \overline{Q}^{k\Delta t}\left(1+\alpha\Delta t\right)^k \Delta t e_i &= \sum_{0\le k\Delta t<t} \left(1-a_i\Delta t\right)^k\left(1+\alpha\Delta t\right)^k \Delta t e_i \\
&= \frac{1-\left(1+\alpha\Delta t\right)^{t/\Delta t}\left(1-a_i\Delta t\right)^{t/\Delta t}}{a_i-\alpha+a_i\alpha\Delta t} e_i.
\end{aligned}$$

If $u = \sum_{i=1}^N u_i e_i$, we get from (1.3.5)

$$\left| \sum_{0 \le k\Delta t < t} \overline{Q}^{k\Delta t} u(1+\alpha\Delta t)^k \Delta t \right|_1^2$$
$$= \sum_{i=1}^{N} b_i u_i^2 \left(\frac{1 - \left(1+\alpha\Delta t\right)^{t/\Delta t} \left(1 - a_i\Delta t\right)^{t/\Delta t}}{a_i - \alpha + a_i\alpha\Delta t} \right)^2,$$

where $b_i = a_i + 1$. Since α is finite, it is easy to check that

$$\frac{1 - \left(1+\alpha\Delta t\right)^{t/\Delta t} \left(1 - a_i\Delta t\right)^{t/\Delta t}}{a_i - \alpha + a_i\alpha\Delta t} \approx 0$$

for all i. Hence, we have

$$\left| \sum_{0 \le k\Delta t < t} \overline{Q}^{k\Delta t} u \left(1+\alpha\Delta t\right)^k \Delta t \right|_1 \approx 0. \tag{1.3.7}$$

We also remark that since $\overline{Q}^{k\Delta t} u \in \mathcal{D}(\overline{\mathcal{E}})$ for all $k\Delta t \ge t$, we must have

$$\sum_{t \le k\Delta t} \overline{Q}^{k\Delta t} u \left(1+\alpha\Delta t\right)^k \Delta t \in \mathcal{D}(\overline{\mathcal{E}}).$$

But by the relation (1.3.7), we have $|\overline{R}_\alpha u - \sum_{t \le k\Delta t} \overline{Q}^{k\Delta t} u \left(1+\alpha\Delta t\right)^k \Delta t|_1 \approx 0$. Hence, we have $\overline{R}_\alpha u \in \mathcal{D}(\overline{\mathcal{E}})$.

(iii) We remark that

$$\begin{aligned} \mathcal{E}_1(-\alpha\overline{G}_\alpha u, -\alpha\overline{G}_\alpha u) &= \sum_{i=1}^{N} b_i u_i^2 \frac{\alpha^2}{(a_i - \alpha)^2} \\ &= \sum_{i=1}^{N} b_i u_i^2 \frac{1}{(a_i/\alpha - 1)^2} \end{aligned} \tag{1.3.8}$$

increases as $\alpha \longrightarrow -\infty$, and is bounded by $\mathcal{E}_1(u,u)$. Applying Proposition 1.2.3 to the sequence $u_n = n\overline{G}_{-n}u$, the lemma follows. □

The next proposition adds two new characterizations of $\mathcal{D}(\overline{\mathcal{E}})$ to the list in Proposition 1.2.2

Proposition 1.3.2. *The following statements are equivalent:*

(i) $u \in \mathcal{D}(\overline{\mathcal{E}})$.
(ii) ${}^\circ\overline{\mathcal{E}}_1(u,u) < \infty$ *and* $\lim_{{}^\circ\alpha \longrightarrow -\infty} {}^\circ\overline{\mathcal{E}}_1(u + \alpha\overline{G}_\alpha u, u + \alpha\overline{G}_\alpha u) = 0$.

(iii) ${}^\circ\overline{\mathcal{E}}_1(u,u) = \lim_{{}^\circ\alpha\longrightarrow-\infty} {}^\circ\overline{\mathcal{E}}_1(-\alpha\overline{G}_\alpha u, -\alpha\overline{G}_\alpha u) < \infty$.

Proof. *(i)* $\Longrightarrow$ *(ii)*. Pick an infinite α such that $-\alpha\overline{G}_\alpha u \in \mathcal{D}(\overline{\mathcal{E}})$. Then $u + \alpha\overline{G}_\alpha u \in \mathcal{D}(\overline{\mathcal{E}}), u + \alpha\overline{G}_\alpha u \approx 0$, and hence $\overline{\mathcal{E}}_1(u+\alpha\overline{G}_\alpha u, u+\alpha\overline{G}_\alpha u) \approx 0$. Part (ii) follows.

(ii) $\Longrightarrow$ *(iii)*. By (1.3.8) and the triangle inequality

$$0 \leq |u|_1 - |-\alpha\overline{G}_\alpha u|_1 \leq |u+\alpha\overline{G}_\alpha u|_1,$$

and multiplying by $|u|_1 + |-\alpha\overline{G}_\alpha u|_1 \leq 2\mathcal{E}_1(u,u)$, we get

$$0 \leq |u|_1^2 - |-\alpha\overline{G}_\alpha u|_1^2 \leq 2\mathcal{E}_1(u,u)\cdot|u+\alpha\overline{G}_\alpha u|_1,$$

which shows that *(ii)* $\Longrightarrow$ *(iii)*.

(iii) $\Longrightarrow$ *(i)*. Pick an infinite α such that $-\alpha\overline{G}_\alpha u \in \mathcal{D}(\overline{\mathcal{E}})$. Then $\| u + \alpha\overline{G}_\alpha u \| \approx 0$ and $\mathcal{E}_1(u,u) \approx \mathcal{E}_1(-\alpha\overline{G}_\alpha u, -\alpha\overline{G}_\alpha u)$, and hence $u \in \mathcal{D}(\overline{\mathcal{E}})$. □

The following results gives a way of reconstructing a form from its resolvent. In the study of singular perturbations in Albeverio et al. [25], Chap. 6, we have found it much easier to control the resolvent of the perturbed form than the form itself. Once we have a good grasp of the resolvent, Theorem 1.3.1 will give us the form.

Theorem 1.3.1. *Let $\mathcal{E}(\cdot,\cdot)$ be a nonnegative hyperfinite form on H, and let $\overline{E}(\cdot,\cdot)$ be its symmetric standard part. For all $x \in {}^\circ H$ and all $v \in x$, we have*

$$\overline{E}(x,x) = -\lim_{{}^\circ\alpha\longrightarrow-\infty} {}^\circ\Big(\alpha^2\langle\overline{G}_\alpha v, v\rangle + \alpha\langle v,v\rangle\Big). \tag{1.3.9}$$

Proof. Notice that since $\overline{G}_\alpha$ is bounded, it does not matter which $v \in x$ we use. We split the proof into two cases.

(i) *x is not in the domain of $\overline{E}(\cdot,\cdot)$:* Let $\{e_i \mid i \leq N\}$ be an orthonormal basis of eigenvectors for $\overline{A}$, and assume that the corresponding eigenvalues $\{a_i\}_{i\leq N}$ are in decreasing order. Pick $v = \sum_{i\leq N} v_i e_i$ in x. An easy calculation shows that

$$-\Big(\langle\alpha^2\overline{G}_\alpha v, v\rangle + \alpha\langle v,v\rangle\Big) = \sum_{i=1}^N a_i v_i^2 \frac{-\alpha}{a_i-\alpha}.$$

Assume for contradiction that the limit in (1.3.9) is finite. Then there is an infinite α such that

$${}^\circ\Big(\sum_{i=1}^N a_i v_i^2 \frac{-\alpha}{a_i-\alpha}\Big) < \infty.$$

If H is the largest integer such that $a_H > |\alpha|$, we have

$$\begin{aligned}\sum_{i=1}^{N} a_i v_i^2 \frac{-\alpha}{a_i-\alpha} &= \sum_{i=1}^{H} a_i v_i^2 \frac{-\alpha}{a_i-\alpha} + \sum_{i=H+1}^{N} a_i v_i^2 \frac{-\alpha}{a_i-\alpha} \\ &= -\alpha \sum_{i=1}^{H} v_i^2 \frac{1}{1-\alpha/a_i} + \sum_{i=H+1}^{N} a_i v_i^2 \frac{1}{1-a_i/\alpha} \\ &\geq -\frac{\alpha}{2}\sum_{i=1}^{H} v_i^2 + \frac{1}{2}\sum_{i=H+1}^{N} a_i v_i^2.\end{aligned}$$

Hence, the last two terms $-\frac{\alpha}{2}\sum_{i=1}^{H} v_i^2$ and $\frac{1}{2}\sum_{i=H+1}^{N} a_i v_i^2$ are finite. But if $-(\alpha/2)\sum_{i=1}^{H} v_i^2$ is finite, v is infinitely close to

$$v' = \sum_{i=H+1}^{N} v_i e_i.$$

In addition, if $\frac{1}{2}\sum_{i=H+1}^{N} a_i v_i^2$ is finite, then ${}^\circ\mathcal{E}_1(v',v') < \infty$. This contradicts the assumption that $x \notin D(\overline{E})$.

(ii) *Assume that* $x \in D(\overline{E})$: Let $v \in x$ be such that

$$^\circ\overline{\mathcal{E}}_1(v,v) < \infty. \tag{1.3.10}$$

Since $\overline{\mathcal{E}}_{-\alpha}(\overline{G}_\alpha u, w) = \langle u, w\rangle$, we have

$$\begin{aligned}\overline{\mathcal{E}}(-\alpha\overline{G}_\alpha v, -\alpha\overline{G}_\alpha v) &= \overline{\mathcal{E}}_{-\alpha}(\alpha\overline{G}_\alpha v, \alpha\overline{G}_\alpha v) + \alpha\langle\alpha\overline{G}_\alpha v, \alpha\overline{G}_\alpha v\rangle \\ &= \alpha^2\langle\overline{G}_\alpha v, v\rangle + \alpha^3\langle\overline{G}_\alpha v, \overline{G}_\alpha v\rangle.\end{aligned}$$

The theorem will follow from Proposition 1.3.2 (iii) if we can prove that for all v satisfying the condition (1.3.10),

$$\overset{\circ}{\lim_{{}^\circ\alpha\to\infty}}\left(\alpha^2\langle\overline{G}_\alpha v, v\rangle + \alpha^3\langle\overline{G}_\alpha v, \overline{G}_\alpha v\rangle + \alpha^2\langle\overline{G}_\alpha v, v\rangle + \alpha\langle v, v\rangle\right) = 0.$$

By simple algebra, this is the same as

$$\overset{\circ}{\lim_{{}^\circ\alpha\to-\infty}}\left(\alpha\|\alpha\overline{G}_\alpha v + v\|^2\right) = 0.$$

Pulling α inside the norm and reformulating the problem in nonstandard terms, we see that what we have to prove is

$$\left|\left||\alpha|^{3/2}\overline{G}_\alpha v - |\alpha|^{\frac{1}{2}}v\right|\right|^2 \approx 0 \tag{1.3.11}$$

for all infinite, negative α of sufficiently small absolute value.

If $v = \sum_{i=1}^{N} v_i e_i$ is the eigenvector expansion of v, we see that

$$\begin{aligned}\left|\left||\alpha|^{3/2}\overline{G}_\alpha v - |\alpha|^{1/2}v\right|\right|^2 &= \sum_{i=1}^{N}\left(\frac{|\alpha|^{3/2}}{a_i-\alpha} - |\alpha|^{1/2}\right)^2 v_i^2 \\ &= \sum_{i=1}^{N}\frac{-\alpha a_i^2}{(a_i-\alpha)^2}v_i^2 \\ &= \sum_{i=1}^{N} a_i v_i^2\Big(\frac{1}{\beta_i+\beta_i^{-1}+2}\Big),\end{aligned} \tag{1.3.12}$$

where $\beta_i = -\alpha/a_i$.

Notice that if a_i is infinitesimal compared to α or α is infinitesimal compared to a_i, then $1/(\beta_i+\beta_i^{-1}+2)$ is infinitesimal. To get the sum on the right hand side of (1.3.12) to be infinitesimal, we only have to choose α such that the contributions from the terms satisfying neither of these requirements are infinitesimal.

Assuming that the eigenvalues $\{a_i\}$ are given in descending order, we define

$$\gamma = \sup\left\{ {}^\circ\Big(\sum_{i=1}^{k} a_i v_i^2\Big) \Big| a_k \text{ is infinite} \right\}.$$

Since ${}^\circ(\sum_{i=1}^{k} a_i v_i^2) \leq {}^\circ\mathcal{E}(v,v)$ is finite by (1.3.10), γ is a real number.

Using saturation[8] on the sets

$$A_n = \left\{ j \in {}^*\mathrm{N} \middle| \sum_{i=1}^{j} a_i v_i^2 > \gamma - \frac{1}{n} \quad\text{and}\quad a_j > n \right\},$$

we find a hyperinteger K such that a_K is infinite and

$$\sum_{i=1}^{K} a_i v_i^2 \approx \gamma.$$

[8] Saturation is also a term of nonstandard analysis, we refer to Albeverio et al. [25].

We choose $|\alpha|$ to be infinitely large, but infinitesimal compared to a_K.

For each $\varepsilon \in \mathbb{R}_+$, let

$$M_\varepsilon = \inf\left\{ k \,\middle|\, \sum_{i=1}^{k} a_i v_i^2 \geq \gamma + \varepsilon \right\}.$$

By our choice of γ, the term a_{M_ε} must be finite. But

$$\sum_{i=1}^{n} a_i v_i^2 \frac{1}{\beta_i + \beta_i^{-1} + 2} \leq \sum_{i=1}^{K} a_i v_i^2 \frac{1}{\beta_i + \beta_i^{-1} + 2} + \sum_{i=K+1}^{M_\varepsilon - 1} a_i v_i^2 + \sum_{i=M_\varepsilon}^{N} a_i v_i^2 \frac{1}{\beta_i + \beta_i^{-1} + 2},$$

where the first term is infinitesimal since each β_i is; the second term is less than 2ε by our choice of M_ε; and the last term is infinitesimal since each β_i is infinite. Since $\varepsilon \in \mathbb{R}_+$ is arbitrary, the sum on the left must be infinitesimal. This proves the approximation (1.3.11). Hence, the theorem is also proved. □

1.4 Weak Coercive Quadratic Forms

In Sect. 1.3, we have discussed the resolvent of the symmetric part $\overline{\mathcal{E}}(\cdot,\cdot)$ of $\mathcal{E}(\cdot,\cdot)$. We have heavily depended on the eigenvectors and eigenvalues of the generator $\overline{A}$. In this section, we shall study the resolvent of $\mathcal{E}(\cdot,\cdot)$ directly. However, the generator A is not symmetric. This forces us to find an alternative way for the discussion.

Let $\mathcal{E}(\cdot,\cdot)$ be a nonnegative quadratic form on a hyperfinite dimensional linear space H. Let A and $\hat{A}$ be the infinitesimal generator and co-generator of $\mathcal{E}(\cdot,\cdot)$, respectively. For $\alpha \in {}^*(-\infty, 0), u \in H$, we have

$$\langle (A - \alpha)u, u \rangle \geq -\alpha \langle u, u \rangle \text{ and } \langle (\hat{A} - \alpha)u, u \rangle \geq -\alpha \langle u, u \rangle. \qquad (1.4.1)$$

Hence, $(A-\alpha)^{-1} = G_\alpha$ and $(\hat{A}-\alpha)^{-1} = \hat{G}_\alpha$ exist for all $\alpha \in {}^*(-\infty, 0)$ by elementary linear algebra. Moreover, we have

Proposition 1.4.1. *Let $\mathcal{E}(\cdot,\cdot)$ be a nonnegative quadratic form on a hyperfinite dimensional linear space H. Then*

(i) $(A-\alpha)^{-1} = G_\alpha$ and $(\hat{A}-\alpha)^{-1} = \hat{G}_\alpha$ exist for all $\alpha \in {}^*(-\infty, 0)$. Moreover, we have $||G_\alpha|| \leq \frac{1}{|\alpha|}$ and $||\hat{G}_\alpha|| \leq \frac{1}{|\alpha|}$.

(ii) For all $\alpha, \beta \in {}^*(-\infty, 0)$, *we have*

$$\begin{aligned} G_\alpha - G_\beta &= (\alpha - \beta) G_\alpha G_\beta \\ &= (\alpha - \beta) G_\beta G_\alpha \end{aligned} \tag{1.4.2}$$

and

$$\begin{aligned} \hat{G}_\alpha - \hat{G}_\beta &= (\alpha - \beta) \hat{G}_\alpha \hat{G}_\beta \\ &= (\alpha - \beta) \hat{G}_\beta \hat{G}_\alpha . \end{aligned} \tag{1.4.3}$$

Proof. (i) From (1.4.1), we know $(A - \alpha)^{-1} = G_\alpha$ and $(\hat{A} - \alpha)^{-1} = \hat{G}_\alpha$ exist for all $\alpha \in {}^*(-\infty, 0)$. Besides, we have

$$\begin{aligned} \langle \alpha G_\alpha u, \alpha G_\alpha u \rangle &\leq -\alpha \mathcal{E}_{-\alpha}(G_\alpha u, G_\alpha u) \\ &= -\alpha \langle u, G_\alpha u \rangle \\ &\leq ||u|| \, ||\alpha G_\alpha u||. \end{aligned}$$

This implies that $||G_\alpha|| \leq \frac{1}{|\alpha|}$. Similarly, we have $||\hat{G}_\alpha|| \leq \frac{1}{|\alpha|}$.

(ii) We notice that

$$\alpha - \beta = (A - \beta) - (A - \alpha).$$

Hence, we have

$$(\alpha - \beta)(A - \alpha)^{-1} = (A - \alpha)^{-1}(A - \beta) - I.$$

This implies

$$(\alpha - \beta)(A - \alpha)^{-1}(A - \beta)^{-1} = (A - \alpha)^{-1} - (A - \beta)^{-1}.$$

Therefore, we have

$$(\alpha - \beta) G_\alpha G_\beta = G_\alpha - G_\beta .$$

Similarly, we can show that

$$(\alpha - \beta) G_\beta G_\alpha = G_\alpha - G_\beta .$$

This proves the relation (1.4.2). In the same manner, we can prove the relation (1.4.3). □

Thereafter, we shall call $\{G_\alpha \mid \alpha < 0\}$ the *resolvent of* $\mathcal{E}(\cdot, \cdot)$, and $\{\hat{G}_\alpha \mid \alpha < 0\}$ the *co-resolvent of* $\mathcal{E}(\cdot, \cdot)$. The relation (1.4.2) will be called the *first resolvent equation*, and the relation (1.4.3) will be called the *first co-resolvent equation.*

For $\alpha < 0$, we define

$$\begin{aligned} {}^{(\alpha)}\mathcal{E}(u,v) &= -\alpha\langle u + \alpha G_\alpha u, v\rangle \\ &= -\alpha\langle v + \alpha\hat{G}_\alpha v, u\rangle, u, v \in H, \end{aligned}$$

and

$$\begin{aligned} {}^{(\alpha)}\hat{\mathcal{E}}(u,v) &= -\alpha\langle v + \alpha G_\alpha v, u\rangle \\ &= -\alpha\langle u + \alpha\hat{G}_\alpha u, v\rangle, u, v \in H. \end{aligned}$$

Then

$$\begin{aligned} {}^{(\alpha)}\mathcal{E}(u, -\alpha G_\alpha u) &= -\alpha\langle u + \alpha G_\alpha u, -\alpha G_\alpha u\rangle \\ &= {}^{(\alpha)}\mathcal{E}(u,u) + \alpha\langle u + \alpha G_\alpha u, u + \alpha G_\alpha u\rangle, u \in H, \end{aligned}$$

and

$$\begin{aligned} {}^{(\alpha)}\mathcal{E}(-\alpha\hat{G}_\alpha u, u) &= -\alpha\langle u + \alpha\hat{G}_\alpha u, -\alpha\hat{G}_\alpha u\rangle \\ &= {}^{(\alpha)}\mathcal{E}(u,u) + \alpha\langle u + \alpha\hat{G}_\alpha u, u + \alpha\hat{G}_\alpha u\rangle, u \in H. \end{aligned}$$

Therefore, we get

$${}^{(\alpha)}\mathcal{E}(u, -\alpha G_\alpha u) \le {}^{(\alpha)}\mathcal{E}(u,u), u \in H, \tag{1.4.4}$$

and

$${}^{(\alpha)}\mathcal{E}(-\alpha\hat{G}_\alpha u, u) \le {}^{(\alpha)}\mathcal{E}(u,u), u \in H.$$

Actually, we have

Lemma 1.4.1. *Let $\mathcal{E}(\cdot,\cdot)$ be a nonnegative quadratic form on a hyperfinite dimensional linear space H. Then*

(i) ${}^{(\alpha)}\mathcal{E}(u,v) = \mathcal{E}(-\alpha G_\alpha u, v)$ *and* ${}^{(\alpha)}\hat{\mathcal{E}}(u,v) = \mathcal{E}(-\alpha\hat{G}_\alpha u, v)$ *for all* $u, v \in H$.
(ii) $\mathcal{E}(-\alpha G_\alpha u, -\alpha G_\alpha u) \le {}^{(\alpha)}\mathcal{E}(u,u)$ *and* $\mathcal{E}(-\alpha\hat{G}_\alpha u, -\alpha\hat{G}_\alpha u) \le {}^{(\alpha)}\mathcal{E}(u,u)$ *for all* $u \in H$.

Proof. (i) We have

$$\begin{aligned} {}^{(\alpha)}\mathcal{E}(u,v) &= -\alpha\langle u, v\rangle - \alpha^2\langle G_\alpha u, v\rangle \\ &= -\alpha\mathcal{E}_{-\alpha}(G_\alpha u, v) - \alpha^2\langle G_\alpha u, v\rangle \\ &= \mathcal{E}(-\alpha G_\alpha u, v). \end{aligned}$$

(ii) It follows from (i) and (1.4.4) that

$$\begin{aligned}\mathcal{E}(-\alpha G_\alpha u, -\alpha G_\alpha u) &= {}^{(\alpha)}\mathcal{E}(u, -\alpha G_\alpha u)\\ &\le {}^{(\alpha)}\mathcal{E}(u,u).\end{aligned}$$

Notice that

$$\mathcal{E}(-\alpha G_\alpha u, -\alpha G_\alpha u) = \mathcal{E}(-\alpha \hat{G}_\alpha u, -\alpha \hat{G}_\alpha u).$$

This implies (ii). □

Definition 1.4.1. Let $\mathcal{E}(\cdot,\cdot)$ be a nonnegative quadratic form on a hyperfinite dimensional linear space H. We call $\mathcal{E}(\cdot,\cdot)$ a *hyperfinite weak coercive quadratic form* if and only if there exists a constant $C \in {}^*\mathbb{R}_+$ with $0 \le {}^\circ C < \infty$ such that

$$|\mathcal{E}_1(u,v)| \le C\sqrt{\mathcal{E}_1(u,u)}\sqrt{\mathcal{E}_1(v,v)}, \forall u, v \in H. \tag{1.4.5}$$

We call (1.4.5) the *hyperfinite weak sector condition* and C a *continuity constant.*

Lemma 1.4.2. *Let $\mathcal{E}(\cdot,\cdot)$ be a nonnegative quadratic form on a hyperfinite dimensional linear space H. Then the following statements are equivalent:*

(i) $\mathcal{E}(\cdot,\cdot)$ satisfies the hyperfinite weak sector condition.
(ii) For $\beta \in {}^*(0,\infty), \infty > {}^\circ\beta > 0$, there exists $C_\beta \in {}^*\mathbb{R}_+$ with $0 \le {}^\circ C_\beta < \infty$ such that

$$|\mathcal{E}_\beta(u,v)| \le C_\beta\sqrt{\mathcal{E}_\beta(u,u)}\sqrt{\mathcal{E}_\beta(v,v)}, \forall u, v \in H.$$

(iii) For $\beta \in {}^*(0,\infty), \infty > {}^\circ\beta > 0$, there exists $C'_\beta \in {}^*\mathbb{R}_+$ with $0 \le {}^\circ C'_\beta < \infty$ such that

$$|\mathcal{E}(u,v)| \le C'_\beta\sqrt{\mathcal{E}_\beta(u,u)}\sqrt{\mathcal{E}_\beta(v,v)}, \forall u, v \in H.$$

Proof. The proof is easy and is left as an exercise. □

Lemma 1.4.3. *Let $\mathcal{E}(\cdot,\cdot)$ be a hyperfinite weak coercive quadratic form on a hyperfinite dimensional linear space H. Then for all $\alpha \in {}^*(-\infty, 0)$*

(i) $|{}^{(\alpha)}\mathcal{E}_1(u,v)| \le (C'_1+1)\sqrt{\mathcal{E}_1(u,u)}\sqrt{{}^{(\alpha)}\mathcal{E}_1(v,v)}$ for all $u, v \in H$.
(ii) $\mathcal{E}_1(-\alpha G_\alpha u, -\alpha G_\alpha u) \le (C'_1+1)^2\mathcal{E}_1(u,u)$ for all $u \in H$.

Proof. (i) By using Lemma 1.4.1 and Lemma 1.4.2, we have

$$\begin{aligned}\left|{}^{(\alpha)}\mathcal{E}(u,v)\right| &= \left|\mathcal{E}(u, -\alpha\hat{G}_\alpha v)\right|\\ &\le C'_1\sqrt{\mathcal{E}_1(u,u)}\sqrt{\mathcal{E}_1(-\alpha\hat{G}_\alpha v, -\alpha\hat{G}_\alpha v)}\\ &\le C'_1\sqrt{\mathcal{E}_1(u,u)}\sqrt{{}^{(\alpha)}\mathcal{E}(v,v)}.\end{aligned}$$

(ii) From (i) and Lemma 1.4.1 (ii), we have

$$\begin{aligned}\mathcal{E}_1(-\alpha G_\alpha u, -\alpha G_\alpha u) &\le {}^{(\alpha)}\mathcal{E}(u,u)\\ &\le (C_1'+1)^2\mathcal{E}_1(u,u).\end{aligned}$$

□

Lemma 1.4.4. *Let $\mathcal{E}(\cdot,\cdot)$ be a hyperfinite weak coercive quadratic form on a hyperfinite dimensional linear space H. If $^\circ\mathcal{E}_1(u,u) < \infty$, then for all infinite $\alpha < 0$, we have $||u + \alpha G_\alpha u|| \approx 0$.*

Proof. From Lemma 1.4.1 (i) and Lemma 1.4.3 (i), we have

$$\begin{aligned}||u+\alpha G_\alpha u||^2 &= \langle u+\alpha G_\alpha u, u+\alpha G_\alpha u\rangle\\ &= -\frac{{}^{(\alpha)}\mathcal{E}(u, u+\alpha G_\alpha u)}{\alpha}\\ &= -\frac{1}{\alpha}\left[{}^{(\alpha)}\mathcal{E}(u,u) + {}^{(\alpha)}\mathcal{E}(u,\alpha G_\alpha u)\right]\\ &= -\frac{1}{\alpha}\left[{}^{(\alpha)}\mathcal{E}(u,u) - \mathcal{E}(\alpha G_\alpha u,\alpha G_\alpha u)\right]\\ &\le -\frac{{}^{(\alpha)}\mathcal{E}(u,u)}{\alpha}\\ &\le -\frac{(C_1'+1)^2}{\alpha}\mathcal{E}_1(u,u)\\ &\approx 0.\end{aligned}$$

□

It is the time to introduce the definition of the domain $\mathcal{D}(\mathcal{E})$ of $\mathcal{E}(\cdot,\cdot)$.

Definition 1.4.2. Let $\mathcal{E}(\cdot,\cdot)$ be a nonnegative quadratic form on a hyperfinite dimensional linear space H. The domain $\mathcal{D}(\mathcal{E})$ of $\mathcal{E}(\cdot,\cdot)$ is the set of all $u \in H$ satisfying

(i) $^\circ\mathcal{E}_1(u,u) < \infty$.
(ii) For all infinite $\alpha < 0$, $\mathcal{E}(u+\alpha G_\alpha u, u+\alpha G_\alpha u) \approx 0$ and $\mathcal{E}(u+\alpha\hat{G}_\alpha u, u+\alpha\hat{G}_\alpha u) \approx 0$.

Proposition 1.4.2. *Let $\mathcal{E}(\cdot,\cdot)$ be a hyperfinite weak coercive quadratic form on a hyperfinite dimensional linear space H. Then the following statements are equivalent:*

(i) $u \in \mathcal{D}(\mathcal{E})$.
(ii) $^\circ\mathcal{E}_1(u,u) < \infty$, *and for all infinite* $\alpha < 0$, $\mathcal{E}(u+\alpha G_\alpha u, u) \approx 0$ *and* $\mathcal{E}(u+\alpha\hat{G}_\alpha u, u) \approx 0$.
(iii) $^\circ\mathcal{E}_1(u,u) < \infty$, *and for all infinite* $\alpha < 0$, ${}^{(\alpha)}\mathcal{E}(u,u) \approx \mathcal{E}(u,u)$ *and* ${}^{(\alpha)}\hat{\mathcal{E}}(u,u) \approx \mathcal{E}(u,u)$.

Proof. $(i) \Longrightarrow (ii)$. This is easily seen from Lemma 1.4.2 (iii) and Lemma 1.4.4.

$(ii) \Longrightarrow (i)$. From Lemma 1.4.1, we have

$$\begin{aligned} 0 &\leq \mathcal{E}(u + \alpha G_\alpha u, u + \alpha G_\alpha u) \\ &= \mathcal{E}(u, u + \alpha G_\alpha u) + \mathcal{E}(\alpha G_\alpha u, u + \alpha G_\alpha u) \\ &= \mathcal{E}(u, u + \alpha G_\alpha u) + \mathcal{E}(-\alpha G_\alpha u, -\alpha G_\alpha u) - {}^{(\alpha)}\mathcal{E}(u, u) \\ &\leq \mathcal{E}(u, u + \alpha G_\alpha u) \\ &= \mathcal{E}(u + \alpha \hat{G}_\alpha u, u) \\ &\approx 0. \end{aligned}$$

Similarly, we can show

$$\mathcal{E}(u + \alpha \hat{G}_\alpha u, u + \alpha \hat{G}_\alpha u) \approx 0.$$

$(ii) \Longleftrightarrow (iii)$. This is easily seen from Lemma 1.4.1. □

Lemma 1.4.5. *Let $\mathcal{E}(\cdot, \cdot)$ be a hyperfinite weak coercive quadratic form on a hyperfinite dimensional linear space H. If ${}^\circ\mathcal{E}(u, u) < \infty$, then for all finite $\beta < 0, {}^\circ\beta \neq 0$, $G_\beta u \in \mathcal{D}(\mathcal{E})$.*

Proof. We have from Lemma 1.4.3 (ii) that

$$\mathcal{E}_1(G_\beta u, G_\beta u) \leq {}^\circ\left[\frac{(C_1' + 1)^2}{\beta^2}\mathcal{E}_1(u, u)\right] < \infty.$$

Hence, it suffices to show that for all infinite $\alpha < 0$ by Proposition 1.4.2 (ii)

$$\mathcal{E}(G_\beta u + \alpha G_\alpha G_\beta u, G_\beta u) \approx 0 \tag{1.4.6}$$

and

$$\mathcal{E}(G_\beta u + \alpha \hat{G}_\alpha G_\beta u, G_\beta u) \approx 0. \tag{1.4.7}$$

Actually, we have

$$\begin{aligned} &\mathcal{E}(G_\beta u + \alpha G_\alpha G_\beta u, G_\beta u) \\ &\quad = \langle u + \alpha G_\alpha u, G_\beta u \rangle + \beta \langle G_\beta u + \alpha G_\alpha G_\beta u, G_\beta u \rangle. \end{aligned} \tag{1.4.8}$$

From Lemma 1.4.4, we see that

$$\langle u + \alpha G_\alpha u, G_\beta u \rangle \approx 0. \tag{1.4.9}$$

From the relation (1.4.2), we have

$$\begin{aligned}\beta\langle G_\beta u+\alpha G_\alpha G_\beta u, G_\beta u\rangle &= \beta\langle G_\beta u+\frac{\alpha}{\alpha-\beta}(G_\alpha-G_\beta)u, G_\beta u\rangle \\ &= \beta\langle (1+\frac{-\alpha}{\alpha-\beta})G_\beta u+\frac{\alpha}{\alpha-\beta}G_\alpha u, G_\beta u\rangle \\ &= \frac{-\beta^2}{\alpha-\beta}\langle G_\beta u, G_\beta u\rangle+\frac{\alpha\beta}{\alpha-\beta}\langle G_\alpha u, G_\beta u\rangle \\ &\approx 0. \end{aligned} \tag{1.4.10}$$

By the relations (1.4.8), (1.4.9), and (1.4.10), we have proved the relation (1.4.6). Similarly, we can prove the relation (1.4.7). □

We recall that the norm $|\cdot|_1$ is defined by $|u|_1 = \sqrt{\mathcal{E}_1(u,u)}$. A subset F of H is called *$\mathcal{E}$-closed* if and only if for all sequences $\{u_n\}_{n\in\mathbb{N}}$ of elements from F such that $^\circ|u_n - u_m|_1 \longrightarrow 0$ as $n, m \longrightarrow \infty$, there exists an element u in F such that $^\circ|u_n - u|_1 \longrightarrow 0$ as $n \longrightarrow \infty$. One may want to notice that a subset F of H is $\mathcal{E}$-closed if and only if it is $\overline{\mathcal{E}}$-closed.

As we just mention above, the reader may already note that the definitions of $\overline{\mathcal{E}}$-closedness and $\mathcal{E}$-closedness are formally identical. However, in the definition of $\overline{\mathcal{E}}$-closedness, the norm $|\cdot|_1$ is defined via $|u|_1 = \sqrt{\overline{\mathcal{E}}(u,u)}$, whereas in the definition of $\mathcal{E}$-closedness, it is defined via $|u|_1 = \sqrt{\mathcal{E}(u,u)}$. It will always be clear from the context which norm we actually mean by $|\cdot|_1$.

Proposition 1.4.3. *Let $\mathcal{E}(\cdot,\cdot)$ be a hyperfinite weak coercive quadratic form on a hyperfinite dimensional linear space H. Then $\mathcal{D}(\mathcal{E})$ is $\mathcal{E}$-closed. Moreover, if $\{u_n\}_{n\in\mathbb{N}}$ is a $|\cdot|_1$ Cauchy sequence from $\mathcal{D}(\mathcal{E})$, and $\{u_n \mid n \in {}^*\mathbb{N}\}$ is an internal extension of $\{u_n\}_{n\in\mathbb{N}}$, then there is a $\gamma \in {}^*\mathbb{N} - \mathbb{N}$ such that $u_\eta \in \mathcal{D}(\mathcal{E})$ for all $\eta \le \gamma$.*

Proof. Let $\{u_n \mid n \in \mathbb{N}\}$ be a $|\cdot|_1$ Cauchy sequence from $\mathcal{D}(\mathcal{E})$, and let $\{u_n \mid n \in {}^*\mathbb{N}\}$ be an internal extension of it. There is an element $\gamma \in {}^*\mathbb{N} - \mathbb{N}$ such that $|u_n - u_m|_1 \approx 0$ whenever n and m are infinite and less than γ. Let $\eta \in {}^*\mathbb{N} - \mathbb{N}, \eta \le \gamma$. By the choice of γ, $^\circ\mathcal{E}_1(u_\eta, u_\eta) < \infty$ and $^\circ|u_n - u_\eta|_1 \longrightarrow 0$ as n approaches infinity in $\mathbb{N}$. All that remains is to prove that $u_\eta \in \mathcal{D}(\mathcal{E})$.

Assuming not, there is an $\varepsilon \in \mathbb{R}_+$ and infinite $\beta < 0$ such that

$$|u_\eta + \beta G_\beta u_\eta|_1 > \varepsilon \tag{1.4.11}$$

or

$$|u_\eta - \beta \hat{G}_\beta u_\eta|_1 > \varepsilon. \tag{1.4.12}$$

(1) Assume that the relation (1.4.11) holds. Choose $m \in \mathbb{N}$ so large that

$$|u_\eta - u_m|_1 < \frac{\varepsilon}{4(C_1'+1)^2}.$$

Then by Lemma 1.4.3 (ii), we have

$$|\beta G_\beta u_\eta - \beta G_\beta u_m|_1 < \frac{\varepsilon}{4}.$$

Combining the inequalities above, we get

$$\begin{aligned}\varepsilon &< |u_\eta + \beta G_\beta u_\eta|_1 \\ &\le |u_\eta - u_m|_1 + |u_m + \beta G_\beta u_m|_1 + |\beta G_\beta u_m - \beta G_\beta u_\eta|_1 \\ &\le \frac{\varepsilon}{2} + |u_m + \beta G_\beta u_m|_1.\end{aligned}$$

However, the last term $|u_m + \beta G_\beta u_m|_1$ is infinitesimal by Lemma 1.4.4 since $u_m \in \mathcal{D}(\mathcal{E})$. We have the contradiction we wanted.

(2) If the relation (1.4.12) holds, then we can get a corresponding contradiction. □

As in Sect. 1.2, we may now sum up our results on $\mathcal{D}(\mathcal{E})$ in one statement.

Theorem 1.4.1. *Let $\mathcal{E}(\cdot,\cdot)$ be a hyperfinite weak coercive quadratic form on a hyperfinite dimensional space H. Then*

(i) If $u, v \in \mathcal{D}(\mathcal{E})$ and α is a finite element of ${}^*\mathbb{R}$, then $\alpha u, u + v \in \mathcal{D}(\mathcal{E})$.
(ii) $\mathcal{D}(\mathcal{E})$ is $\mathcal{E}$-closed.
(iii) If ${}^\circ\mathcal{E}_1(u,u) < \infty$, then there exists a $v \in \mathcal{D}(\mathcal{E})$ with $||u - v|| \approx 0$. Moreover, we have

$$\begin{aligned}{}^\circ\mathcal{E}(v,v) &= \lim_{\substack{{}^\circ\alpha\downarrow-\infty \\ {}^\circ\alpha\neq-\infty}} {}^\circ\mathcal{E}(\alpha G_\alpha u, \alpha G_\alpha u) \\ &= \lim_{\substack{{}^\circ\alpha\downarrow-\infty \\ {}^\circ\alpha\neq-\infty}} {}^\circ[{}^{(\alpha)}\mathcal{E}(u,u)].\end{aligned}$$

(iv) If $u, v \in \mathcal{D}(\mathcal{E})$ and $u \approx v$, then $\mathcal{E}(u,u) \approx \mathcal{E}(v,v)$.
(v) If $u \in \mathcal{D}(\mathcal{E})$, then ${}^\circ\mathcal{E}(u,u) = \inf\{{}^\circ\mathcal{E}(v,v) \mid v \approx u\}$.
(vi) If ${}^\circ\mathcal{E}(u,u) = \inf\{{}^\circ\mathcal{E}(v,v) \mid v \approx u\} < \infty$, then $u \in \mathcal{D}(\mathcal{E})$.

Proof. (i) We can prove this in the same way as the proof of Corollary 1.2.1.

(ii) It is proved in Proposition 1.4.3.

(iii) Since ${}^\circ\mathcal{E}_1(u,u) < \infty$, we have from Lemma 1.4.3 (ii) that for all $\alpha \in {}^*(-\infty, 0)$

$${}^\circ[\mathcal{E}_1(-\alpha G_\alpha u, -\alpha G_\alpha u)] \le {}^\circ\left[(C_1' + 1)^2 \mathcal{E}_1(u,u)\right] < \infty.$$

Let $\{-\alpha_n G_{\alpha_n}\}$ be a $|\cdot|_1$ Cauchy sequence, ${}^\circ\alpha \downarrow -\infty$. Let $\{-\alpha_n G_{\alpha_n} \mid n \in {}^*\mathbb{N}\}$ be an internal extension of it. There exists an infinite $\eta \in {}^*\mathbb{N}$ such that $-\alpha_n G_{\alpha_n} u \in \mathcal{D}(\mathcal{E}), n \leq \eta$. By Lemma 1.4.4, we have $u \approx -\alpha_\eta G_{\alpha_\eta} u$. Hence by letting $v = -\alpha_\eta G_{\alpha_\eta} u$, we have gotten (iii).

(iv) It is obviously enough to show that if $u \in \mathcal{D}(\mathcal{E})$ and $u \approx 0$, then $\mathcal{E}(u,u) \approx 0$. But if $u \in \mathcal{D}(\mathcal{E})$, we know from (iii):

$$^\circ\mathcal{E}(u,u) = \lim_{\substack{^\circ\alpha \downarrow -\infty \\ ^\circ\alpha \neq -\infty}} {}^{(\alpha)}[^\circ\mathcal{E}(u,u)]. \tag{1.4.13}$$

Also

$$\begin{aligned} {}^{(\alpha)}\mathcal{E}(u,u) &= -\alpha\langle u + \alpha G_\alpha u, u\rangle \\ &= -\alpha\langle u,u\rangle - \alpha^2\langle G_\alpha u, u\rangle \\ &\leq -\alpha||u||^2, \end{aligned}$$

which is infinitesimal for $\alpha \not\approx -\infty$. Combining this with (1.4.13), (iv) follows.

(v) This follows in the same way as for the proof of Theorem 1.2.1 (v).

(vi) For simplicity, we assume that $u \approx 0$. Then $\mathcal{E}(u,u) \approx 0$. From Lemma 1.4.3 (i), we have for all $\alpha \in {}^*(-\infty, 0)$

$$\begin{aligned} 0 &\leq |{}^{(\alpha)}\mathcal{E}_1(u,u)| \\ &\leq (C_1' + 1)^2 \mathcal{E}_1(u,u) \\ &\approx 0. \end{aligned}$$

Hence, ${}^{(\alpha)}\mathcal{E}_1(u,u) \approx 0, \forall \alpha \in {}^*(-\infty,0)$. Similarly, we can show that ${}^{(\alpha)}\hat{\mathcal{E}}_1(u,u) \approx 0$. By Proposition 1.4.2, we have $u \in \mathcal{D}(\mathcal{E})$. □

The following definition now makes sense.

Definition 1.4.3. Let $\mathcal{E}(\cdot,\cdot)$ be a hyperfinite weak coercive quadratic form on a hyperfinite dimensional space H. The *standard part of* $\mathcal{E}(\cdot,\cdot)$ is the quadratic form $E(\cdot,\cdot)$ on $^\circ H$ defined by:

(i) The domain $D(E)$ of $E(\cdot,\cdot)$ is the set of all equivalence classes $^\circ u \in {}^\circ H$ such that $\inf\{^\circ\mathcal{E}_1(v,v) \mid v \in {}^\circ u\} < \infty$.
(ii) If $x, y \in {}^\circ H$ are in the domain of $E(\cdot,\cdot)$, let $E(x,y) = {}^\circ\mathcal{E}(u,v)$, where $u \in x, v \in y$ are in $\mathcal{D}(\mathcal{E})$.

For $\alpha \in [0,\infty)$, let us set

$$E_\alpha(\cdot,\cdot) = E(\cdot,\cdot) + \alpha(\cdot,\cdot).$$

We recall that $(\cdot,\cdot)$ is the inner product of $^\circ H$.

An E_1-*Cauchy sequence* is a sequence $\{x_n\}$ of elements from $D(E)$ such that $E_1(u_n - u_m, u_n - u_m) \longrightarrow 0$ as $n, m \longrightarrow \infty$. We say that $E(\cdot,\cdot)$ is *closed* if all E_1-Cauchy sequences converge in E_1-norm to an element in $D(E)$. The next proposition follows immediately from Theorem 1.4.1 and the definition of $E(\cdot,\cdot)$.

Proposition 1.4.4. *Let $\mathcal{E}(\cdot,\cdot)$ be a hyperfinite weak coercive quadratic form on a hyperfinite dimensional space H. Let $E(\cdot,\cdot)$ be the standard part of $\mathcal{E}(\cdot,\cdot)$. Then $E(\cdot,\cdot)$ is closed, and for all $x \in {}^\circ H$*

$$E(x,x) = \inf\{{}^\circ\mathcal{E}(u,u) \mid u \in x\}, \tag{1.4.14}$$

where we take the value ∞ on the right to mean that the expression on the left is undefined.

Proof. If ${}^\circ\mathcal{E}_1(u,u) = \infty$ for all $u \in x$, it is easy to see (1.4.14) holds. Assume that $\inf\{{}^\circ\mathcal{E}(u,u) \mid u \in x\} < \infty$, then $E(x,x) = {}^\circ\mathcal{E}(v,v)$ for some $v \in \mathcal{D}(\mathcal{E}), v \in x$. Hence, we only need to show ${}^\circ\mathcal{E}(v,v) = \inf\{{}^\circ\mathcal{E}(u,u) \mid u \in x\}$. But this is implied by Theorem 1.4.1 (iv). □

In Sect. 1.2, we have gotten the symmetric standard part $\overline{E}(\cdot,\cdot)$ of $\overline{\mathcal{E}}(\cdot,\cdot)$. We have studied the domain $\mathcal{D}(\overline{\mathcal{E}})$ of $\overline{\mathcal{E}}(\cdot,\cdot)$ also. Now it is very natural to discuss the relation between the results of Sect. 1.2 and those of this section. Actually, we have

Theorem 1.4.2. *Let $\mathcal{E}(\cdot,\cdot)$ be a hyperfinite weak coercive quadratic form on a hyperfinite dimensional space H. Then $\mathcal{D}(\mathcal{E}) = \mathcal{D}(\overline{\mathcal{E}})$ and $D(E) = D(\overline{E})$. Moreover, we have*

$$\overline{E}(x,y) = \frac{1}{2}\Big(E(x,y) + E(y,x)\Big). \tag{1.4.15}$$

Proof. From Theorem 1.4.1 (v) and (vi), we know that $v \in \mathcal{D}(\mathcal{E})$ if and only if ${}^\circ\mathcal{E}(v,v) = \inf\{{}^\circ\mathcal{E}(u,u) \mid v \approx u\}$ and ${}^\circ\mathcal{E}(v,v) < \infty$. This statement is also true for the elements in $\mathcal{D}(\overline{\mathcal{E}})$ from Theorem 1.2.1 (v) and (vi). Hence, $\mathcal{D}(\mathcal{E}) = \mathcal{D}(\overline{\mathcal{E}})$. The relation (1.4.15) follows immediately. □

Remark 1.4.1. Let $\mathcal{E}(\cdot,\cdot)$ be a hyperfinite weak coercive quadratic form on a hyperfinite dimensional space H. It is easy to see that $(E(\cdot,\cdot), D(E))$ satisfies the following *weak sector condition*

$$|E_1(x,y)| \le C\sqrt{E_1(x,x)}\sqrt{E_1(y,y)} \text{ for all } x, y \in D(E),$$

where $C \in \mathbb{R}_+$ is a positive real number. Hence, $(E(\cdot,\cdot), D(E))$ is a coercive closed form on ${}^\circ H$ (the definition of coercive closed form will be given in Sect. 1.6).

1.5 Hyperfinite Dirichlet Forms

It is time to discuss the hyperfinite forms associated with Markov processes, i.e., the Dirichlet forms. The aim is to give a reasonably detailed account of the relationship between the properties of these forms and the behavior of the associated processes.

Consider a particle which can be in $N+1$ different states $s_0, s_1, s_2, \cdots, s_N$. Assume that if the particle is in state s_i at some instant t, then – independently of what its past history may be – the probability that it will be in state s_j at the next instant $t + \Delta t$ is given by a fixed number q_{ij}. This is the familiar setting for the theory of stationary Markov chains with finite state space, see, e.g., Chung [119], Dynkin [147], and Dynkin and Yushkevich [148]. We shall be interested in the case where $S = \{s_0, s_1, \cdots, s_N\}$ is a hyperfinite set, and $T = \{k\Delta t \mid k \in {}^*\mathbb{N}_0\}$ is a hyperfinite time line with $\Delta t \approx 0$ (for technical reasons it is convenient to have an *-infinite time line to work with). The idea is to use the hyperfinite setup to reduce the highly sophisticated theory of continuous parameter Markov processes taking values in topological spaces (see [96, 175, 193, 270, 330, 333, 334]) to the much simpler theory of finite Markov chains.

Let Y be a *Hausdorff space*, i.e., Y is a topological space such that for each pair x, y of distinct points in Y, there are open sets U, V satisfying $x \in U, y \in V$, and $U \cap V = \emptyset$ [121]. Here $\emptyset$ denotes the empty set. Let ${}^*Y$ be the nonstandard extension of Y. Let $S = \{s_0, s_1, \cdots, s_N\}$ be an S-dense subset of ${}^*Y$ for some $N \in {}^*\mathbb{N} - \mathbb{N}$ and m be a hyperfinite measure on S. Denote by $\mathcal{S}$ the internal algebra of subsets of S. Assume that $Q = \{q_{ij}\}$ is an $(N+1) \times (N+1)$ matrix with nonnegative entries, and assume that

$$\sum_{j=0}^{N} q_{ij} = 1 \quad \text{for all} \quad i = 0, 1, \cdots, N, \tag{1.5.1}$$

and the state s_0 is a trap, i.e.,

$$q_{0i} = 0 \quad \text{for all} \quad i \neq 0. \tag{1.5.2}$$

In the sequel, we shall write m_i for $m(\{s_i\})$ and q_{ij} for $q_{s_i s_j}$ respectively, whenever it is convenient.

If (Ω, P) is an internal measure space, and $X : \Omega \times T \longrightarrow S$ is an internal process, let

$$[\omega]_t = \{\omega' \in \Omega \mid X(\omega', s) = X(\omega, s) \text{ for all } s \leq t\}. \tag{1.5.3}$$

For each $t \in T$, let $\mathcal{F}_t$ be the internal algebra on Ω generated by the sets $[\omega]_t$.

If for all $\omega \in \Omega$

$$P([\omega]_0) = m\{X(\omega, 0)\}, \tag{1.5.4}$$

and whenever $X(\omega, t) = s_i$,

$$P\{\omega' \in [\omega]_t \mid X(t + \Delta t, \omega') = s_j\} = q_{ij} P([\omega]_t), \tag{1.5.5}$$

then we call X a *hyperfinite Markov chain* with initial distribution m and transition matrix Q. Notice that we do not assume that m and P are probability measures. $P(\Omega)$ could be an infinite, hyperfinite number.

In fact, given m and Q, it is easy to construct an associated Markov chain $X(\omega, t)$. Let Ω be the set of all internal functions $\omega : T \longrightarrow S$. Denote by X the coordinate function $X(\omega, t) = \omega(t)$. Let P be the measure defined by

$$P\Big([\omega]_{k\Delta t}\Big) = m(\{\omega(0)\}) \prod_{n=0}^{k-1} q_{\omega(n\Delta t), \omega((n+1)\Delta t)}. \tag{1.5.6}$$

In particular, we define a family $(\Omega, \mathcal{F}_t, P_i, i \in S)$ of internal probability spaces by

$$P_i\Big([\omega]_{k\Delta t}\Big) = \delta_{i\omega(0)} \prod_{n=0}^{k-1} q_{\omega(n\Delta t), \omega((n+1)\Delta t)} \tag{1.5.7}$$

for each $i \in S$, where δ_{ij} is the Kronecker symbol.

Similarly, let $\hat{Q} = \{\hat{q}_{ij}\}$ be an $(N+1) \times (N+1)$ matrix with nonnegative entries, and assume that

$$\sum_{j=0}^{N} \hat{q}_{ij} = 1 \quad \text{for all} \quad i = 0, 1, \cdots, N, \tag{1.5.8}$$

and the state s_0 is a trap, i.e.,

$$\hat{q}_{0i} = 0 \quad \text{for all} \quad i \neq 0. \tag{1.5.9}$$

In the same manner as above, we shall write $\hat{q}_{ij}$ for $\hat{q}_{s_i s_j}$, respectively, whenever it is convenient.

If $(\hat{\Omega}, \hat{P})$ is an internal measure space, and $\hat{X} : \hat{\Omega} \times T \longrightarrow S$ is an internal process, let

$$[\hat{\omega}]_t = \{\hat{\omega}' \in \hat{\Omega} \mid \hat{X}(\hat{\omega}', s) = \hat{X}(\hat{\omega}, s) \text{ for all } s \leq t\}.$$

For each $t \in T$, let $\hat{\mathcal{F}}_t$ be the internal algebra on $\hat{\Omega}$ generated by the sets $[\hat{\omega}]_t$.

If for all $\hat{\omega} \in \hat{\Omega}$

$$\hat{P}([\hat{\omega}]_0) = m\{\hat{X}(\hat{\omega}, 0)\}, \tag{1.5.10}$$

and whenever $\hat{X}(\hat{\omega}, t) = s_i$,

$$\hat{P}\{\hat{\omega}' \in [\hat{\omega}]_t \mid \hat{X}(t + \Delta t, \hat{\omega}') = s_j\} = \hat{q}_{ij}\hat{P}([\hat{\omega}]_t), \tag{1.5.11}$$

then we call $\hat{X}$ a *hyperfinite Markov chain* with initial distribution m and transition matrix $\hat{Q}$.

Given m and $\hat{Q}$, we can construct an associated Markov chain $\hat{X}(\hat{\omega}, t)$ as that of (1.5.6). Moreover, we define a family $(\hat{\Omega}, \hat{\mathcal{F}}_t, \hat{P}_i, i \in S)$ of internal probability spaces by

$$\hat{P}_i\Big([\hat{\omega}]_{k\Delta t}\Big) = \delta_{i\hat{\omega}(0)} \prod_{n=0}^{k-1} \hat{q}_{\hat{\omega}(n\Delta t),\hat{\omega}((n+1)\Delta t)} \tag{1.5.12}$$

for each $i \in S$. It is easy to see that we can take

$$\Omega = \hat{\Omega}, \quad X = \hat{X}, \quad [\omega]_t = [\hat{\omega}]_t, \quad \mathcal{F}_t = \hat{\mathcal{F}}_t.$$

However, $\hat{Q}$, $\hat{P}$, and $\hat{P}_i$ are different from Q, P, and P_i.

Now let us introduce some regularity conditions. We assume that the measure m and the transition matrices Q and $\hat{Q}$ satisfy the dual conditions

$$m_i q_{ij} = m_j \hat{q}_{ji} \quad \text{for all} \quad i \neq 0, j \neq 0. \tag{1.5.13}$$

Besides, we assume that

$$m_i \neq 0 \quad \text{for at least one} \quad i \neq 0. \tag{1.5.14}$$

It is easy to find examples of transition matrices Q and $\hat{Q}$ such that no m satisfies the conditions (1.5.13) and (1.5.14). Thus, these assumptions may be regarded as conditions on Q and $\hat{Q}$.

Notice that for most i, the transition probabilities q_{i0} and $\hat{q}_{i0}$ should be of order of magnitude Δt, since if not the process will die in infinitesimal time.

Given m, Q and $\hat{Q}$ which satisfy conditions (1.5.1), (1.5.2), (1.5.4), (1.5.5), (1.5.8), (1.5.9), (1.5.10), (1.5.11), (1.5.13), and (1.5.14), the processes X and $\hat{X}$ as above are called *dual hyperfinite Markov chains.* These are the processes we will study in detail. We shall first associate to a given process a hyperfinite quadratic form.

If

$$S_0 = \{s_1, s_2, \cdots, s_N\}$$

is the state space S without the trap s_o, let us set $\mathcal{S}_0 = \mathcal{S} \cap S_0$. Let H be the linear space of all internal functions $u : S_0 \longrightarrow {}^*\mathbb{R}$ with the inner product

$$\begin{aligned} \int_{S_0} uv\, dm &= \langle u, v \rangle \\ &= \sum_{i=1}^{N} u(s_i)v(s_i)m(s_i). \end{aligned} \tag{1.5.15}$$

Just as we usually write m_i for $m(s_i)$, we shall write $u(i)$ or u_i for $u(s_i)$. And we shall identify H with the set of all internal functions $u : S \longrightarrow {}^*\mathbb{R}$ such that $u(s_0) = 0$.

Our convention of letting the trap s_0 be the zeroth element is notationally convenient, but we call attention of the reader to the fact that she/he should distinguish between sums of the forms $\sum_{i=0}^{N}$ and $\sum_{i=1}^{N}$, e.g.

For $t \in T$ and $u \in H$, we define new functions $Q^t u, \hat{Q}^t u \in H$ by

$$\begin{aligned} Q^t u(i) &= E_i u(X(t)), \\ \hat{Q}^t u(i) &= \hat{E}_i u(X(t)), \end{aligned} \tag{1.5.16}$$

where E_i and $\hat{E}_i$ are the expectations with respect to the measures P_i and $\hat{P}_i$ defined in (1.5.7) and (1.5.12), respectively. Intuitively, $Q^t u(i)$ and $\hat{Q}^t u(i)$ are the expected values of $u(X(t))$ for a particle starting in state s_i. Notice that

$$\begin{aligned} Q^{\Delta t} u(i) &= (Q \cdot u)(i) \\ &= \sum_{j=1}^{N} u(j) q_{ij}, \\ \hat{Q}^{\Delta t} u(i) &= (\hat{Q} \cdot u)(i) \\ &= \sum_{j=1}^{N} u(j) \hat{q}_{ij}, \end{aligned}$$

where $\cdot$ are the matrix multiplications in the middle terms. Since

$$q_{ij}^{(t+s)} = \sum_{k=1}^{N} q_{ik}^{(t)} q_{kj}^{(s)},$$
$$\hat{q}_{ij}^{(t+s)} = \sum_{k=1}^{N} \hat{q}_{ik}^{(t)} \hat{q}_{kj}^{(s)},$$

we must have

$$Q^{t+s} = Q^t \cdot Q^s,$$
$$\hat{Q}^{t+s} = \hat{Q}^t \cdot \hat{Q}^s,$$

where $q_{ij}^{(t)}$ and $\hat{q}_{ij}^{(t)}$ are the transition probabilities given by operator Q^t and $\hat{Q}^t$, respectively. Hence, the families $\{Q^t \mid t \in T\}$ and $\{\hat{Q}^t \mid t \in T\}$ are semigroups of operators on H. Actually, $\{\hat{Q}^t \mid t \in T\}$ is the co-semigroup of $\{Q^t \mid t \in T\}$.

The *infinitesimal generator* A of the semigroup $\{Q^t \mid t \in T\}$ is given by

$$Au(i) = \frac{1}{\Delta t}\Big(u(i) - \sum_{j=1}^{N} u(j) q_{ij}\Big). \tag{1.5.17}$$

The *hyperfinite quadratic form* associated with Q and m is defined to be

$$\begin{aligned}\mathcal{E}(u,v) &= \langle Au, v\rangle \\ &= \sum_{i=1}^{N} Au(i)v(i)m(i).\end{aligned} \tag{1.5.18}$$

Combining (1.5.17) and (1.5.18), we get

$$\mathcal{E}(u,v) = \frac{1}{\Delta t}\sum_{i=1}^{N}\left[u(i)v(i)m(i) - \sum_{j=1}^{N} u(j)v(i)q_{ij}m(i)\right]. \tag{1.5.19}$$

The *infinitesimal co-generator* $\hat{A}$ of the co-semigroup $\{\hat{Q}^t \mid t \in T\}$ is given by

$$\hat{A}u(i) = \frac{1}{\Delta t}\Big(u(i) - \sum_{j=1}^{N} u(j) \hat{q}_{ij}\Big). \tag{1.5.20}$$

The *hyperfinite quadratic co-form* associated with $\hat{Q}$ and m is defined to be

$$\begin{aligned}\hat{\mathcal{E}}(u,v) &= \mathcal{E}(v,u) \\ &= \langle Av, u\rangle \\ &= \sum_{i=1}^{N} Av(i)u(i)m(i). \end{aligned} \tag{1.5.21}$$

Combining (1.5.13), (1.5.17), (1.5.20), and (1.5.21), we get

$$\begin{aligned}\hat{\mathcal{E}}(u,v) &= \frac{1}{\Delta t}\sum_{i=1}^{N}\left[u(i)v(i)m(i) - \sum_{j=1}^{N} u(i)v(j)q_{ij}m(i)\right] \\ &= \frac{1}{\Delta t}\sum_{i=1}^{N}\left[u(i)v(i)m(i) - \sum_{j=1}^{N} u(i)v(j)\hat{q}_{ji}m(j)\right] \\ &= \langle \hat{A}u, v\rangle. \end{aligned} \tag{1.5.22}$$

Now let us look at the symmetric part $\overline{\mathcal{E}}(\cdot,\cdot)$ of $\mathcal{E}(\cdot,\cdot)$. Let $\{\overline{Q}^t \mid t \in T\}$ be the semigroup and $\overline{A}$ be the generator of $\overline{\mathcal{E}}(\cdot,\cdot)$. Then, we have

$$\overline{q}_{ij} = \frac{1}{2}(q_{ij} + \hat{q}_{ij}),$$

where $(\overline{q}_{ij}) = \overline{Q} = \overline{Q}^{\Delta t}$. It is easy to see that

$$m_i\overline{q}_{ij} = m_j\overline{q}_{ji}.$$

Moreover, the generator $\overline{A}$ of $\overline{\mathcal{E}}(\cdot,\cdot)$ is given by

$$\begin{aligned}\overline{A}u(i) &= \frac{1}{2}\Big(Au(i) + \hat{A}u(i)\Big) \\ &= \frac{1}{\Delta t}\Big(u(i) - \sum_{j=1}^{N} u(j)\frac{1}{2}(q_{ij} + \hat{q}_{ij})\Big) \\ &= \frac{1}{\Delta t}\Big(u(i) - \sum_{j=1}^{N} u(j)\overline{q}_{ij}\Big). \end{aligned}$$

The *symmetric hyperfinite quadratic form* $\overline{\mathcal{E}}(\cdot,\cdot)$ is given by

$$\overline{\mathcal{E}}(u,v) = \sum_{i=1}^{N} \overline{A}u(i)v(i)m(i)$$
$$= \frac{1}{\Delta t}\sum_{i=1}^{N}\left[u(i)v(i)m(i) - \sum_{j=1}^{N} u(j)v(i)\overline{q}_{ij}m(i)\right]. \qquad (1.5.23)$$

Our first result in the following lemma gives alternative ways of expressing hyperfinite quadratic forms in term of m_i, q_{ij}, $\hat{q}_{ij}$ and $\overline{q}_{ij}$. They are nonstandard versions of the Beurling–Deny formulae [87, 88], and are often more useful than the expressions (1.5.19), (1.5.22), and (1.5.23).

Lemma 1.5.1. *Let $\mathcal{E}(\cdot,\cdot)$ be the hyperfinite quadratic form as above. Then, we have*

$$(i) \quad \mathcal{E}(u,v) = \frac{1}{\Delta t}\Big[\sum_{1\le i,j\le N}\Big(u(i)-u(j)\Big)v(i)q_{ij}m_i + \sum_{i=1}^{N} u(i)v(i)q_{i0}m_i\Big]. \qquad (1.5.24)$$

$$(ii) \quad \mathcal{E}(u,v) = \frac{1}{\Delta t}\Big[\sum_{1\le i,j\le N}\Big(v(i)-v(j)\Big)u(i)\hat{q}_{ij}m_i + \sum_{i=1}^{N} u(i)v(i)\hat{q}_{i0}m_i\Big]. \qquad (1.5.25)$$

$$(iii) \quad \overline{\mathcal{E}}(u,v) = \frac{1}{\Delta t}\Big[\sum_{1\le i<j\le N}\Big(u(i)-u(j)\Big)\Big(v(i)-v(j)\Big)\overline{q}_{ij}m(i) + \sum_{i=1}^{N} u(i)v(i)\overline{q}_{i0}m(i)\Big]. \qquad (1.5.26)$$

Proof. We only prove (iii) since the other proofs are similar. Notice that

$$\overline{\mathcal{E}}(u,v) = \frac{1}{\Delta t}\left[\sum_{i=1}^{N} u(i)v(i)m_i - \sum_{i=1}^{N}\sum_{j=1}^{N} u(j)v(i)\overline{q}_{ij}m_i\right]$$
$$= \frac{1}{\Delta t}\left[\sum_{i=1}^{N}\sum_{j=0}^{N} u(i)v(i)\overline{q}_{ij}m_i - \sum_{i=1}^{N}\sum_{j=1}^{N} u(j)v(i)\overline{q}_{ij}m_i\right]$$
$$= \frac{1}{\Delta t}\left[\sum_{1\le i,j\le N}\Big(u(i)-u(j)\Big)v(i)\overline{q}_{ij}m_i + \sum_{i=1}^{N} u(i)v(i)\overline{q}_{i0}m_i\right],$$

where the first line is a trivial modification of the expression (1.5.23), the second line follows from the first since $\sum_{j=0}^{N}\overline{q}_{ij} = 1$, and the last line is just a rearrangement of the second line.

Fix a pair (i,j) and consider the terms in the last expression above involving both i and j. If $i=j$, there is only one such term, and that term is zero. If $i \neq j$, there are two terms to consider, i.e.,

$$(u(i)-u(j))v(i)\overline{q}_{ij}m_i \quad \text{and} \quad (u(j)-u(i))v(j)\overline{q}_{ji}m_j.$$

Since $\overline{q}_{ij}m_i = \overline{q}_{ji}m_j$, the sum of the two terms equals

$$(u(i)-u(j))(v(i)-v(j))\overline{q}_{ij}m_i.$$

Summing over all pairs (i,j), the relation (1.5.26) follows. □

As an immediate consequence, we have:

Corollary 1.5.1. *The hyperfinite quadratic form $\mathcal{E}(\cdot,\cdot)$ associated with Q and m as above is nonnegative.*

In order to get some further properties of the hyperfinite quadratic form $\mathcal{E}(\cdot,\cdot)$, let us introduce the following definitions.

For $u \in H, v \in H$, we define

$$(u \vee v)(s) = \max\{u(s), v(s)\} \text{ and } (u \wedge v)(s) = \min\{u(s), v(s)\}, s \in S_0.$$

Moreover, let $u^+ = u \vee 0$ and $u^- = -u \wedge 0$.

If $u \in H$, the function $\tilde{u} = (0 \vee u) \wedge 1$ is called the *unit contraction* of u. A quadratic form $\mathcal{E}(\cdot,\cdot)$ is said to have the *Markov property* if for all u

$$\mathcal{E}(\tilde{u},\tilde{u}) \leq \mathcal{E}(u,u).$$

Corollary 1.5.2. *Let $\mathcal{E}(\cdot,\cdot)$ be the hyperfinite quadratic form as above. Then*
(1) For all $u \in H, \alpha \in {}^*\mathbb{R}, \alpha \geq 0$, we have

$$\mathcal{E}(u \wedge \alpha, u - u \wedge \alpha) \geq 0,$$
$$\mathcal{E}(u - u \wedge \alpha, u \wedge \alpha) \geq 0.$$

(2) $\mathcal{E}(\tilde{u},\tilde{u}) \leq \mathcal{E}(u,u)$.

Proof. (1) By the expression (1.5.24), we have

$$\mathcal{E}(u \wedge \alpha, u - u \wedge \alpha) = \frac{1}{\Delta t}\Big[\sum_{1\leq i,j\leq N}\Big(u(i)\wedge\alpha - u(j)\wedge\alpha\Big)\Big(u - u\wedge\alpha\Big)(i)q_{ij}m_i$$
$$+\sum_{i=1}^{N}(u\wedge\alpha)(i)\Big(u(i) - u(i)\wedge\alpha\Big)q_{i0}m_i\Big]. \qquad (1.5.27)$$

If $u(i) \geq \alpha$, then

$$\Big(u(i) \wedge \alpha - u(j) \wedge \alpha\Big)\Big(u - u \wedge \alpha\Big)(i) = \Big(\alpha - u(j) \wedge \alpha\Big)\Big(u(i) - \alpha\Big) \geq 0 \tag{1.5.28}$$

and

$$\Big(u \wedge \alpha\Big)(i)\Big(u(i) - u(i) \wedge \alpha\Big) = \alpha\Big(u(i) - \alpha\Big) \geq 0. \tag{1.5.29}$$

If $u(i) < \alpha$, then

$$\Big(u(i) \wedge \alpha - u(j) \wedge \alpha\Big)\Big(u - u \wedge \alpha\Big)(i) = (u \wedge \alpha)(i)\Big(u - u \wedge \alpha\Big)(i) = 0. \tag{1.5.30}$$

It follows from the relations (1.5.27), (1.5.28), (1.5.29), and (1.5.30) that $\mathcal{E}(u \wedge \alpha, u - u \wedge \alpha) \geq 0$. Similarly, we can show that $\mathcal{E}(u - u \wedge \alpha, u \wedge \alpha) \geq 0$ by using the relation (1.5.25).

(2) From the relation (1.5.26), we have $\overline{\mathcal{E}}(\tilde{u}, \tilde{u}) \leq \overline{\mathcal{E}}(u, u)$. This implies the conclusion (2), since $\mathcal{E}(u, u) = \overline{\mathcal{E}}(u, u)$ for all $u \in H$. □

We call a map $T : H \longrightarrow H$ a *Markov operator*, if it maps nonnegative functions into nonnegative functions, and never increases the supremum norm, i.e.,

$$||Tu||_\infty \leq ||u||_\infty$$

for all $u \in H$, where $||u||_\infty = \max_{1 \leq i \leq N} |u(i)|$.

The next result gives us four ways of deciding whether a given quadratic form is a hyperfinite quadratic form associated with some Q and m without actually constructing an associated Markov process.

Proposition 1.5.1. *Let*

$$\mathcal{E}(u, v) = \sum_{i,j=1}^{N} b_{ij} u(i) v(j)$$

be a nonnegative quadratic form which is non-zero. The following statements are equivalent:

(*i*) $\mathcal{E}(\cdot,\cdot)$ *is the hyperfinite quadratic form of some* Q, $\hat{Q}$, *and* m *which satisfy the conditions (1.5.1), (1.5.2), (1.5.4), (1.5.5), (1.5.8), (1.5.9), (1.5.10), (1.5.11), (1.5.13), and (1.5.14).*

(*ii*) *There exist a hyperfinite measure* m *on* S_0 *and two Markov operators* $Q^{\Delta t}, \hat{Q}^{\Delta t} : H \longrightarrow H$ *such that*

$$\langle Q^{\Delta t}u, v\rangle = \langle u, \hat{Q}^{\Delta t}v\rangle$$

and

$$\begin{aligned}\mathcal{E}(u,v) &= \frac{1}{\Delta t}\langle (I - Q^{\Delta t})u, v\rangle \\ &= \frac{1}{\Delta t}\langle u, (I - \hat{Q}^{\Delta t})v\rangle,\end{aligned}$$

where $(H, \langle\cdot,\cdot\rangle)$ *is defined through (1.5.15) by* m, *and* Δt *is an infinitesimal.*

(*iii*) *For all* $u \in H, \alpha \in {}^*\mathbb{R}, \alpha \geq 0$, *we have*

$$\begin{aligned}\mathcal{E}(u \wedge \alpha, u - u \wedge \alpha) \geq 0,\\ \mathcal{E}(u - u \wedge \alpha, u \wedge \alpha) \geq 0.\end{aligned}$$

(*iv*) $\mathcal{E}(\cdot,\cdot)$ *satisfies* $\mathcal{E}(\tilde{u}, u - \tilde{u}) \geq 0$ *and* $\mathcal{E}(u - \tilde{u}, \tilde{u}) \geq 0, \forall u \in H$, *where* $\tilde{u}$ *is the unit contraction of* u.

(*v*) *Whenever* $i \neq j, b_{ij} \leq 0$; *but* $b_{ii} \geq -\sum_{j\neq i} b_{ij}$ *and* $b_{ii} \geq -\sum_{j\neq i} b_{ji}$ *for all* i.

Proof. We shall prove (*i*) $\Longrightarrow$ (*iii*) $\Longrightarrow$ (*iv*) $\Longrightarrow$ (*v*) $\Longrightarrow$ (*i*) $\Longrightarrow$ (*ii*) $\Longrightarrow$ (*v*).

(*i*) $\Longrightarrow$ (*iii*). This follows immediately from Corollary 1.5.2.

(*iii*) $\Longrightarrow$ (*iv*). We have

$$\begin{aligned}\mathcal{E}(\tilde{u}, u - \tilde{u}) &= \mathcal{E}(\tilde{u}, (u \vee 0) - \tilde{u}) + \mathcal{E}(\tilde{u}, u - (u \vee 0)) \\ &= \mathcal{E}((u \vee 0) \wedge 1, (u \vee 0) - (u \vee 0) \wedge 1) + \mathcal{E}(\tilde{u}, u - u \vee 0) \\ &\geq \mathcal{E}(\tilde{u}, u - u \vee 0) \\ &= \mathcal{E}((u \wedge 1)^+, -(u \wedge 1)^-).\end{aligned}$$

Let $v = u \wedge 1$, then

$$\begin{aligned}\mathcal{E}(\tilde{u}, u - \tilde{u}) &\geq \mathcal{E}((u \wedge 1)^+, -(u \wedge 1)^-) \\ &= -\mathcal{E}(v^+, v^-) \\ &= -\mathcal{E}(v^+, v^+ - v) \\ &= \mathcal{E}((-v) \wedge 0, (-v) - (-v) \wedge 0) \\ &\geq 0.\end{aligned}$$

Similarly, we can show

$$\mathcal{E}(u - \tilde{u}, \tilde{u}) \geq 0.$$

$(iv) \Longrightarrow (v)$. Fix $k, l \in S_0, k \neq l$, define u by

$$u(i) = \begin{cases} -1 & \text{if } i = k, \\ 1 & \text{if } i = l, \\ 0 & \text{otherwise.} \end{cases}$$

If $\tilde{u}$ is the unit contraction of u, we have

$$\begin{aligned} 0 &\leq \mathcal{E}(\tilde{u}, u) - \mathcal{E}(\tilde{u}, \tilde{u}) \\ &= b_{ll} - b_{lk} - b_{ll} \\ &= -b_{lk}. \end{aligned}$$

Therefore, $b_{lk} \leq 0$.

To get the second half of (v), we fix a $k \in S_0$ and consider the function

$$v(i) = \begin{cases} 2 & \text{if } i = k, \\ 1 & \text{otherwise} \end{cases}$$

If $\tilde{v}$ is the unit contraction of v, then

$$\begin{aligned} 0 &\leq \mathcal{E}(\tilde{v}, v - \tilde{v}) \\ &= \sum_{i=1}^{N} b_{ik} \end{aligned}$$

and

$$\begin{aligned} 0 &\leq \mathcal{E}(v - \tilde{v}, \tilde{v}) \\ &= \sum_{j=1}^{N} b_{kj}. \end{aligned}$$

$(v) \Longrightarrow (i)$. We first observe

$$\begin{aligned} \sum_{1 \leq i,j \leq N} (v(i) - v(j))u(i)b_{ij} &= \sum_{1 \leq i,j \leq N} u(i)v(i)b_{ij} - \sum_{1 \leq i,j \leq N} u(i)v(j)b_{ij} \\ &= \sum_{1 \leq i,j \leq N} u(i)v(i)b_{ij} - \mathcal{E}(u, v). \end{aligned}$$

Thus, we have

$$\mathcal{E}(u,v) = -\sum_{1\le i,j\le N} (v(i)-v(j))u(i)b_{ij} + \sum_{1\le i,j\le N} u(i)v(i)b_{ij}. \quad (1.5.31)$$

Similarly, we have

$$\begin{aligned}\sum_{1\le i,j\le N} (u(i)-u(j))v(i)b_{ji} &= \sum_{1\le i,j\le N} u(i)v(i)b_{ji} - \sum_{1\le i,j\le N} u(j)v(i)b_{ji} \\ &= \sum_{1\le i,j\le N} u(i)v(i)b_{ji} - \mathcal{E}(u,v).\end{aligned}$$

Hence, we have

$$\mathcal{E}(u,v) = -\sum_{1\le i,j\le N} (u(i)-u(j))v(i)b_{ji} + \sum_{1\le i,j\le N} u(i)v(i)b_{ji}. \quad (1.5.32)$$

In the following, we shall define Q and m by marching the two expressions (1.5.25) and (1.5.31), (1.5.24) and (1.5.32) term by term. It turns out that we have one degree of freedom for each i. We may choose

$$0 \le q_{ii} < 1, \quad q_{ii} = \hat{q}_{ii}.$$

Comparing (1.5.25) and (1.5.31), (1.5.24) and (1.5.32), we notice that we must have

$$\begin{aligned}\frac{1}{\Delta t} m_i \hat{q}_{ij} &= -b_{ij} \\ &= \frac{1}{\Delta t} q_{ji} m_j \quad \text{for} \quad j \ne 0, i \end{aligned} \quad (1.5.33)$$

and

$$\begin{aligned}\frac{1}{\Delta t} m_i \hat{q}_{i0} &= \sum_{j=1}^{N} b_{ij}, \\ \frac{1}{\Delta t} m_i q_{i0} &= \sum_{j=1}^{N} b_{ji}. \end{aligned} \quad (1.5.34)$$

In order to have $\sum_{j=0}^{N} \hat{q}_{ij} = 1$ satisfied, we must have

$$\begin{aligned}
m_i &= \sum_{j=0}^{N} m_i \hat{q}_{ij} \\
&= m_i \hat{q}_{ii} - \Delta t \sum_{j \neq i,0} b_{ij} + \Delta t \sum_{j=1}^{N} b_{ij} \\
&= m_i \hat{q}_{ii} + b_{ii} \Delta t.
\end{aligned}$$

Hence, this gives us

$$\begin{aligned}
m_i &= \frac{b_{ii}\Delta t}{1 - q_{ii}} \\
&= \frac{b_{ii}\Delta t}{1 - \hat{q}_{ii}}.
\end{aligned}$$

Combining this with (1.5.33), we get

$$\begin{aligned}
q_{ij} &= -\frac{b_{ji}(1 - q_{ii})}{b_{ii}}, \\
\hat{q}_{ij} &= -\frac{b_{ij}(1 - q_{ii})}{b_{ii}} \quad \text{for} \quad j \neq 0, i,
\end{aligned}$$

and from (1.5.34), we have

$$\begin{aligned}
\hat{q}_{i0} &= \frac{\sum_{j=1}^{N} b_{ij}(1 - \hat{q}_{ii})}{b_{ii}}, \\
q_{i0} &= \frac{\sum_{j=1}^{N} b_{ji}(1 - q_{ii})}{b_{ii}}.
\end{aligned}$$

This construction breaks down when $b_{ii} = 0$, but in that case $b_{ij} = 0$ for all j, and we get $m_i = 0$ and can choose the q_{ij}'s arbitrarily.

We finally observe that (1.5.33) implies (1.5.13), and that the non-triviality of $\mathcal{E}(\cdot, \cdot)$ implies (1.5.14).

$(i) \Longrightarrow (ii)$. This follows immediately from (1.5.16).

$(ii) \Longrightarrow (v)$. Notice that

$$Q^{\Delta t} u(i) = u(i) - \Delta t \sum_{j=1}^{N} b_{ji} u(j) \frac{1}{m(i)},$$

for $m(i) \neq 0$. For $m(i) = 0$, we simplify here $Q^{\Delta t} u(i) = u(i)$. For each $j \in S_0$, let u_j be given by $u_j(i) = \delta_{ij}$. Since $Q^{\Delta t}$ is Markov, we have for all $i \neq j$

$$\begin{aligned} 0 &\leq Q^{\Delta t} u_j(i) \\ &= -\Delta t b_{ji} \frac{1}{m(i)}, \end{aligned}$$

and thus $b_{ji} \leq 0$.

On the other hand, applying $Q^{\Delta t}$ to the function which is constant one and using that $Q^{\Delta t}$ cannot increase the supremum norm, we get

$$\begin{aligned} 1 &\geq Q^{\Delta t} 1(i) \\ &= 1 - \Delta t \sum_{j=1}^{N} b_{ji} \frac{1}{m(i)}, \text{ if } m(i) \neq 0, \end{aligned}$$

and $Q^{\Delta t} 1(i) = 1$ if $m(i) = 0$. It follows that $b_{ii} \geq -\sum_{j \neq i} b_{ji}$.

Besides, we have

$$\hat{Q}^{\Delta t} u(i) = u(i) - \Delta t \sum_{j=1}^{N} b_{ij} u(j) \frac{1}{m(i)}.$$

Hence, we have

$$\begin{aligned} 1 &\geq \hat{Q}^{\Delta t} 1(i) \\ &= 1 - \Delta t \sum_{j=1}^{N} b_{ij} \frac{1}{m(i)}. \end{aligned}$$

From this, we have $b_{ii} \geq -\sum_{j \neq i} b_{ij}$. □

Now it is the time to introduce the following:

Definition 1.5.1. The hyperfinite quadratic form $\mathcal{E}(\cdot, \cdot)$ associated with Q and m is called a *hyperfinite Dirichlet form* if it satisfies the hyperfinite weak sector condition in the sense of Definition 1.4.1.

If $\delta \in T$, let T_δ be the sub-line

$$T_\delta = \{k\delta \mid k \in {}^*\mathbb{N}_0\}.$$

We write $X^{(\delta)}$ for the restriction $X|T_\delta$. For each $t \in T_\delta$, let $\mathcal{F}_t^{(\delta)}$ be the internal algebra on Ω generated by the sets

$$[\omega]_t^{(\delta)} = \left\{ \omega' \in \Omega \middle| X^{(\delta)}(\omega', s) = X^{(\delta)}(\omega, s) \text{ for all } s \in T_\delta, s \leq t \right\}. \tag{1.5.35}$$

It is easy to verify the following for all $k \in {}^*\mathbb{N}$,

$$P\Big([\omega]^{(\delta)}_{k\delta}\Big) = m(\omega(0)) \prod_{n=0}^{k-1} q^{(\delta)}_{\omega(n\delta),\omega((n+1)\delta)}$$

and

$$P_i\Big([\omega]^{(\delta)}_{k\delta}\Big) = \delta_{i\omega(0)} \prod_{n=0}^{k-1} q^{(\delta)}_{\omega(n\delta),\omega((n+1)\delta)}.$$

Therefore, for any $t \in T_\delta$ and $u \in H$, we have

$$\begin{aligned} E_i u(X^{(\delta)}(t)) &= Q^t u(i) \\ &= E_i u(X(t)). \end{aligned}$$

In particular, we get

$$\begin{aligned} E_i u(X^{(\delta)}(\delta)) &= Q^\delta u(i) \\ &= \sum_{j=1}^{N} q^{(\delta)}_{ij} u(j). \end{aligned}$$

This implies that the semigroup of $X^{(\delta)}$ is $\{Q^t \mid t \in T_\delta\}$. The infinitesimal generator $A^{(\delta)}$ of $X^{(\delta)}$ is given by

$$A^{(\delta)} u(i) = \frac{1}{\delta}\Big(u(i) - \sum_{j=1}^{N} u(j) q^{(\delta)}_{ij}\Big).$$

Similarly, we can get $\hat{X}^{(\delta)}, \hat{\mathcal{F}}^{(\delta)}_t, [\hat{\omega}]^{(\delta)}_t, \{\hat{Q}^t \mid t \in T_\delta\}, \hat{A}^{(\delta)}$. It is easy to see that $A^{(\delta)}$ and $\hat{A}^{(\delta)}$ are nonnegative, i.e.,

$$\langle A^{(\delta)} u, u\rangle \geq 0 \text{ and } \langle \hat{A}^{(\delta)} u, u\rangle \geq 0 \text{ for all } u \in H.$$

Moreover, we can easily show that

$$\sum_{j=0}^{N} q^{(\delta)}_{ij} = 1 \text{ and } \sum_{j=0}^{N} \hat{q}^{(\delta)}_{ij} = 1 \text{ for all } i = 0, 1, 2 \cdots, N,$$

$$q^{(\delta)}_{0i} = 0 \text{ and } \hat{q}^{(\delta)}_{0i} = 0 \qquad \text{for all} \quad i \neq 0$$

and

$$m_i q_{ij}^{(\delta)} = m_j \hat{q}_{ji}^{(\delta)} \quad \text{for all} \quad i \neq 0, j \neq 0.$$

The hyperfinite quadratic form associated with $Q^{(\delta)}$ and m is defined to be

$$\begin{aligned}\mathcal{E}^{(\delta)}(u,v) &= \langle A^{(\delta)}u, v\rangle \\ &= \sum_{i=1}^{N} A^{(\delta)}u(i)v(i)m(i).\end{aligned}$$

Therefore, we have

$$\begin{aligned}\mathcal{E}^{(\delta)}(u,v) &= \frac{1}{\delta}\sum_{i=1}^{N}\left[u(i)v(i)m(i) - \sum_{j=1}^{N} u(j)v(i)q_{ij}^{(\delta)}m(i)\right] \\ &= \frac{1}{\delta}\left[\sum_{1\le i,j\le N}\Big(u(i)-u(j)\Big)v(i)q_{ij}^{(\delta)}m(i) + \sum_{i=1}^{N} u(i)v(i)q_{i0}^{(\delta)}m(i)\right].\end{aligned}$$

As in Sect. 1.4, let $G_\alpha^{(\delta)} = (A^{(\delta)} - \alpha)^{-1}$ and $\hat{G}_\alpha^{(\delta)} = (\hat{A}^{(\delta)} - \alpha)^{-1}$ be the resolvents of $A^{(\delta)}$ and $\hat{A}^{(\delta)}, \alpha \in {}^*(-\infty, 0)$, respectively. Similarly, the domain $\mathcal{D}(\mathcal{E}^{(\delta)})$ of $\mathcal{E}^{(\delta)}(\cdot,\cdot)$ is the set of all $u \in H$ such that

(i) ${}^\circ\mathcal{E}_1^{(\delta)}(u,u) = \overset{\circ}{\left[\mathcal{E}^{(\delta)}(u,u) + \langle u,u\rangle\right]} < \infty$.

(ii) For all infinitesimal $\alpha < 0$, we have ${}^{(\alpha)}\mathcal{E}(u+\alpha G_\alpha^{(\delta)}u, u+\alpha G_\alpha^{(\delta)}u) \approx 0$ and ${}^{(\alpha)}\mathcal{E}(u+\alpha \hat{G}_\alpha^{(\delta)}u, u+\alpha \hat{G}_\alpha^{(\delta)}u) \approx 0$.

Moreover, we can define $\hat{\mathcal{E}}^{(\delta)}(\cdot,\cdot)$ and $\overline{\mathcal{E}}^{(\delta)}(\cdot,\cdot)$, $\overline{A}^{(\delta)}$, $\{\overline{Q}^t \mid t \in T_\delta\}$ and $\mathcal{D}(\overline{\mathcal{E}}^{(\alpha)})$ as in Sects. 1.1 and 1.2. We can get similar results of the Beurling–Deny formulae for these hyperfinite quadratic forms; in addition, we may get similar results as Proposition 1.5.1 for $\mathcal{E}^{(\delta)}(\cdot,\cdot)$.

Let us try to illustrate the theory by a simple example.

Example 1.5.1. (Brownian motion on a circle). Let $N = (\Delta t)^{-1/2}$ be an even hyperfinite integer, and let $S_0 = \{s_1, \cdots, s_N\}$ be uniformly distributed on a circle of circumference one. If $i, j \in \{1, 2, \cdots, N\}$, let the transition probability q_{ij} be $\frac{1}{2}$ if s_i and s_j are neighbors, and 0 otherwise. The semigroup $\{Q^t\}$ is given by

$$Q^{\Delta t}u(i) = \frac{1}{2}u(i+1) + \frac{1}{2}u(i-1),$$

where the addition is modulo N inside the u's. The infinitesimal generator is

$$Au(i) = -\frac{u(i+1) - 2u(i) + u(i-1)}{2\Delta t}. \qquad (1.5.36)$$

If $m_i = \frac{1}{N}$ for all i, the associated hyperfinite quadratic form $\mathcal{E}(\cdot,\cdot)$ is given by

$$\mathcal{E}(u,v) = -\frac{N}{2}\sum_{i=1}^{N}\Big(u(i+1) - 2u(i) + u(i-1)\Big)v(i),$$

or – in the Beurling–Deny formulation –

$$\mathcal{E}(u,v) = \frac{N}{2}\sum_{i=0}^{N}\Big(u(i+1) - u(i)\Big)\Big(v(i+1) - v(i)\Big). \qquad (1.5.37)$$

Notice that by (1.5.36), the infinitesimal generator A is a nonstandard version of the operator

$$\tilde{A}f = -\frac{1}{2}f'',$$

and by (1.5.37), the form $\mathcal{E}(\cdot,\cdot)$ is a representation of

$$E(f,g) = \frac{1}{2}\int_C f'g'dm,$$

where m is the Lebesgue measure on the circle C, and all derivatives are taken along the circle. Passing from the original expression for $\mathcal{E}(\cdot,\cdot)$ to the Beurling–Deny version amounts to an integration by parts.

1.6 Hyperfinite Representations

In this section, we will relate the results we have obtained so far to the standard Dirichlet space theory. We have proved that a hyperfinite weak coercive quadratic form on a hyperfinite dimensional space H induces a closed form on the hull $^\circ H$ of H. For most applications, the space $^\circ H$ is too large. What we really want is a form defined on a Hilbert space K given in advance of the nonstandard construction. If K can be identified with a subspace of $^\circ H$, we get the desired form by restricting the form on $^\circ H$ to K. Noticing that since K is a closed subspace of $^\circ H$, the restricted form is also closed. The result we shall prove in this section states that any coercive closed form on any Hilbert space K can be obtained from a nonnegative quadratic form in this way. There are two reasons for proving such a representation theorem. The first and most important is to obtain the "correct" relationship between the already developed standard theory on the one hand, and the new hyperfinite theory on the other. In fact, we shall utilize this relation to solve the problem in classical theory, referring to Theorem 4.1.1 in Chap. 4. The second, closely related reason is to show that no generality is lost by working within the nonstandard framework.

Let K be a standard Hilbert space with an inner product $(\cdot,\cdot)$. A hyperfinite dimensional subspace H of ${}^*K$ is called *S-dense* in ${}^*K$ if for all $x \in K$, there is a $y \in H$ such that $||x-y|| \approx 0$. We recall that $\approx$ is the equivalent relation on H given by $u \approx v$ if and only if $||u-v|| \approx 0$, and that ${}^\circ u$ denotes the equivalence class of u under $\approx$. If H is S-dense in ${}^*K$, we can identify K with a subspace of ${}^\circ H$ by identifying x and ${}^\circ u$ whenever $||x-u|| \approx 0$.

If $\mathcal{E}(\cdot,\cdot)$ is a hyperfinite weak coercive quadratic form on H, we let $E(\cdot,\cdot)$ denote the standard part of $\mathcal{E}(\cdot,\cdot)$ as defined in Definition 1.4.3, and we let $E_K(\cdot,\cdot)$ be the restriction of $E(\cdot,\cdot)$ to K. As mentioned above, our goal is to show that all coercive closed forms can be obtained in this way. Before we prove this, we shall recall a few results from the standard theory of quadratic forms ([270] is a convenient reference).

Let $D(F)$ be a linear subspace of K, and let $F : D(F) \times D(F) \longrightarrow \mathbb{R}$ be a bilinear map which is nonnegative, i.e., $F(x,x) \geq 0$ for all $x \in D(F)$. For $\alpha \in \mathbb{R}_+$, we set

$$F_\alpha(\cdot,\cdot) = F(\cdot,\cdot) + \alpha(\cdot,\cdot).$$

$(F(\cdot,\cdot), D(F))$ is said to satisfy the *weak sector condition* if

$$|F_1(x,y)| \leq C\sqrt{F_1(x,x)}\sqrt{F_1(y,y)} \text{ for all } x,y \in D(F),$$

where C is positive real number which is called a *continuity constant.*

$(F(\cdot,\cdot), D(F))$ is said to be *closed* if $D(F)$ is complete with respect to the norm $\sqrt{F_1(\cdot,\cdot)}$.

We say that $(F(\cdot,\cdot), D(F))$ is a *coercive closed form* on K if and only if

(1) $D(F)$ is a dense subspace of K.
(2) $(F(\cdot,\cdot), D(F))$ satisfies the weak sector condition.
(3) $(F(\cdot,\cdot), D(F))$ is a closed form.

Proposition 1.6.1. *Let $(F(\cdot,\cdot), D(F))$ be a coercive closed form on K with continuity constant C. Then there exist unique strongly continuous contraction resolvent $\{R_\alpha \mid \alpha \in (-\infty,0)\}$ and co-resolvent $\{\hat{R}_\alpha \mid \alpha \in (-\infty,0)\}$ on K such that*

$$R_\alpha(K), \hat{R}_\alpha(K) \subset D(F) \textit{ and } F_{-\alpha}(R_\alpha f, u) = (f,u) = F_{-\alpha}(u, \hat{R}_\alpha f) \tag{1.6.1}$$

for all $f \in K, u \in D(F), \alpha \in (-\infty, 0)$. *In particular, we have*

$$(R_\alpha f, g) = (f, \hat{R}_\alpha g) \text{ for all } f, g \in K,$$

i.e., $\hat{R}_\alpha$ *is the adjoint of* R_α *for all* $\alpha \in (-\infty, 0)$.

Proof. We refer to Ma and Röckner [270], Chap. I, Theorem 2.8 and use negative value of α for the resolvents. □

We assume that $(F(\cdot,\cdot), D(F))$ is a coercive closed form on K with continuity constant C. Let $(L, D(L))$ and $(\hat{L}, D(\hat{L}))$ be the generators of $\{R_\alpha \mid \alpha \in (-\infty, 0)\}$ and $\{\hat{R}_\alpha \mid \alpha \in (-\infty, 0)\}$, respectively. Then we have from Ma and Röckner [270], Chap. I, Corollary 2.10,

$$R_\alpha(K) = D(L) \subset D(F) \text{ and } F(u, v) = (Lu, v), \forall u \in D(L), v \in D(F)$$

and

$$\hat{R}_\alpha(K) = D(\hat{L}) \subset D(F) \text{ and } F(v, u) = (\hat{L}u, v), \forall u \in D(\hat{L}), v \in D(F).$$

Define for $\alpha \in (-\infty, 0), u, v \in K$,

$$\begin{aligned} {}^{(\alpha)}F(u, v) &= -\alpha(u + \alpha R_\alpha u, v) \\ &= -\alpha(v + \alpha \hat{R}_\alpha v, u) \end{aligned}$$

and

$$\begin{aligned} {}^{(\alpha)}\hat{F}(u, v) &= -\alpha(v + \alpha R_\alpha v, u) \\ &= -\alpha(u + \alpha \hat{R}_\alpha u, v). \end{aligned}$$

Then for all $\alpha \in (-\infty, 0), u \in K$, we have

$${}^{(\alpha)}F(u, -\alpha R_\alpha u) = {}^{(\alpha)}F(u, u) + \alpha(u + \alpha R_\alpha u, u + \alpha R_\alpha u)$$

and

$${}^{(\alpha)}F(-\alpha \hat{R}_\alpha u, u) = {}^{(\alpha)}F(u, u) + \alpha(u + \alpha \hat{R}_\alpha u, u + \alpha \hat{R}_\alpha u).$$

Therefore, we have for all $\alpha \in (-\infty, 0), u \in K$,

$$\begin{aligned} {}^{(\alpha)}F(u, -\alpha R_\alpha u) &\le {}^{(\alpha)}F(u, u), \\ {}^{(\alpha)}F(-\alpha \hat{R}_\alpha u, u) &\le {}^{(\alpha)}F(u, u). \end{aligned}$$

Proposition 1.6.2. *Let* $(F(\cdot,\cdot), D(F))$ *be a coercive closed form on* K *with continuity constant* $C > 0$. *Then, we have*

(i) ${}^{(\alpha)}F(u,v) = F(-\alpha R_\alpha u, v)$ *and* ${}^{(\alpha)}\hat{F}(u,v) = F(-\alpha \hat{R}_\alpha u, v)$ *for all* $u \in K, v \in D(F)$.
(ii) $F(\alpha R_\alpha u, \alpha R_\alpha u) \le {}^{(\alpha)}F(u,u)$ *and* $F(\alpha \hat{R}_\alpha u, \alpha \hat{R}_\alpha u) \le {}^{(\alpha)}F(u,u)$ *for all* $u \in K$ *and* $\alpha \in (-\infty, 0)$.
(iii) $|{}^{(\alpha)}F_1(u,v)| \le (C+1)\sqrt{F_1(u,u)}\sqrt{{}^{(\alpha)}F_1(v,v)}$ *for all* $u \in D(F), v \in K$.
(iv) $F_1(\alpha R_\alpha u, \alpha R_\alpha u) \le (C+1)^2 F_1(u,u)$ *for all* $u \in D(F)$.

Proof. We refer to Ma and Röckner [270], Chap. I, Lemma 2.11. □

Proposition 1.6.3. *Let* $(F(\cdot,\cdot), D(F))$ *be a coercive closed form on* K *with continuity constant* $C > 0$. *Then*

(i) *Let* $u \in K$. *Then* $u \in D(F)$ *if and only if* $\sup_{\alpha \in (-\infty,0)} {}^{(\alpha)}F(u,u) < \infty$.
(ii) $D(L) = R_\alpha(K)$ *is dense in* $D(F)$ *with respect to* $\sqrt{F_1(\cdot,\cdot)}$. *Moreover, we have for all* $u \in D(F)$

$$\lim_{\alpha \longrightarrow -\infty} F_1(\alpha R_\alpha u + u, \alpha R_\alpha u + u) = 0.$$

In particular, $(F(\cdot,\cdot), D(F))$ *is determined by* $\{R_\alpha \mid \alpha \in (-\infty, 0)\}$ *via (1.6.1) uniquely.*
(iii) $\lim_{\alpha \longrightarrow -\infty} {}^{(\alpha)}F(u,v) = F(u,v)$ *for all* $u, v \in D(F)$.

Proof. We refer to Ma and Röckner [270], Chap. I, Theorem 2.13. □

Let $(F(\cdot,\cdot), D(F))$ be a coercive closed form on K with continuity constant $C > 0$. Denote by $\{T_t \mid t \in [0,\infty)\}$ and $\{\hat{T}_t \mid t \in [0,\infty)\}$ the strongly continuous contraction semigroups corresponding to $\{R_\alpha \mid \alpha \in (-\infty,0)\}$ and $\{\hat{R}_\alpha \mid \alpha \in (-\infty,0)\}$, respectively. Then, we have from Ma and Röckner [270], Chap. I, Theorem 2.8 that

$$(T_t f, g) = (f, \hat{T}_t g) \text{ for all } f, g \in K, t > 0.$$

Thereafter, $\{R_\alpha \mid \alpha \in (-\infty,0)\}$ and $\{\hat{R}_\alpha \mid \alpha \in (-\infty,0)\}$ shall be called *resolvent* and *co-resolvent* of $(F(\cdot,\cdot), D(F))$, respectively. Similarly, $\{T_t \mid t \in [0,\infty)\}$ and $\{\hat{T}_t \mid t \in [0,\infty)\}$ shall be called *semigroup* and *co-semigroup of* $(F(\cdot,\cdot), D(F))$, respectively. For $t \in (0,\infty)$, we define

$$F^{(t)}(u,v) = \frac{1}{t}(u - T_t u, v), u, v \in K.$$

As in Proposition 1.6.3, we have

Proposition 1.6.4. *Let* $(F(\cdot,\cdot), D(F))$ *be a coercive closed form on* K *with continuity constant* $C > 0$. *Then*

(i) Let $u \in K$*. Then* $u \in D(F)$ *if and only if* $\sup_{t>0} F^{(t)}(u,u) < \infty$*.*
(ii) For all $u, v \in D(F)$*, we have* $\lim_{t \downarrow 0} F^{(t)}(u,v) = F(u,v)$*.*
(iii) $\lim_{t \downarrow 0} F_1(u - T_t u, u - T_t u) = 0$ *for all* $u \in D(F)$*.*

Proof. We refer to Albeverio et al. [9], Theorem 3.4. □

Proposition 1.6.5. *Let* $(F(\cdot,\cdot), D(F))$ *be a coercive closed form on* K *with continuity constant* $C > 0$*. Then, we have for* $n \in \mathbb{N}$ *and* $\forall f \in K$

$$\begin{aligned} T_t f &= \lim_{n \longrightarrow \infty} \left(I + \frac{t}{n} L \right)^{-n} f \\ &= \lim_{n \longrightarrow \infty} \left[\frac{n}{t} R_{-n/t} \right]^n f \end{aligned}$$

and

$$\begin{aligned} \hat{T}_t f &= \lim_{n \longrightarrow \infty} \left(I + \frac{t}{n} \hat{L} \right)^{-n} f \\ &= \lim_{n \longrightarrow \infty} \left[\frac{n}{t} \hat{R}_{-n/t} \right]^n f. \end{aligned}$$

Proof. We refer to Pazy [299], Theorem 8.3, Chap. I, page 33. □

Let st_K be the standard part map from $^*K$ to K.

Theorem 1.6.1. *Let* $(F(\cdot,\cdot), D(F))$ *be a coercive closed form on a Hilbert space* K*, and let* H *be an* S*-dense, hyperfinite dimensional subspace of* $^*K$*. Then, there exists a nonnegative quadratic form* $\mathcal{E}(\cdot,\cdot)$ *on* H *– associated with an internal time line* $T = \{0, \Delta t, 2\Delta t, \cdots, k\Delta t, \cdots\} = \{k\Delta t \mid k \in {}^*\mathbb{N}_0\}$ *with an infinitesimal* $\Delta t > 0$ *– such that*

$$F(\cdot,\cdot) = E_K(\cdot,\cdot). \tag{1.6.2}$$

Moreover, if $\{G_\alpha \mid \alpha \in {}^*(-\infty, 0)\}$ *and* $\{\hat{G}_\alpha \mid \alpha \in {}^*(-\infty, 0)\}$ *are the resolvent and co-resolvent generated by* $\mathcal{E}(\cdot,\cdot)$*, then for all* $\beta \in (-\infty, 0), \alpha \in {}^*(-\infty, 0), u \in K, v \in H$ *such that* $\beta = {}^\circ\alpha, u = st_K(v)$*, we have*

$$st_K G_\alpha v = R_\beta u \text{ and } st_K \hat{G}_\alpha v = \hat{R}_\beta u. \tag{1.6.3}$$

On the other hand, if $\{Q^s \mid s \in T\}$ *and* $\{\hat{Q}^s \mid s \in T\}$ *are the semigroup and co-semigroup generated by* $\mathcal{E}(\cdot,\cdot)$*, then for all* $t \in [0, \infty), s \in T, u \in K, v \in H$ *such that* $t = {}^\circ s, u = st_K(v)$*, we have*

$$st_K Q^s v = T_t u \text{ and } st_K \hat{Q}^s v = \hat{T}_t u. \tag{1.6.4}$$

Proof. Let P be the projection of ${}^*K$ on H. Let us write $\{{}^*R_\alpha\}_{\alpha\in{}^*(-\infty,0)}$ for ${}^*(\{R_\alpha\}_{\alpha\in(-\infty,0)})$, and $\{{}^*\hat{R}_\alpha\}_{\alpha\in{}^*(-\infty,0)}$ for ${}^*(\{\hat{R}_\alpha\}_{\alpha\in(-\infty,0)})$. Our plan is first to define an internal semigroup by putting $Q^{\Delta t} = -P^*(\gamma R_\gamma)$ for a carefully chosen infinitesimal $\Delta t = -\frac{1}{\gamma}$, and then let

$$\begin{aligned}\mathcal{E}(u,v) &= \frac{1}{\Delta t}\langle (I - Q^{\Delta t})u, v\rangle \\ &= -\gamma\langle (I + P^*(\gamma R_\gamma))u, v\rangle, \end{aligned} \tag{1.6.5}$$

by using Proposition 1.6.3. Notice that if $u \in H$, then

$$\begin{aligned}\langle Q^{\Delta t}u, u\rangle &= -\langle P^*(\gamma R_\gamma)u, u\rangle \\ &= -\langle {}^*(\gamma R_\gamma)u, u\rangle \\ &\geq 0.\end{aligned}$$

This shows that $Q^{\Delta t}$ is positive on H. Also, since the operator norm of ${}^*(\gamma R_\gamma)$ is less than or equal to one, so is the norm of $Q^{\Delta t}$. Hence, the conditions in Sect. 1.1 are satisfied.

We shall now choose Δt such that the relations (1.6.2), (1.6.3), and (1.6.4) hold. If $u \in H$ is nearstandard and $-{}^\circ\alpha < \infty$, then

$$||P^*(\alpha R_\alpha)u - {}^*(\alpha R_\alpha)u|| \approx 0 \text{ and } ||P^*(\alpha \hat{R}_\alpha)u - {}^*(\alpha \hat{R}_\alpha)u|| \approx 0$$

because ${}^*(\alpha R_\alpha)$ and ${}^*(\alpha\hat{R}_\alpha)$ take the nearstandard elements to nearstandard elements, and because H is S-dense in ${}^*K$. By induction, we get

$$||[P^*(\alpha R_\alpha)]^n u - {}^*(\alpha R_\alpha)^n u|| \approx 0 \text{ and } ||[P^*(\alpha \hat{R}_\alpha)]^n u - {}^*(\alpha \hat{R}_\alpha)^n u|| \approx 0$$

for all $n \in \mathbb{N}$. For each $u \in K$, let $v_u = Pu$. Then $v_u \in H$ and $||u - v_u|| \approx 0$. We consider the set

$$\begin{aligned}A_u = \Big\{ n \in {}^*\mathbb{N} \,\Big|\, \forall k \leq 2^{2n} \Big(&||[P^*(2^n R_{-2^n})]^k v_u - [{}^*(2^n R_{-2^n})]^k u|| \leq \frac{1}{n} \text{ and} \\ &||[P^*(2^n \hat{R}_{-2^n})]^k v_u - [{}^*(2^n \hat{R}_{-2^n})]^k u|| \leq \frac{1}{n}\Big)\Big\}.\end{aligned} \tag{1.6.6}$$

For each $u \in K$, this set contains $\mathbb{N}$, and hence an internal segment $\{n \in {}^*\mathbb{N} \mid n \leq n_u\}$. By saturation, there is an infinite n smaller than all the n_u's, $u \in K$.

Next, we consider

$$B_u = \Big\{ m \in {}^*\mathbb{N} \Big| \forall k \leq 2^{2m} \Big(|k\langle (I + kR_{-k})v_u, v_u \rangle$$
$$-k\langle (I + P^*(kR_{-k}))v_u, v_u\rangle| \leq \frac{1}{m} \Big) \Big\}.$$

For each $u \in K$, this set contains $\mathbb{N}$, and hence an internal segment $\{m \in {}^*\mathbb{N} \mid m \leq m_u\}$. By saturation, there is an infinite m smaller than all the m_u's, $u \in K$.

We now take Δt to be the infinitesimal $2^{-m} \vee 2^{-n}$. That is, $\Delta t = 2^{-m} \vee 2^{-n}$, or $\gamma = -2^m \wedge 2^n$. Since $R_\alpha = (L - \alpha)^{-1}$, we have $L = \alpha + R_\alpha^{-1}$ for all $\alpha \in (-\infty, 0)$. Hence, we have for all $u \in D(L) = R_1(K)$,

$$\begin{aligned} -\alpha(I + \alpha R_\alpha)u &= -\alpha R_\alpha(R_\alpha^{-1} + \alpha)u \\ &= -\alpha R_\alpha(Lu) \\ &\longrightarrow Lu, \alpha \longrightarrow -\infty, \alpha \in (-\infty, 0). \end{aligned}$$

Therefore, we get for all infinite $\alpha \in {}^*(-\infty, 0)$

$$\begin{aligned} 0 &\approx ||Lu + \alpha R_\alpha(Lu)|| \\ &= ||Lu + \alpha(I + \alpha R_\alpha)u||, \forall u \in D(L). \end{aligned} \tag{1.6.7}$$

From the relations (1.6.5) and (1.6.7), we have for all $u \in D(L) = R_1(K)$ that

$$^\circ\mathcal{E}(Pu, Pu) = F(u, u).$$

Because of the closedness of $F(\cdot, \cdot)$, we have got the following equation for all $u \in D(F)$

$$\begin{aligned} ^\circ\mathcal{E}(Pu, Pu) &= E_K(u, u) \\ &= F(u, u). \end{aligned}$$

This is (1.6.2). From (1.6.6) and Proposition 1.6.5, we have the results (1.6.4).

In the following, we shall show that $\mathcal{E}(\cdot, \cdot)$ satisfies (1.6.3). For all $\beta \in (-\infty, 0)$, $\alpha \in {}^*(-\infty, 0), u \in K, v \in H$ such that $\beta = {}^\circ\alpha, u = st_K(v)$, we have

$$\mathcal{E}_{-\alpha}(G_\alpha v, G_\alpha v - R_\alpha v_u) = \langle v, G_\alpha v - R_\alpha v_u \rangle$$

and

$$\begin{aligned}
&\mathcal{E}_{-\alpha}(R_\alpha v_u, G_\alpha v - R_\alpha v_u) \\
&\quad = -\gamma\langle (I + P^*(\gamma R_\gamma))R_\alpha v_u, G_\alpha v - R_\alpha v_u\rangle - \alpha\langle R_\alpha v_u, G_\alpha v - R_\alpha v_u\rangle \\
&\quad \approx -\gamma\langle (I + \gamma R_\gamma))R_\alpha v_u, G_\alpha v - R_\alpha v_u\rangle - \alpha\langle R_\alpha v_u, G_\alpha v - R_\alpha v_u\rangle \\
&\quad \approx \langle L R_\alpha v_u, G_\alpha v - R_\alpha v_u\rangle - \alpha\langle R_\alpha v_u, G_\alpha v - R_\alpha v_u\rangle \\
&\quad \approx \langle v, G_\alpha v - R_\alpha v_u\rangle.
\end{aligned}$$

Hence, we have

$$\begin{aligned}
|\alpha| \cdot ||G_\alpha v - R_\alpha v_u|| &\le \mathcal{E}_{-\alpha}(G_\alpha v - R_\alpha v_u, G_\alpha v - R_\alpha v_u) \\
&\approx 0.
\end{aligned}$$

This implies the relations (1.6.3). □

Now let us make a few comments on Theorem 1.6.1.

Remark 1.6.1. In the proof of Theorem 1.6.1, we apply Proposition 1.6.3 by using the resolvent $\{R_\alpha \mid \alpha \in (-\infty, 0)\}$ to construct $Q^{\Delta t}$. An alternative way is to apply Proposition 1.6.4 by using the semigroup $\{T_t \mid t \in (0,\infty)\}$ to construct $Q^{\Delta t}$, and the readers may refer to the proof of 5.2.1. Proposition, Albeverio et al. [25] for the details.

Remark 1.6.2. The assumption that $(F(\cdot,\cdot), D(F))$ is densely defined is for convenience only. If it is not satisfied, we just apply the proposition to the closure of $D(F)$. If $F(\cdot,\cdot)$ is not closed, we obviously cannot obtain $F(\cdot,\cdot)$ as $E_K(\cdot,\cdot)$ for any hyperfinite form $\mathcal{E}(\cdot,\cdot)$ since we need the closedness of $F(\cdot,\cdot)$ in the proof of Theorem 1.6.1. However, if $F(\cdot,\cdot)$ is closable (i.e., there exists a closed form extending $F(\cdot,\cdot)$), all closed extensions of $F(\cdot,\cdot)$ can be represented as standard parts of hyperfinite forms. A natural representation for a closed form $F(\cdot,\cdot)$ would be a representation of its smallest closed extension – the Friedrichs extension. If $F(\cdot,\cdot)$ is not closable, no hyperfinite representation (in our sense) is possible. Any representation we try will change some $F(\cdot,\cdot)$ values, and restrict and extend $D(F)$ in different directions in order to turn $F(\cdot,\cdot)$ into a closed form. As we commented in Sect. 1.2, the fact that non-closable forms do not have hyperfinite representation is more than a curse. In the standard theory, a lot of effort goes into showing that the forms one constructs are closable. In the hyperfinite theory, this is an immediate consequence of the construction.

Remark 1.6.3. In Theorem 1.6.1, the space H was just any S-dense, hyperfinite dimensional subspace of ${}^*K$. In applications, we often want to choose special kind of subspaces which are appropriate for the problems we have in mind. In Chap. 4, we shall study the case $K = L^2(Y, \nu)$ for some Hausdorff space Y.

Remark 1.6.4. The lifting $\mathcal{E}(\cdot,\cdot)$ for a coercive closed form $F(\cdot,\cdot)$ in Theorem 1.6.1 needs not satisfy the hyperfinite weak sector condition. Hence, $\mathcal{E}(\cdot,\cdot)$ is not necessarily a hyperfinite weak coercive quadratic form. This makes it impossible for us to use the results in Sect. 1.4. However, we do have some good property about $\mathcal{E}(\cdot,\cdot)$ which will be very useful in Chap. 4. Actually, we have the following from Proposition 1.6.2:

Corollary 1.6.1. *Let $(F(\cdot,\cdot), D(F))$ be a coercive closed form on K with continuity constant $C > 0$. For $\alpha \in {}^*(-\infty, 0)$, we define for $u, v \in {}^*K$*

$$\begin{aligned} {}^{(\alpha)}F(u,v) &= -\alpha(u + \alpha R_\alpha u, v) \\ &= -\alpha(v + \alpha \hat{R}_\alpha v, u) \end{aligned}$$

and

$$\begin{aligned} {}^{(\alpha)}\hat{F}(u,v) &= -\alpha(v + \alpha R_\alpha v, u) \\ &= -\alpha(u + \alpha \hat{R}_\alpha u, v). \end{aligned}$$

Then

(i) ${}^{(\alpha)}F(u,v) = F(-\alpha R_\alpha u, v)$ *and* ${}^{(\alpha)}\hat{F}(u,v) = F(-\alpha \hat{R}_\alpha u, v)$ *for all* $u \in {}^*K, v \in {}^*(D(F))$ *and* $\alpha \in {}^*(-\infty, 0)$.
(ii) $F(\alpha R_\alpha u, \alpha R_\alpha u) \le {}^{(\alpha)}F(u,u)$ *and* $F(\alpha \hat{R}_\alpha u, \alpha \hat{R}_\alpha u) \le {}^{(\alpha)}F(u,u)$ *for all* $u \in {}^*K$ *and* $\alpha \in {}^*(-\infty, 0)$.
(iii) $|{}^{(\alpha)}F_1(u,v)| \le (C+1)\sqrt{F_1(u,u)}\sqrt{{}^{(\alpha)}F_1(v,v)}$ *for all* $u \in {}^*(D(F))$, $v \in {}^*K$ *and* $\alpha \in {}^*(-\infty, 0)$.
(iv) $F_1(\alpha R_\alpha u, \alpha R_\alpha u) \le (C+1)^2 F_1(u,u)$ *for all* $u \in {}^*(D(F))$ *and* $\alpha \in {}^*(-\infty, 0)$.

Proof. The proof follows from Proposition 1.6.2. □

When are the standard forms generated by two hyperfinite forms different? The last result in this section we shall prove shows that to answer this question, it is enough to check whether the forms have the same resolvents. We recall that in Theorem 1.4.1 we found a way to construct a form from its resolvent. This representation will be helpful to solve our problem.

Theorem 1.6.2. *Let K be a Hilbert space and H be an S-dense, hyperfinite dimensional subspace of ${}^*K$. Let $\mathcal{E}(\cdot,\cdot)$ and $\breve{\mathcal{E}}(\cdot,\cdot)$ be two hyperfinite weak coercive quadratic forms on H inducing $E_K(\cdot,\cdot)$ and $\breve{E}_K(\cdot,\cdot)$ on K, respectively. Let $\{G_\alpha\}$ and $\{\breve{G}_\alpha\}$ be the resolvents of $\mathcal{E}(\cdot,\cdot)$ and $\breve{\mathcal{E}}(\cdot,\cdot)$. Assume that for some finite, non-infinitesimal $\alpha \in {}^*(-\infty, 0)$, there is a $u \in H$ with ${}^\circ\mathcal{E}_1(u,u) < \infty$ such that $v = G_\alpha u$, $w = \breve{G}_\alpha u$ are both nearstandard, but ${}^\circ||v - w|| \ne 0$. Then $E_K(\cdot,\cdot) \ne \breve{E}_K(\cdot,\cdot)$.*

Proof. Assume for contradiction that $E_K(\cdot,\cdot) = \breve{E}_K(\cdot,\cdot)$. Pick $\tilde{v} \approx v$, $\tilde{w} \approx w$ such that $\tilde{v} \in \mathcal{D}(\breve{\mathcal{E}}), \tilde{w} \in \mathcal{D}(\mathcal{E})$. Notice that by Lemma 1.4.5, $v \in \mathcal{D}(\mathcal{E}), w \in \mathcal{D}(\breve{\mathcal{E}})$. We have

$$\begin{aligned}\langle u, v - w\rangle &\approx \langle u, v - \tilde{w}\rangle \\ &= \mathcal{E}_{-\alpha}(v, v - \tilde{w}).\end{aligned} \tag{1.6.8}$$

Since v, w are nearstandard and $E_K(\cdot,\cdot) = \breve{E}_K(\cdot,\cdot)$, we have

$$\begin{aligned}{}^\circ\mathcal{E}_{-\alpha}(v, v - \tilde{w}) &= E_K({}^\circ v, {}^\circ v - {}^\circ w) - ({}^\circ\alpha)({}^\circ v, {}^\circ v - {}^\circ w) \\ &= \breve{E}_K({}^\circ v, {}^\circ v - {}^\circ w) - ({}^\circ\alpha)({}^\circ v, {}^\circ v - {}^\circ w) \\ &= {}^\circ\breve{\mathcal{E}}_{-\alpha}(\tilde{v}, \tilde{v} - w).\end{aligned} \tag{1.6.9}$$

On the other hand, we have

$$\begin{aligned}\langle u, v - w\rangle &\approx \langle u, \tilde{v} - w\rangle \\ &= \breve{\mathcal{E}}_{-\alpha}(w, \tilde{v} - w).\end{aligned} \tag{1.6.10}$$

Combining the relations (1.6.8), (1.6.9), and (1.6.10), we see that

$$\begin{aligned}0 &= {}^\circ\breve{\mathcal{E}}_{-\alpha}(\tilde{v} - w, \tilde{v} - w) \\ &\geq {}^\circ|\alpha|^\circ||v - w||^2 \\ &> 0.\end{aligned}$$

The theorem is proved. □

1.7 Weak Coercive Quadratic Forms, Revisited

Let $\mathcal{E}(\cdot,\cdot)$ be a hyperfinite weak coercive quadratic form on a hyperfinite dimensional space H. Let $\{G_\alpha \mid \alpha \in {}^*(-\infty, 0)\}$ be the resolvent of $\mathcal{E}(\cdot,\cdot)$, and let $\{\hat{G}_\alpha \mid \alpha \in {}^*(-\infty, 0)\}$ be the co-resolvent of $\mathcal{E}(\cdot,\cdot)$, respectively. Let us still denote by A the infinitesimal generator of $\mathcal{E}(\cdot,\cdot)$, and by $\hat{A}$ the infinitesimal co-generator of $\mathcal{E}(\cdot,\cdot)$. Such as in Sect. 1.1, we fix an infinitesimal Δt, and we define new operators $Q^{\Delta t}$ and $\hat{Q}^{\Delta t}$ by

$$\begin{aligned}Q^{\Delta t} &= I - \Delta t A, \\ \hat{Q}^{\Delta t} &= I - \Delta t \hat{A}.\end{aligned}$$

Introduce a nonstandard time line T by $T = \{k\Delta t \mid k \in {}^*\mathbb{N}_0\}$. For each element $t = k\Delta t \in T$, define the semigroup Q^t and the co-semigroup $\hat{Q}^t$ to be the families of operators

$$\begin{aligned}Q^t &= (Q^{\Delta t})^k, \\ \hat{Q}^t &= (\hat{Q}^{\Delta t})^k, t \in T.\end{aligned}$$

For each $t \in T$, we may define approximations $A^{(t)}$ of A and $\hat{A}^{(t)}$ of $\hat{A}$ by

$$A^{(t)} = \frac{1}{t}\left(I - Q^t\right),$$
$$\hat{A}^{(t)} = \frac{1}{t}\left(I - \hat{Q}^t\right).$$

From $A^{(t)}$ and $\hat{A}^{(t)}$, we get the forms $\mathcal{E}^{(t)}(u,v) = \langle A^{(t)}u, v\rangle = \langle u, \hat{A}^{(t)}v\rangle$.

Let $E(\cdot,\cdot)$ be the standard part of $\mathcal{E}(\cdot,\cdot)$. Then $E(\cdot,\cdot)$ is closed. In addition, $(E(\cdot,\cdot), D(E))$ satisfies the weak sector condition by Remark 1.4.1. Hence, $(E(\cdot,\cdot), D(E))$ is a coercive closed form on ${}^\circ H$. Let $\{R_\beta \mid \beta \in (-\infty, 0)\}$ and $\{\hat{R}_\beta \mid \beta \in (-\infty, 0)\}$ be the resolvent and co-resolvent of $(E(\cdot,\cdot), D(E))$, respectively. Similarly, let $\{T_t \mid t \in [0,\infty)\}$ and $\{\hat{T}_t \mid t \in [0,\infty)\}$ be the semigroup and co-semigroup of $(E(\cdot,\cdot), D(E))$, respectively. For $t \in (0,\infty)$, we define

$$E^{(t)}(x,y) = \frac{1}{t}(x - T_t x, y), x, y \in {}^\circ H.$$

By Albeverio et al. [9], Theorem 3.4 (or referring to Proposition 1.6.4), we have

Lemma 1.7.1. *(i) Let $x \in {}^\circ H$. Then $x \in D(E)$ if and only if $\sup_{t>0} E^{(t)}(x,x) < \infty$.*

(ii) For all $x, y \in D(E)$, we have

$$\lim_{t \downarrow 0} E^{(t)}(x,y) = E(x,y).$$

(iii) For all $x \in D(E)$, we have

$$\lim_{t \downarrow 0} E_1(x - T_t x, x - T_t x) = 0.$$

By applying Lemma 1.7.1, we can see from the proof of Theorem 1.6.1 that there exists an infinitesimal $\delta \in T$ such that $(E(\cdot,\cdot), D(E))$ is the standard part of $\mathcal{E}^{(\delta)}(u,v)$ (referring to Remark 1.6.1, and replacing $-\gamma R_\gamma$ by Q^δ in the proof of Theorem 1.6.1). In addition, for any $x \in {}^\circ H, u \in x$, and for all $t \in [0,\infty), s \in \{k\delta \mid k \in {}^*\mathbb{N}_0\}, t = {}^\circ s$, we have that $Q^s u \in T_t x$. Since $Q^{s_1+s_2} \approx Q^{s_1}$ if $s_2 \approx 0$, we have that $Q^s u \in T_t x$ for all $t \in [0,\infty), s \in T = \{k\Delta t \mid k \in {}^*\mathbb{N}_0\}, t = {}^\circ s$. Notice that $(E(\cdot,\cdot), D(E))$ is the standard part of both $\mathcal{E}(u,v)$ and $\mathcal{E}^{(\delta)}(u,v)$, and so the resolvent of $\mathcal{E}(u,v)$ is almost the same as that of $\mathcal{E}^{(\delta)}(u,v)$ by Theorem 1.6.2.

Summarizing above results, we have

Theorem 1.7.1. *Let $\mathcal{E}(\cdot,\cdot)$ be a hyperfinite weak coercive quadratic form on a hyperfinite dimensional space H. Then, we have*

(i) Let $u \in H$. *Then* $u \in \mathcal{D}(\mathcal{E})$ *if and only if* $\sup_s {}^\circ\mathcal{E}^{(s)}(u,u) < \infty$.
(ii) For all $u, v \in \mathcal{D}(\mathcal{E})$, *we have*

$$\lim_{^\circ s \downarrow 0} \mathcal{E}^{(s)}(u,v) = \mathcal{E}(u,v).$$

(iii) For all $u \in \mathcal{D}(\mathcal{E})$, *we have*

$$\lim_{^\circ s \downarrow 0} \mathcal{E}_1(u - Q^s u, u - Q^s u) \approx 0.$$

Proof. Let $x = {}^\circ u$. Then for all $t \in [0,\infty), s \in T = \{k\Delta t \mid k \in {}^*\mathbb{N}_0\}$, $t = {}^\circ s$, we have that $Q^s u \in T_t x$. By Lemma 1.7.1, the theorem follows easily. □

Proposition 1.7.1. *Let* $\mathcal{E}(\cdot,\cdot)$ *be a hyperfinite weak coercive quadratic form on a hyperfinite dimensional linear space* H. *If* ${}^\circ\mathcal{E}(u,u) < \infty$, *then for all finite* $s > 0, s \in T$, *and* ${}^\circ s \neq 0$, *we have* ${}^\circ[Q^s u] \in D(E)$.

Proof. Let $x = {}^\circ u$. For $t = {}^\circ s$, we have that $Q^s u \in T_t x$. Therefore, we have ${}^\circ[Q^s u] = T_t x \in D(E)$. □

Chapter 2
Potential Theory of Hyperfinite Dirichlet Forms

Probabilistic potential theory has been a very important component in the study of hyperfinite Dirichlet space theory. It provides a probabilistic interpretation of potential theory; and, more generally, it establishes a beautiful bridge between functional analysis and the theory of Markov processes. There are many applications of this theory, especially in the area of infinite dimensional stochastic analysis and mathematical physics. Our purpose in this chapter is to develop the probabilistic potential theory associated with hyperfinite Dirichlet forms and the related Markov chains. The motivation is twofold. On the one hand, we want to establish a relationship between the standard Dirichlet space theory and the hyperfinite counterpart. On the other hand, we want to provide new methods for the theory of hyperfinite Dirichlet forms itself. Infinite dimensional stochastic analysis has been developed extensively in the last decades. We hope to convince the reader that nonstandard analysis can provide a new tool to deal with problems in this exciting area, see particularly Chap. 4, for example.

The arrangement of the present chapter is as follows. In Sect. 2.1, we shall define exceptional sets for non-symmetric hyperfinite Markov chains. Sect. 2.2 will discuss excessive functions and equilibrium potentials. Moreover, we introduce a capacity theory for hyperfinite quadratic forms and show that it is a Choquet capacity in Sect. 2.3. Furthermore, we establish a relation between the family of exceptional sets and the family of zero capacity sets in Sect. 2.4. In Sect. 2.5, we consider positive measures of hyperfinite energy integrals and the associated theory. That is, we establish connections among hyperfinite excessive functions and hyperfinite potentials. Zero capacity subsets will be characterized by positive measures of hyperfinite energy integrals. In Sect. 2.6, we introduce internal additive functionals. The relationship between hyperfinite measures and additive functionals will be considered. Moreover, we shall obtain a positive hyperfinite measure $\mu_{<u>}$ associated with an internal function u. In Sect. 2.7, we get Fukushima's decomposition theorem under individual probability measures. This extends the work of Albeverio et al. [25] and Fan [166]. In Sect. 2.8, we shall discuss the properties of internal multiplicative functionals, subordinate semigroups, subprocesses,

S. Albeverio et al., *Hyperfinite Dirichlet Forms and Stochastic Processes*, Lecture Notes of the Unione Matematica Italiana 10,
DOI 10.1007/978-3-642-19659-1_2,

and a Feynman-Kac formula. The motivation for this work is given by the corresponding standard theory developed by Blumenthal and Getoor [96]. This serves as a basis for the development of a perturbation theory of hyperfinite Dirichlet forms characterized by internal additive functionals. The content of Sect. 2.9 is the nonstandard version of time change of standard Dirichlet forms. In standard Dirichlet space theory, the question how to change a non-conservative symmetric Markov process into a conservative one has received an answer in Fukushima and Takeda [184] in terms of the Girsanov transformation. In Sect. 2.10, we show that this problem is quite simple in the hyperfinite setting (Theorem 2.10.1).

In this chapter, we shall use the notations developed in Sect. 1.5. However, we need not assume all the conditions required in Sect. 1.5. In Sect. 2.1, we will define exceptional sets for a hyperfinite Markov chain X. We do not need anything about the dual hyperfinite Markov chain of X. Starting from Sect. 2.2, we shall work under the setting of Sect. 1.5. That is, we shall assume the conditions (1.5.1), (1.5.2), (1.5.4), (1.5.5), (1.5.8), (1.5.9), (1.5.10), (1.5.11), (1.5.13), and (1.5.14) in Sect. 1.5, and the related notations.

2.1 Exceptional Sets

Albeverio et al. [25] has given us a definition of exceptional sets in the framework of the theory of symmetric hyperfinite Dirichlet forms, which is too restrictive for certain cases. In Fan [166], we extended this concept in the same symmetric setting by removing a lot of unnecessary assumptions. In this section, we shall define exceptional sets for non-symmetric hyperfinite Markov chains.

2.1.1 Exceptional Sets

We take an infinitesimal Δt such that $\Delta t > 0$. Set

$$T = \{k\Delta t \mid k \in {}^*\mathbb{N}_0\}\,. \tag{2.1.1}$$

Let Y be a Hausdorff space and let ${}^*Y$ be the nonstandard extension of Y. Let $S = \{s_0, s_1, \ldots, s_N\}$ be an S-dense subset of ${}^*Y$ for some $N \in {}^*\mathbb{N} - \mathbb{N}$ and let m be a hyperfinite measure on S. Denote by $\mathcal{S}$ the internal algebra of subsets of S. Let $Q = \{q_{ij}\}$ be an $(N+1) \times (N+1)$ matrix with nonnegative entries. Assume that

$$\sum_{j=0}^{N} q_{ij} = 1 \quad \text{for all} \quad i = 0, 1, \ldots, N, \tag{2.1.2}$$

and assume that the state s_0 is a trap, i.e.,

$$q_{0i} = 0 \quad \text{for all} \quad i \neq 0. \tag{2.1.3}$$

If (Ω, P) is an internal measure space, and $X : \Omega \times T \longrightarrow S$ is an internal process, let

$$[\omega]_t = \{\omega' \in \Omega \mid X(\omega', s) = X(\omega, s) \text{ for all } s \leq t\}. \tag{2.1.4}$$

For each $t \in T$, let $\mathcal{F}_t$ be the internal algebra on Ω generated by the sets $[\omega]_t$.

Assume that for all $\omega \in \Omega$, we have

$$P([\omega]_0) = m\{X(\omega, 0)\}, \tag{2.1.5}$$

and whenever $X(\omega, t) = s_i$, we have

$$P\{\omega' \in [\omega]_t \mid X(t + \Delta t, \omega') = s_j\} = q_{ij} P([\omega]_t). \tag{2.1.6}$$

That is, X is a hyperfinite Markov chain with the initial distribution m and transition matrix Q. The family $(\Omega, \mathcal{F}_t, P_i, i \in S)$ of internal probability spaces is defined by

$$P_i\Big([\omega]_{k\Delta t}\Big) = \delta_{i\omega(0)} \prod_{n=0}^{k-1} q_{\omega(n\Delta t), \omega((n+1)\Delta t)}, \tag{2.1.7}$$

for each $i \in S$, where δ_{ij} is the Kronecker symbol as in Chap. 1.

As in Sect. 1.5, let $\delta \in T$ and let the sub-line

$$T_\delta = \{k\delta \mid k \in {}^*\mathbb{N}_0\}.$$

Set

$$\begin{aligned} T^{\text{fin}} &= \{t \in T \mid t \text{ is finite}\}, \\ T_\delta^{\text{fin}} &= \{t \in T_\delta \mid t \text{ is finite}\}. \end{aligned}$$

In addition, for $r \in T$, we let

$$\begin{aligned} T^r &= \{t \in T \mid t \leq r\}, \\ T_\delta^r &= \{t \in T_\delta \mid t \leq r\}. \end{aligned}$$

Moreover, we know that $X^{(\delta)}$ is the restriction $X|T_\delta$.

For every $y \in Y$, let us define the *monad* $\mu(y)$ of y by

$$\mu(y) = \bigcap \{{}^*O \mid O \text{ is open such that } \quad y \in O\}.$$

We call a point $y \in {}^*Y$ *nearstandard* if and only if $y \in \mu(x)$ for some $x \in Y$. Denote by $Ns({}^*Y)$ the set of all nearstandard points in ${}^*Y$. Since Y is a Hausdorff topological space, each element $y \in Ns({}^*Y)$ is nearstandard to exactly one element x in Y (we refer to page 48, 2.1.6. Proposition, [25]). We call x the *standard part* of y and denote it by ${}^\circ y$ or $\mathrm{st}(y)$. In particular, we can take $Y = \mathbb{R}$ and use this notation also.

Definition 2.1.1. (i) A subset B of S_0 is called *δ-exceptional for X* if

$$L(P)\left\{\omega \middle| \exists t \in T_\delta^{\text{fin}}\Big(X(\omega,t) \in B\Big)\right\} = 0, \tag{2.1.8}$$

where $L(P)$ is the Loeb measure of P.

(ii) A subset B of S_0 is called *exceptional* for X if it is δ-exceptional for some infinitesimal $\delta \in T$.

Remark 2.1.1. For symmetric hyperfinite Markov chains, δ-exceptional sets are defined in Albeverio et al. [25] in the following way

$$L(P)\left\{\omega \middle| (X(\omega,0) \in \overline{S}_0) \wedge \Big(\exists t \in T_\delta^{\text{fin}}\Big(X(\omega,t) \in B\Big)\Big)\right\} = 0, \tag{2.1.9}$$

where $\overline{S}_0 = S_0 \cap Ns({}^*Y)$. Therefore, if a subset B is δ-exceptional in the sense of our Definition 2.1.1, it is δ-exceptional in the sense of (2.1.9). We have noticed this result in Fan [166] for the symmetric case. Here we deal with the general hyperfinite non-symmetric Markov chain $X(t)$.

Remark 2.1.2. From (2.1.8), we see that for every exceptional set B

$$L(P)\{\omega \mid X(\omega,0) \in B\} = 0.$$

This implies that $L(m)(B) = 0$, where $L(m)$ is the Loeb measure of m.

We have the following lemma, whose proof is easy and therefore will be omitted.

Lemma 2.1.1. *(i) All internal subsets $B \subset S_0$ with $m(B) \approx 0$ are exceptional.*

(ii) The families of exceptional and δ-exceptional sets are closed under countable unions.

Definition 2.1.2. (i) A δ-exceptional subset A of S_0 is called *properly δ-exceptional* for X if there is a family $\{B_{m,n} \mid m, n \in \mathbb{N}\}$ of internal subsets such that

$$A = \bigcup_{m \in \mathbb{N}} \bigcap_{n \in \mathbb{N}} B_{m,n}$$

and for all $s_i \notin A$,

$$L(P_i)\left\{\omega \middle| \exists t \in T_\delta^{\text{fin}}\Big(X(\omega, t) \in A\Big)\right\} = 0,$$

where $L(P_i)$ is the Loeb measure of P_i.

(ii) A subset A of S_0 is called *properly exceptional* for X if it is properly δ-exceptional for some $\delta \approx 0$, $\delta \in T$.

Proposition 2.1.1. *If $A \subset S_0$ is a δ-exceptional set, there is a properly δ-exceptional set $B \supset A$.*

Proof. Since A is δ-exceptional, there is an internal subset B_{mn} for each pair (m, n) of natural numbers such that

$$A \subset B_{mn}, \quad L(P)\left\{\omega \middle| \exists t \in T_\delta^{\text{fin}}\Big(X(\omega, t) \in B_{mn}\Big)\right\} \le \frac{1}{n^2 m}. \tag{2.1.10}$$

Define

$$C_{mn} = \left\{i \in S \middle| P_i\left\{\exists t \in T_\delta^m\Big(X(t) \in B_{mn}\Big)\right\} \ge \frac{1}{nm}\right\}$$

and

$$\bar{A} = \bigcup_{m \in \mathbb{N}} \bigcap_{n \in \mathbb{N}} C_{mn}.$$

Let $\sigma(\omega)$ be a stopping time defined by

$$\sigma(\omega) = \min\left\{t \in T_\delta \middle| P_{X(\omega,t)}\left\{\exists s \in T_\delta^m\Big(X(s) \in B_{mn}\Big)\right\} \ge \frac{1}{mn}\right\}.$$

Then, we have

$$\begin{aligned}
&L(P)\left\{\omega \middle| \exists t \in T_\delta^{\text{fin}}\Big(X(\omega, t) \in C_{mn}\Big)\right\} \frac{1}{mn} \\
&\quad = L(P)\left\{\omega \middle| \exists t \in T_\delta^{\text{fin}}\left(P_{X(\omega,t)}\left\{\exists s \in T_\delta^m\Big(X(s) \in B_{mn}\Big)\right\} \ge \frac{1}{mn}\right)\right\} \frac{1}{mn}
\end{aligned}$$

$$
\begin{aligned}
&= L(P)\left\{1_{(^\circ\sigma(\omega)<\infty)}\right\}\frac{1}{mn} \\
&\leq L(P)\left\{1_{(^\circ\sigma(\omega)<\infty)}P_{X(\sigma(\omega))}\left(\exists s \in T_\delta^m\left(X(s) \in B_{mn}\right)\right)\right\} \\
&= L(P)\left\{\left(^\circ\sigma(\omega) < \infty\right)\bigcap\left(\exists s \in T_\delta^m\left(X(\omega, s+\sigma(\omega)) \in B_{mn}\right)\right)\right\} \\
&\leq L(P)\left\{\omega\middle|\exists s \in T_\delta^{\text{fin}}\left(X(\omega, s) \in B_{mn}\right)\right\}.
\end{aligned} \tag{2.1.11}
$$

Therefore, we get from the relations (2.1.10) and (2.1.11) that

$$
L(P)\left\{\omega\middle|\exists t \in T_\delta^{\text{fin}}\left(X(\omega,t) \in C_{mn}\right)\right\} \leq \frac{1}{n}.
$$

This implies that

$$
L(P)\left\{\omega\middle|\exists t \in T_\delta^{\text{fin}}\left(X(\omega,t) \in \bigcap_{n\in\mathbb{N}} C_{mn}\right)\right\} = 0.
$$

Therefore, $\bigcap_{n\in\mathbb{N}} C_{mn}$ is δ-exceptional. By Lemma 2.1.1 (ii), we know that

$$
\bar{A} = \bigcup_{m\in\mathbb{N}}\bigcap_{n\in\mathbb{N}} C_{mn}
$$

is also δ-exceptional. From the definition of the family $\{C_{mn} \mid n, m \in \mathbb{N}\}$, we see that for any $s_i \notin \bar{A}$,

$$
L(P_i)\left\{\omega\middle|\exists t \in T_\delta^{\text{fin}}\left(X(\omega,t) \in A\right)\right\} = 0. \tag{2.1.12}
$$

By using the principle of mathematical induction, we obtain an increasing sequence

$$
A_0 \subset A_1 \subset A_2 \subset \cdots
$$

of δ-exceptional sets, where $A_0 = A$ and $A_{i+1} = \bar{A}_i$ for every i. Put $B = \bigcup_{i\in\mathbb{N}} A_i$. Then B is a δ-exceptional set of the form

$$
B = \bigcup_{i\in\mathbb{N}}\bigcup_{m\in\mathbb{N}}\bigcap_{n\in\mathbb{N}} C_{mn}^{(i)}
$$

for some internal subsets $C_{mn}^{(i)}$. From (2.1.12), we know that if $s_i \notin B$, then

$$L(P_i)\left\{\omega\middle|\exists t\in T_\delta^{\text{fin}}\Big(X(\omega,t)\in B\Big)\right\}=0.$$

We have completed the proof of Proposition 2.1.1. □

Remark 2.1.3. The proof of Proposition 2.1.1 is similar to Albeverio et al. [25] 5.4.7 Lemma. For a similar result, see also Fan [166].

2.1.2 Co-Exceptional Sets

Let $\hat{Q}=\{\hat{q}_{ij}\}$ be an $(N+1)\times(N+1)$ matrix with nonnegative entries. Assume that

$$\sum_{j=0}^{N}\hat{q}_{ij}=1 \quad \text{for all} \quad i=0,1,\ldots,N, \tag{2.1.13}$$

and assume that the state s_0 is a trap, i.e.,

$$\hat{q}_{0i}=0 \quad \text{for all} \quad i\neq 0. \tag{2.1.14}$$

In the same manner as above, we shall write $\hat{q}_{ij}$ for $\hat{q}_{s_is_j}$, respectively, whenever it is convenient.

If $(\hat{\Omega},\hat{P})$ is an internal measure space, and $\hat{X}:\hat{\Omega}\times T\longrightarrow S$ is an internal process, let

$$[\hat{\omega}]_t=\{\hat{\omega}'\in\hat{\Omega}\mid\hat{X}(\hat{\omega}',s)=\hat{X}(\hat{\omega},s)\text{ for all }s\leq t\}. \tag{2.1.15}$$

For each $t\in T$, let $\hat{\mathcal{F}}_t$ be the internal algebra on $\hat{\Omega}$ generated by the sets $[\hat{\omega}]_t$.

Assume that for all $\hat{\omega}\in\hat{\Omega}$

$$\hat{P}([\hat{\omega}]_0)=m\{\hat{X}(\hat{\omega},0)\}, \tag{2.1.16}$$

and whenever $\hat{X}(\hat{\omega},t)=s_i$, we have

$$\hat{P}\{\hat{\omega}'\in[\hat{\omega}]_t\mid\hat{X}(t+\Delta t,\hat{\omega}')=s_j\}=\hat{q}_{ij}\hat{P}([\hat{\omega}]_t). \tag{2.1.17}$$

That is, $\hat{X}$ is a hyperfinite Markov chain with initial distribution m and transition matrix $\hat{Q}$. The family $(\hat{\Omega},\hat{\mathcal{F}}_t,\hat{P}_i,i\in S)$ of internal probability spaces is defined by

$$\hat{P}_i\Big([\hat{\omega}]_{k\Delta t}\Big) = \delta_{i\hat{\omega}(0)} \prod_{n=0}^{k-1} \hat{q}_{\hat{\omega}(n\Delta t),\hat{\omega}((n+1)\Delta t)}, \qquad (2.1.18)$$

for each $i \in S$.

Similarly as in Sect. 2.1.1, we have the following results about $\hat{X}$. Here, we remark that the dual condition (1.5.13) and the regular condition (1.5.14) of Sect. 1.5 of Chap. 1 are not necessary for the results in this section.

Definition 2.1.3. (i) A subset B of S_0 is called *δ-co-exceptional* for $\hat{X}$ if

$$L(\hat{P})\left\{\omega \middle| \exists t \in T_\delta^{\text{fin}}\Big(\hat{X}(\omega,t) \in B\Big)\right\} = 0,$$

where $L(\hat{P})$ is the Loeb measure of $\hat{P}$.

(ii) A subset B of S_0 is called *co-exceptional* for $\hat{X}$ if it is δ-co-exceptional for some infinitesimal $\delta \in T$.

Lemma 2.1.2. *(i) All internal subsets $B \subset S_0$ with $m(B) \approx 0$ are co-exceptional.*

(ii) The families of co-exceptional and δ-co-exceptional sets are closed under countable unions.

Definition 2.1.4. (i) A δ-co-exceptional subset A of S_0 is called *properly δ-co-exceptional* for $\hat{X}$ if there is a family $\{B_{m,n} \mid m, n \in \mathbb{N}\}$ of internal subsets such that

$$A = \bigcup_{m\in\mathbb{N}} \bigcap_{n\in\mathbb{N}} B_{m,n}$$

and for all $s_i \notin A$,

$$L(\hat{P}_i)\left\{\omega \middle| \exists t \in T_\delta^{\text{fin}}\Big(\hat{X}(\omega,t) \in A\Big)\right\} = 0,$$

where $L(\hat{P}_i)$ is the Loeb measure of $\hat{P}_i$.

(ii) A subset A of S_0 is called *properly co-exceptional* for $\hat{X}$ if it is properly δ-co-exceptional for some $\delta \approx 0$, $\delta \in T$.

Proposition 2.1.2. *If $A \subset S_0$ is a δ-co-exceptional set, there is a properly δ-co-exceptional set $B \supset A$.*

2.2 Excessive Functions and Equilibrium Potentials

Let Y be a Hausdorff space and $^*Y$ be the nonstandard extension of Y. Let $S = \{s_0, s_1, \ldots, s_N\}$ be an S-dense subset of $^*Y$ for some $N \in {}^*\mathbb{N} - \mathbb{N}$ and let m be a hyperfinite measure on S. Denote by $\mathcal{S}$ the internal algebra of subsets of S. Let X be a hyperfinite Markov chain with the initial distribution m and transition matrix Q, which is introduced in Sect. 2.1 by the conditions (2.1.1), (2.1.2), (2.1.3), (2.1.4), (2.1.5), (2.1.6), and (2.1.7). Moreover, let $\hat{X}$ be a hyperfinite Markov chain with the initial distribution m and transition matrix $\hat{Q}$, which is introduced in Sect. 2.1 by the conditions (2.1.13), (2.1.14), (2.1.15), (2.1.16), (2.1.17), and (2.1.18).

We assume the measure m and the transition matrices Q and $\hat{Q}$ satisfy the dual conditions

$$m_i q_{ij} = m_j \hat{q}_{ji} \quad \text{for all} \quad i \neq 0, j \neq 0. \tag{2.2.1}$$

Besides, we assume that

$$m_i \neq 0 \quad \text{for at least one} \quad i \neq 0. \tag{2.2.2}$$

Hence X and $\hat{X}$ are dual hyperfinite Markov chains.

We have defined an inner product $\langle \cdot, \cdot \rangle$ on the hyperfinite dimensional space H by

$$\begin{aligned} \int_{S_0} uv \, dm &= \langle u, v \rangle \\ &= \sum_{i=1}^{N} u(s_i) v(s_i) m(s_i). \end{aligned} \tag{2.2.3}$$

Let $\|\cdot\|$ be the norm generated by $\langle \cdot, \cdot \rangle$. Let $\{Q^t \mid t \in T\}$, A and $\mathcal{E}(\cdot, \cdot)$ be the semigroup, infinitesimal generator and hyperfinite quadratic form of X. Similarly, let $\{\hat{Q}^t \mid t \in T\}$, $\hat{A}$, and $\hat{\mathcal{E}}(\cdot, \cdot)$ be the co-semigroup, infinitesimal co-generator, and hyperfinite quadratic co-form of $\hat{X}$. We remark here that the reader may refer to Sect. 1.5 of Chap. 1 for the other notations.

Besides, we have for $\alpha \in {}^*\mathbb{R}$, $\alpha \geq 0$ and $\delta \in T$ that

$$\begin{aligned} \mathcal{E}_\alpha^{(\delta)}(u, v) &= \mathcal{E}^{(\delta)}(u, v) + \alpha \langle u, v \rangle, \\ \hat{\mathcal{E}}_\alpha^{(\delta)}(u, v) &= \hat{\mathcal{E}}^{(\delta)}(u, v) + \alpha \langle u, v \rangle. \end{aligned}$$

Each of these forms generates a norm (possibly a semi-norm in the case $\alpha = 0$):

$$\begin{aligned}|u|_\alpha^{(\delta)} &= \left[\mathcal{E}_\alpha^{(\delta)}(u,u)\right]^{\frac{1}{2}}\\ &= \left[\hat{\mathcal{E}}_\alpha^{(\delta)}(u,u)\right]^{\frac{1}{2}}.\end{aligned}$$

Definition 2.2.1. Fix $\alpha \in {}^*\mathbb{R}, \alpha \geq 0$ and $\delta \in T$.

(i) An element $u \in H$ is called *hyperfinite α-excessive* associated with $\mathcal{E}^{(\delta)}(\cdot,\cdot)$, if

$$u(i) \geq 0, \quad Q^\delta u(i) \leq (1+\alpha\delta)u(i)$$

for every $i \in S_0$.

(ii) An element $u \in H$ is called *hyperfinite α-co-excessive* associated with $\hat{\mathcal{E}}^{(\delta)}(\cdot,\cdot)$, if

$$u(i) \geq 0, \quad \hat{Q}^\delta u(i) \leq (1+\alpha\delta)u(i)$$

for every $i \in S_0$.

Lemma 2.2.1. *The following results hold:*

(i) Let $u \in H$ be hyperfinite α-excessive associated with $\mathcal{E}^{(\delta)}(\cdot,\cdot)$. Then

$$Q^{k\delta}u(i) \leq (1+\alpha\delta)^k u(i), \quad \forall k \in {}^*\mathbb{N}, \tag{2.2.4}$$

for every $i \in S_0$.

(ii) Let $u \in H$ be hyperfinite α-co-excessive associated with $\hat{\mathcal{E}}^{(\delta)}(\cdot,\cdot)$. Then

$$\hat{Q}^{k\delta}u(i) \leq (1+\alpha\delta)^k u(i), \quad \forall k \in {}^*\mathbb{N}, \tag{2.2.5}$$

for every $i \in S_0$.

Proof. (i) We have from the Markov property of X that

$$\begin{aligned}Q^{2\delta}u(i) &= E_i u(X(2\delta))\\ &= E_i E_{X(\delta)} u(X(\delta))\\ &\leq (1+\alpha\delta)E_i u(X(\delta))\\ &\leq (1+\alpha\delta)^2 u(i).\end{aligned}$$

Now it is easy to prove (2.2.4) by mathematical induction.

(ii) In the same way as above, we get (2.2.5). □

Remark 2.2.1. The results of Lemma 2.2.1 are true for all hyperfinite Markov chains X and $\hat{X}$ without dual condition (2.2.1) and regular condition (2.2.2).

Let A be an internal subset of S_0, and let $\delta \in T$. Define a stopping time $\sigma_A^{(\delta)}$ by

$$\sigma_A^{(\delta)}(\omega) = \min\left\{t \in T_\delta \mid X^{(\delta)}(\omega, t) \in A\right\}.$$

Let $f : A \longrightarrow {}^*\mathbb{R}$ be an internal function defined on A. For $\alpha \in {}^*\mathbb{R}, \alpha \geq 0$, set

$$e_\alpha^{(\delta)}(f)(i) = E_i\left[(1 + \alpha\delta)^{-\sigma_A^{(\delta)}/\delta} f\left(X^{(\delta)}(\sigma_A^{(\delta)})\right)\right].$$

Similarly, we define a stopping time $\hat{\sigma}_A^{(\delta)}$ by

$$\hat{\sigma}_A^{(\delta)}(\hat{\omega}) = \min\left\{t \in T_\delta \mid \hat{X}^{(\delta)}(\hat{\omega}, t) \in A\right\}.$$

For $\alpha \in {}^*\mathbb{R}, \alpha \geq 0$, set

$$\hat{e}_\alpha^{(\delta)}(f)(i) = \hat{E}_i\left[(1 + \alpha\delta)^{-\hat{\sigma}_A^{(\delta)}/\delta} f\left(\hat{X}^{(\delta)}(\hat{\sigma}_A^{(\delta)})\right)\right].$$

Then, we have

Proposition 2.2.1. *Assume that $\mathcal{E}(\cdot,\cdot)$ and $\hat{\mathcal{E}}(\cdot,\cdot)$ are the hyperfinite quadratic form and co-form associated with the dual hyperfinite Markov chains X and $\hat{X}$, respectively. Let $f : A \longrightarrow {}^*\mathbb{R}$ be an internal function defined on an internal subset A of S_0. Then $\mathcal{E}_\alpha^{(\delta)}\left(e_\alpha^{(\delta)}(f), u\right) = 0$ and $\mathcal{E}_\alpha^{(\delta)}\left(u, \hat{e}_\alpha^{(\delta)}(f)\right) = 0$ for all u which are zero on A.*

Proof. We notice that if $i \notin A$, then

$$(1 + \alpha\delta)\, e_\alpha^{(\delta)}(f)(i) = \sum_{j=1}^{N} e_\alpha^{(\delta)}(f)(j) q_{ij}^{(\delta)}.$$

If u is zero on A, then

$$\begin{aligned}
\mathcal{E}_\alpha^{(\delta)}\left(e_\alpha^{(\delta)}(f), u\right) &= \frac{1}{\delta}\sum_{i=1}^{N}\left[e_\alpha^{(\delta)}(f)(i) - \sum_{j=1}^{N} e_\alpha^{(\delta)}(f)(j) q_{ij}^{(\delta)}\right] u(i) m_i \\
&\quad + \alpha \sum_{i=1}^{N} e_\alpha^{(\delta)}(f)(i) u(i) m_i \\
&= \frac{1}{\delta}\sum_{i=1}^{N}\left[(1 + \alpha\delta) e_\alpha^{(\delta)}(f)(i) - \sum_{j=1}^{N} e_\alpha^{(\delta)}(f)(j) q_{ij}^{(\delta)}\right] u(i) m_i \\
&= 0,
\end{aligned}$$

since the expression in the bracket is zero when $i \notin A$, and $u(i) = 0$ when $i \in A$. Similarly, we can prove $\mathcal{E}_\alpha^{(\delta)}\left(u, \hat{e}_\alpha^{(\delta)}(f)\right) = 0$. Hence, Proposition 2.2.1 is proved. □

Corollary 2.2.1. *Assume that $\mathcal{E}(\cdot,\cdot)$ and $\hat{\mathcal{E}}(\cdot,\cdot)$ are the hyperfinite quadratic form and co-form associated with the dual hyperfinite Markov chains X and $\hat{X}$, respectively. Let $f : A \longrightarrow {}^*\mathbb{R}$ be an internal function defined on an internal subset A of S_0. Set*

$$\mathcal{M}(f) = \left\{ g \middle| g : S_0 \longrightarrow {}^*\mathbb{R} \quad \textit{is internal and} \quad g\,|_A = f \right\}.$$

Then, we have for all $g \in \mathcal{M}(f)$

$$\begin{aligned}\mathcal{E}_\alpha^{(\delta)}\left(e_\alpha^{(\delta)}(f), e_\alpha^{(\delta)}(f)\right) &= \mathcal{E}_\alpha^{(\delta)}\left(e_\alpha^{(\delta)}(f), g\right) \\ &= \mathcal{E}_\alpha^{(\delta)}\left(g, \hat{e}_\alpha^{(\delta)}(f)\right) \\ &= \mathcal{E}_\alpha^{(\delta)}\left(\hat{e}_\alpha^{(\delta)}(f), \hat{e}_\alpha^{(\delta)}(f)\right). \end{aligned} \tag{2.2.6}$$

Proof. It follows from Proposition 2.2.1 that for all $g \in \mathcal{M}(f)$

$$\mathcal{E}_\alpha^{(\delta)}\left(e_\alpha^{(\delta)}(f), e_\alpha^{(\delta)}(f)\right) = \mathcal{E}_\alpha^{(\delta)}\left(e_\alpha^{(\delta)}(f), g\right) \tag{2.2.7}$$

and

$$\mathcal{E}_\alpha^{(\delta)}\left(g, \hat{e}_\alpha^{(\delta)}(f)\right) = \mathcal{E}_\alpha^{(\delta)}\left(\hat{e}_\alpha^{(\delta)}(f), \hat{e}_\alpha^{(\delta)}(f)\right). \tag{2.2.8}$$

Since $e_\alpha^{(\delta)}(f), \hat{e}_\alpha^{(\delta)}(f) \in \mathcal{M}(f)$, we have from the relations (2.2.7) and (2.2.8) that

$$\begin{aligned}\mathcal{E}_\alpha^{(\delta)}\left(e_\alpha^{(\delta)}(f), e_\alpha^{(\delta)}(f)\right) &= \mathcal{E}_\alpha^{(\delta)}\left(e_\alpha^{(\delta)}(f), \hat{e}_\alpha^{(\delta)}(f)\right) \\ &= \mathcal{E}_\alpha^{(\delta)}\left(\hat{e}_\alpha^{(\delta)}(f), \hat{e}_\alpha^{(\delta)}(f)\right).\end{aligned}$$

Hence, (2.2.6) is proved. □

Corollary 2.2.2. *Assume that $\mathcal{E}(\cdot,\cdot)$ is a symmetric hyperfinite quadratic form, i.e., $\mathcal{E}(\cdot,\cdot) = \hat{\mathcal{E}}(\cdot,\cdot)$. Let $f : A \longrightarrow {}^*\mathbb{R}$ be an internal function defined on an internal subset A of S_0. Then, we have*

$$\mathcal{E}_\alpha^{(\delta)}\left(e_\alpha^{(\delta)}(f), e_\alpha^{(\delta)}(f)\right) = \min\left\{\mathcal{E}_\alpha^{(\delta)}(g,g) \mid g \in \mathcal{M}(f)\right\}.$$

Proof. First, we observe that $e_\alpha^{(\delta)}(f) = \hat{e}_\alpha^{(\delta)}(f)$ satisfies

$$\begin{aligned}\mathcal{E}_\alpha^{(\delta)}\left(e_\alpha^{(\delta)}(f), u\right) &= \mathcal{E}_\alpha^{(\delta)}\left(u, e_\alpha^{(\delta)}(f)\right) \\ &= 0 \end{aligned} \tag{2.2.9}$$

for all u which are zero on A. In fact, this is implied by Proposition 2.2.1 and the symmetry of $\mathcal{E}(\cdot,\cdot)$.

To prove our corollary, let w be another extension of f, i.e., $w \in \mathcal{M}(f)$. Then $u = w - e_\alpha(f)$ is zero on A, and by (2.2.9)

$$\begin{aligned}\mathcal{E}_\alpha^{(\delta)}(w,w) &= \mathcal{E}_\alpha^{(\delta)}\left(e_\alpha^{(\delta)}(f)+u, e_\alpha^{(\delta)}(f)+u\right) \\ &= \mathcal{E}_\alpha^{(\delta)}\left(e_\alpha^{(\delta)}(f), e_\alpha^{(\delta)}(f)\right) + 2\mathcal{E}_\alpha^{(\delta)}\left(e_\alpha^{(\delta)}(f), u\right) + \mathcal{E}_\alpha^{(\delta)}(u,u) \\ &= \mathcal{E}_\alpha^{(\delta)}\left(e_\alpha^{(\delta)}(f), e_\alpha^{(\delta)}(f)\right) + \mathcal{E}_\alpha^{(\delta)}(u,u).\end{aligned}$$

□

Definition 2.2.2. Let $f : A \longrightarrow {}^*\mathbb{R}$ be an internal function defined on an internal subset A of S_0. The function $e_\alpha^{(\delta)}(f)$ is called the *equilibrium α-potential* of f, and $\hat{e}_\alpha^{(\delta)}(f)$ is called the *co-equilibrium α-potential* of f associated with $\mathcal{E}^{(\delta)}(\cdot,\cdot)$.

Lemma 2.2.2. *Let A be an internal subset of S_0 and 1_A be the indicator function of A. Then for all $\alpha \in {}^*\mathbb{R}_+$ and $\delta \in T$, we have*

(i) $e_\alpha^{(\delta)}(1_A)(i)$ is hyperfinite α-excessive.
(ii) $\hat{e}_\alpha^{(\delta)}(1_A)(i)$ is hyperfinite α-co-excessive.

Proof. (i) First of all, we notice that if $i \notin A$,

$$\begin{aligned}(1+\alpha\delta)\, e_\alpha^{(\delta)}(1_A)(i) &= \sum_{j=1}^{N} e_\alpha^{(\delta)}(1_A)(j) q_{ij}^{(\delta)} \\ &= Q^\delta e_\alpha^{(\delta)}(1_A)(i).\end{aligned} \tag{2.2.10}$$

For $i \in A$, we have

$$\begin{aligned}Q^\delta e_\alpha^{(\delta)}(1_A)(i) &\le 1 \\ &= e_\alpha^{(\delta)}(1_A)(i) \\ &\le (1+\alpha\delta) e_\alpha^{(\delta)}(1_A)(i).\end{aligned} \tag{2.2.11}$$

From the relations (2.2.10) and (2.2.11), we have got (i).

(ii) The proof of this is similar to the one for (i). □

Theorem 2.2.1. *Assume that $\mathcal{E}(\cdot,\cdot)$ and $\hat{\mathcal{E}}(\cdot,\cdot)$ are the hyperfinite quadratic form and co-form associated with the dual hyperfinite Markov chains X and $\hat{X}$, respectively. Let A be an internal subset of S_0 and let 1_A be the indicator function of A. Set*

$$\mathcal{L}(1_A) = \left\{ g \middle| g : S_0 \longrightarrow {}^*\mathbb{R} \quad \textit{is internal and} \quad g(i) \geq 1, \forall i \in A \right\}.$$

Then for all $\alpha \in {}^*\mathbb{R}_+, \delta \in T, u \in \mathcal{L}(1_A)$, we have

$$\mathcal{E}_\alpha^{(\delta)}\left(\mathrm{e}_\alpha^{(\delta)}(1_A), u\right) \geq \mathcal{E}_\alpha^{(\delta)}\left(\mathrm{e}_\alpha^{(\delta)}(1_A), \mathrm{e}_\alpha^{(\delta)}(1_A)\right),$$
$$\mathcal{E}_\alpha^{(\delta)}\left(u, \hat{\mathrm{e}}_\alpha^{(\delta)}(1_A)\right) \geq \mathcal{E}_\alpha^{(\delta)}\left(\hat{\mathrm{e}}_\alpha^{(\delta)}(1_A), \hat{\mathrm{e}}_\alpha^{(\delta)}(1_A)\right).$$

Proof. In the same way as the proof of Proposition 2.2.1, we get our theorem by Lemma 2.2.2. □

Corollary 2.2.3. *Assume that $\mathcal{E}^{(\delta)}(\cdot,\cdot)$ is a hyperfinite weak coercive quadratic form with continuity constant C. Let A be an internal subset of S_0. Then*

(1) For all $u \in \mathcal{L}(1_A)$, we have

$$\mathcal{E}_1^{(\delta)}\left(e_1^{(\delta)}(1_A), e_1^{(\delta)}(1_A)\right) \leq C^2 \mathcal{E}_1^{(\delta)}(u,u),$$
$$\mathcal{E}_1^{(\delta)}\left(\hat{e}_1^{(\delta)}(1_A), \hat{e}_1^{(\delta)}(1_A)\right) \leq C^2 \mathcal{E}_1^{(\delta)}(u,u).$$

(2) If $C = 1$, we have

$$\begin{aligned}\mathcal{E}_1^{(\delta)}\left(e_1^{(\delta)}(1_A), e_1^{(\delta)}(1_A)\right) &= \mathcal{E}_1^{(\delta)}\left(\hat{e}_1^{(\delta)}(1_A), \hat{e}_1^{(\delta)}(1_A)\right) \\ &= \min\{\mathcal{E}_1^{(\delta)}(u,u) \big| u \in \mathcal{L}(1_A)\}.\end{aligned}$$

Proof. The proof is very easy and will therefore be omitted! □

2.3 Capacity Theory

In this section, we assume that all conditions in Sect. 2.2 are satisfied, and we shall use the same notations there. That is, X and $\hat{X}$ are dual hyperfinite Markov chains, and $\mathcal{E}(\cdot,\cdot)$ and $\hat{\mathcal{E}}(\cdot,\cdot)$ are the hyperfinite quadratic form and co-form associated with X and $\hat{X}$, respectively. Let A be an internal subset of S_0. For $\alpha \in {}^*\mathbb{R}_+, \delta \in T$, set

$$\mathrm{Cap}_{\alpha}^{(\delta)}(A) = \mathcal{E}_{\alpha}^{(\delta)}\left(e_{\alpha}^{(\delta)}(1_A), e_{\alpha}^{(\delta)}(1_A)\right).$$

We call $\mathrm{Cap}_{\alpha}^{(\delta)}(A)$ the *α-capacity* of A associated with $\mathcal{E}^{(\delta)}(\cdot,\cdot)$. An internal subset A has *finite energy* of $\mathcal{E}^{(\delta)}(\cdot,\cdot)$ if ${}^{\circ}\mathcal{E}_1^{(\delta)}\left(e_1^{(\delta)}(1_A), e_1^{(\delta)}(1_A)\right) < \infty$. We shall write $e_{\alpha}^{(\delta)}(A)$ for $e_{\alpha}^{(\delta)}(1_A)$. Moreover, we call $e_{\alpha}^{(\delta)}(A)$ the *equilibrium α-potential* of A associated with $\mathcal{E}^{(\delta)}(\cdot,\cdot)$. Whenever $\delta = \Delta t$, we abbreviate $e_{\alpha}^{(\delta)}(f)$ and $\mathrm{Cap}_{\alpha}^{(\delta)}(f)$ by $e_{\alpha}(f)$ and $\mathrm{Cap}_{\alpha}(f)$, respectively.

By Corollary 2.2.1, we have

$$\mathcal{E}_{\alpha}^{(\delta)}\left(e_{\alpha}^{(\delta)}(A), e_{\alpha}^{(\delta)}(A)\right) = \mathcal{E}_{\alpha}\left(e_{\alpha}(A), 1\right).$$

Applying the Beurling–Deny formulae, Lemma 1.5.1, to $\mathcal{E}_{\alpha}^{(\delta)}(e_{\alpha}^{(\delta)}(A), 1)$, we get

$$\mathcal{E}_{\alpha}^{(\delta)}\left(e_{\alpha}^{(\delta)}(A), e_{\alpha}^{(\delta)}(A)\right) = \alpha \sum_{i=1}^{N} e_{\alpha}^{(\delta)}(A)(i) m_i + \sum_{i=1}^{N} e_{\alpha}^{(\delta)}(A)(i) \frac{\hat{q}_{i0}}{\delta} m_i. \tag{2.3.1}$$

The equilibrium potentials will serve as bridgeheads in our campaign to unify the analytic theory of Dirichlet forms and the probabilistic theory of Markov processes. The key to their importance is the observation that they have natural interpretations both in analytic and in probabilistic terms. On the one hand, they minimize the forms $\mathcal{E}_{\alpha}^{(\delta)}(\cdot,\cdot)$ under suitable side conditions. On the other they describe how the process X hits subsets of S_0. For a fuller understanding of the probabilistic description, we point out that for finite α

$${}^{\circ}e_{\alpha}^{(\delta)}(A)(i) = E_i({}^{\circ}e^{-\alpha\sigma_A^{(\delta)}}),$$

and that the function $\alpha \mapsto E_i({}^{\circ}e^{-\alpha\sigma_A^{(\delta)}})$ is the Laplace transform of $\sigma_A^{(\delta)}$. Hence, the distribution of σ_A can be completely recovered from the functions $e_{\alpha}^{(\delta)}(A)$. This fact will be of great importance in Chap. 3.

Lemma 2.3.1. *Let $\delta \in T, \alpha \geq 0$ and $\alpha \in {}^*\mathbb{R}$, then the following results hold:*

(i) If A and B are two internal subsets of S_0, $A \subset B$, then we have

$$\mathrm{Cap}_{\alpha}^{(\delta)}(A) \leq \mathrm{Cap}_{\alpha}^{(\delta)}(B).$$

(ii) If A and B are internal subsets of S_0, then we have

$$\mathrm{Cap}_{\alpha}^{(\delta)}(A \cup B) + \mathrm{Cap}_{\alpha}^{(\delta)}(A \cap B) \leq \mathrm{Cap}_{\alpha}^{(\delta)}(A) + \mathrm{Cap}_{\alpha}^{(\delta)}(B).$$

Proof. (i) Since $e_\alpha^{(\delta)}(B) - e_\alpha^{(\delta)}(A) \geq 0$ on B, we have from Corollary 2.2.1 and Theorem 2.2.1

$$\begin{aligned}\mathrm{Cap}_\alpha^{(\delta)}(A) &= \mathcal{E}_\alpha^{(\delta)}\left(e_\alpha^{(\delta)}(A), \hat{e}_\alpha^{(\delta)}(A)\right)\\ &= \mathcal{E}_\alpha^{(\delta)}\left(e_\alpha^{(\delta)}(A), \hat{e}_\alpha^{(\delta)}(B)\right)\\ &\leq \mathcal{E}_\alpha^{(\delta)}\left(e_\alpha^{(\delta)}(B), \hat{e}_\alpha^{(\delta)}(B)\right)\\ &= \mathrm{Cap}_\alpha^{(\delta)}(B).\end{aligned}$$

(ii) Since

$$e_\alpha^{(\delta)}(A) + e_\alpha^{(\delta)}(B) - e_\alpha^{(\delta)}(A \cup B) - e_\alpha^{(\delta)}(A \cap B) \geq 0 \text{ on } A \cup B,$$

we have the following from Theorem 2.2.1

$$\begin{aligned}&\mathrm{Cap}_\alpha^{(\delta)}(A) + \mathrm{Cap}_\alpha^{(\delta)}(B) - \mathrm{Cap}_\alpha^{(\delta)}(A \cup B) - \mathrm{Cap}_\alpha^{(\delta)}(A \bigcap B)\\ &= \mathcal{E}_\alpha^{(\delta)}\left(e_\alpha^{(\delta)}(A) + e_\alpha^{(\delta)}(B) - e_\alpha^{(\delta)}(A \cup B) - e_\alpha^{(\delta)}(A \cap B), \hat{e}_\alpha^{(\delta)}(A \cup B)\right)\\ &\geq 0.\end{aligned}$$

□

Now let us generalize our capacity theory. For every subset A of S_0, we define

$$\mathrm{Cap}_\alpha^{(\delta)}(A) = \inf\left\{\mathrm{Cap}_\alpha^{(\delta)}(B) \middle| A \subset B, B \in \mathcal{S}_0\right\}.$$

We call $\mathrm{Cap}_\alpha^{(\delta)}(A)$ the *α-capacity* of A associated with $\mathcal{E}^{(\delta)}(\cdot,\cdot)$.

Lemma 2.3.2. *Let $\delta \in T, \alpha \geq 0$ and $\alpha \in {}^*\mathbb{R}$, then the following results hold:*

(i) If $\{A_n \mid n \in \mathbb{N}\}$ is an increasing sequence of internal subsets of S_0, then we have

$$^\circ\mathrm{Cap}_\alpha^{(\delta)}\left(\bigcup_{n\in\mathbb{N}} A_n\right) = \sup\left\{^\circ\mathrm{Cap}_\alpha^{(\delta)}(A_n) \middle| n \in \mathbb{N}\right\}.$$

(ii) If $\{A_n \mid n \in \mathbb{N}\}$ is a decreasing sequence of internal subsets of S_0, then we have

$$^\circ\mathrm{Cap}_\alpha^{(\delta)}\left(\bigcap_{n\in\mathbb{N}} A_n\right) = \inf\left\{^\circ\mathrm{Cap}_\alpha^{(\delta)}(A_n) \middle| n \in \mathbb{N}\right\}.$$

Proof. (i) Set a $= \sup\left\{{}^\circ\mathrm{Cap}_\alpha^{(\delta)}(A_n)\middle| n\in\mathbb{N}\right\}$. Obviously, we have

$$a \le {}^\circ\mathrm{Cap}_\alpha^{(\delta)}\left(\bigcup_{n\in\mathbb{N}} A_n\right).$$

Hence, we can suppose that $a<\infty$. Let $\{A_n \mid n\in {}^*\mathbb{N}\}$ be an increasing internal extension of $\{A_n \mid n\in\mathbb{N}\}$. Given $\varepsilon>0$, we consider the following internal set

$$\left\{n\in {}^*\mathbb{N}\middle| \bigcup_{l=1}^{n} A_l = A_n \quad \text{is internal,} \quad \mathrm{Cap}_\alpha^{(\delta)}\left(\bigcup_{l=1}^{n} A_l\right) = \mathrm{Cap}_\alpha^{(\delta)}(A_n) \le a+\varepsilon\right\}.$$

It is easy to see that $\mathbb{N}$ is contained in above internal set. By saturation, there is an infinite member M belonging to it. Therefore, we have

$$\begin{aligned}\mathrm{Cap}_\alpha^{(\delta)}\left(\bigcup_{n\in\mathbb{N}} A_n\right) &\le \mathrm{Cap}_\alpha^{(\delta)}(A_M)\\ &\le a+\varepsilon.\end{aligned}$$

By letting $\varepsilon\downarrow 0$, we get ${}^\circ\mathrm{Cap}_\alpha^{(\delta)}\left(\bigcup_{n\in\mathbb{N}} A_n\right)\le a$. Therefore, we have

$${}^\circ\mathrm{Cap}_\alpha^{(\delta)}\left(\bigcup_{n\in\mathbb{N}} A_n\right) = \sup\left\{{}^\circ\mathrm{Cap}_\alpha^{(\delta)}(A_n)\middle| n\in\mathbb{N}\right\}.$$

(ii) We extend $\{A_n \mid n\in\mathbb{N}\}$ to be a decreasing sequence $\{A_n \mid n\in {}^*\mathbb{N}\}$. When $\inf\left\{{}^\circ\mathrm{Cap}_\alpha^{(\delta)}(A_l)\middle| l\in\mathbb{N}\right\}<\infty$, consider the following internal subset for $\varepsilon>0$

$$\left\{n\in {}^*\mathbb{N}\middle| \mathrm{Cap}_\alpha^{(\delta)}(A_n)\ge \inf\left\{{}^\circ\mathrm{Cap}_\alpha^{(\delta)}(A_l)\middle| l\in\mathbb{N}\right\}-\varepsilon\right\}.$$

By saturation, there is an infinite M belonging to the above internal set. Hence, we have

$$\begin{aligned}{}^\circ\mathrm{Cap}_\alpha^{(\delta)}\left(\bigcap_{l=1}^{\infty} A_l\right) &\ge {}^\circ\mathrm{Cap}_\alpha^{(\delta)}(A_M)\\ &\ge \inf\left\{{}^\circ\mathrm{Cap}_\alpha^{(\delta)}(A_l)\middle| l\in\mathbb{N}\right\}-\varepsilon.\end{aligned}$$

By letting $\varepsilon \downarrow 0$, we get

$$^\circ\mathrm{Cap}_\alpha^{(\delta)}\left(\bigcap_{l=1}^{\infty} A_l\right) \geq \inf\left\{ ^\circ\mathrm{Cap}_\alpha^{(\delta)}(A_l) \middle| l \in \mathbb{N}\right\}.$$

On the other hand, it is easy to see that

$$^\circ\mathrm{Cap}_\alpha^{(\delta)}\left(\bigcap_{l=1}^{\infty} A_l\right) \leq \inf_{l\in\mathbb{N}}\left\{ ^\circ\mathrm{Cap}_\alpha^{(\delta)}(A_l)\right\}.$$

If $\inf\left\{ ^\circ\mathrm{Cap}_\alpha^{(\delta)}(A_l) \middle| l \in \mathbb{N}\right\} = \infty$, we consider the following internal subset for $N_0 \in \mathbb{N}$

$$\left\{ n \in {}^*\mathbb{N} \middle| \mathrm{Cap}_\alpha^{(\delta)}(A_n) \geq N_0 \right\}.$$

By saturation and letting $N_0 \uparrow \infty$, we can show that (ii) holds. □

Lemma 2.3.3. *If $\{A_n \mid n \in \mathbb{N}\}$ is a sequence of internal subsets of S_0, then we have*

$$^\circ\mathrm{Cap}_\alpha^{(\delta)}\left(\bigcup_{n\in\mathbb{N}} A_n\right) \leq \sum_{n\in\mathbb{N}} {}^\circ\mathrm{Cap}_\alpha^{(\delta)}(A_n), \tag{2.3.2}$$

for all $\delta \in T, \alpha \geq 0$.

Proof. Set $b = \sum_{n\in\mathbb{N}} {}^\circ\mathrm{Cap}_\alpha^{(\delta)}(A_n)$. If $b = \infty$, the inequality (2.3.2) holds. In the following proof, we shall assume $b < \infty$. Let $\{A_n \mid n \in {}^*\mathbb{N}\}$ be an internal extension of $\{A_n \mid n \in \mathbb{N}\}$. For every $\varepsilon > 0$, it follows from Lemma 2.3.1 (ii) that

$$\mathrm{Cap}_\alpha^{(\delta)}\left(\bigcup_{l=1}^{n} A_l\right) \leq \sum_{l=1}^{n} \mathrm{Cap}_\alpha^{(\delta)}(A_l) \leq \varepsilon + b \quad \text{for all} \quad n \in \mathbb{N}.$$

Consider the following internal set

$$\left\{ n \in {}^*\mathbb{N} \middle| \bigcup_{l=1}^{n} A_l \text{ is internal and } \mathrm{Cap}_\alpha^{(\delta)}\left(\bigcup_{l=1}^{n} A_l\right) \leq \sum_{l=1}^{n} \mathrm{Cap}_\alpha^{(\delta)}(A_l) \leq b + \varepsilon \right\}.$$

By saturation, there is an infinite element $M = M(\varepsilon)$ belonging to the above internal set. Hence, we obtain

$$\begin{aligned}\mathrm{Cap}_{\alpha}^{(\delta)}\left(\bigcup_{l\in\mathbb{N}} A_l\right) &\le \mathrm{Cap}_{\alpha}^{(\delta)}\left(\bigcup_{l=1}^{M} A_l\right)\\ &\le \sum_{l=1}^{M} \mathrm{Cap}_{\alpha}^{(\delta)}(A_l)\\ &\le b+\varepsilon.\end{aligned}$$

By letting $\varepsilon \downarrow 0$, we have proved the inequality (2.3.2). □

Proposition 2.3.1. *For all $\delta \in T$, $\alpha \ge 0$, we have*

(i) If A and B are two subsets of S_0, $A \subset B$, then

$$\mathrm{Cap}_{\alpha}^{(\delta)}(A) \le \mathrm{Cap}_{\alpha}^{(\delta)}(B) \tag{2.3.3}$$

(ii) Let $\{A_n \mid n \in \mathbb{N}\}$ be a sequence of subsets of S_0, then

$$^{\circ}\mathrm{Cap}_{\alpha}^{(\delta)}\left(\bigcup_{n\in\mathbb{N}} A_n\right) \le \sum_{n\in\mathbb{N}} {}^{\circ}\mathrm{Cap}_{\alpha}^{(\delta)}(A_n). \tag{2.3.4}$$

(iii) Let $\{A_n \mid n \in \mathbb{N}\}$ be an increasing sequence of subsets of S_0, then

$$^{\circ}\mathrm{Cap}_{\alpha}^{(\delta)}\left(\bigcup_{n\in\mathbb{N}} A_n\right) = \sup\left\{ {}^{\circ}\mathrm{Cap}_{\alpha}^{(\delta)}(A_n) \middle| n \in \mathbb{N}\right\}.$$

Proof. (i) The proof is immediate, using the definition.

(ii) Set $b = \sum_{n\in\mathbb{N}} {}^{\circ}\mathrm{Cap}_{\alpha}^{(\delta)}(A_n)$. We can assume that $b < \infty$. Given $\varepsilon > 0$, for every $n \in \mathbb{N}$, let us take an internal subset B_n such that $A_n \subset B_n$ and

$$\begin{aligned}\mathrm{Cap}_{\alpha}^{(\delta)}(A_n) &\le \mathrm{Cap}_{\alpha}^{(\delta)}(B_n)\\ &\le \mathrm{Cap}_{\alpha}^{(\delta)}(A_n) + \frac{\varepsilon}{2^{n+1}}.\end{aligned}$$

Therefore, we have from (i) and Lemma 2.3.3 that

$$\begin{aligned}{}^{\circ}\mathrm{Cap}_{\alpha}^{(\delta)}\left(\bigcup_{n\in\mathbb{N}} A_n\right) &\le {}^{\circ}\mathrm{Cap}_{\alpha}^{(\delta)}\left(\bigcup_{n\in\mathbb{N}} B_n\right)\\ &\le \sum_{n\in\mathbb{N}} {}^{\circ}\mathrm{Cap}_{\alpha}^{(\delta)}(B_n)\\ &\le \sum_{n\in\mathbb{N}} {}^{\circ}\mathrm{Cap}_{\alpha}^{(\delta)}(A_n) + \varepsilon.\end{aligned}$$

By letting $\varepsilon \downarrow 0$, we get the inequality (2.3.4).

(iii) We may assume that for all $n \in \mathbb{N}$, ${}^\circ\mathrm{Cap}_\alpha^{(\delta)}(A_n) < \infty$. Given $\varepsilon > 0$, for each $n \in \mathbb{N}$, let B_n be an internal subset of S_0 such that

$$A_n \subset B_n, \quad \mathrm{Cap}_\alpha^{(\delta)}(B_n) \le \mathrm{Cap}_\alpha^{(\delta)}(A_n) + \varepsilon.$$

Then, we have from (2.3.3) and Lemma 2.3.2 (i) that

$$\begin{aligned} {}^\circ\mathrm{Cap}_\alpha^{(\delta)}\left(\bigcup_{n\in\mathbb{N}} A_n\right) &\le {}^\circ\mathrm{Cap}_\alpha^{(\delta)}\left(\bigcup_{n\in\mathbb{N}} B_n\right) \\ &= \sup_n \left\{{}^\circ\mathrm{Cap}_\alpha^{(\delta)}(B_n)\right\} \\ &\le \sup_n \left\{{}^\circ\left(\mathrm{Cap}_\alpha^{(\delta)}(A_n) + \varepsilon\right)\right\} \\ &\le \sup_n \left\{{}^\circ\mathrm{Cap}_\alpha^{(\delta)}(A_n)\right\} + \varepsilon. \end{aligned}$$

By letting $\varepsilon \downarrow 0$, we get

$$ {}^\circ\mathrm{Cap}_\alpha^{(\delta)}\left(\bigcup_{n\in\mathbb{N}} A_n\right) \le \sup\left\{{}^\circ\mathrm{Cap}_\alpha^{(\delta)}(A_n)\,\middle|\, n \in \mathbb{N}\right\}. \tag{2.3.5}$$

On the other hand, it is easy to see that

$$ {}^\circ\mathrm{Cap}_\alpha^{(\delta)}\left(\bigcup_{n\in\mathbb{N}} A_n\right) \ge \sup\left\{{}^\circ\mathrm{Cap}_\alpha^{(\delta)}(A_n)\,\middle|\, n \in \mathbb{N}\right\}. \tag{2.3.6}$$

From the inequalities (2.3.5) and (2.3.6), we have proved Proposition 2.3.1 (iii). □

For the purpose of explaining our Theorem 2.3.1 in the following, we first introduce some notations in capacity theory (referring to, e.g., [282]). Let G be a set, $\mathcal{G}$ be a family of some subsets of G. Denote by $\mathcal{G}_\sigma$ (respectively, $\mathcal{G}_\delta$) the closure of a collection of subsets of G under countable union (respectively, countable intersection). That is,

$$\mathcal{G}_\sigma = \left\{\bigcup_{n=1}^{\infty} A_n \,\middle|\, A_n \in \mathcal{G}\right\}, \quad \mathcal{G}_\delta = \left\{\bigcap_{n=1}^{\infty} A_n \,\middle|\, A_n \in \mathcal{G}\right\}.$$

Moreover, we shall write $\mathcal{G}_{\sigma\delta} = (\mathcal{G}_\sigma)_\delta$.

Definition 2.3.1. Let G be a set. A *paving* $\mathcal{G}$ on G is a family of subsets of G such that the empty set $\emptyset$ is contained in $\mathcal{G}$. The pair $(G, \mathcal{G})$ consisting of a set G and a paving $\mathcal{G}$ on G is called a *paved set*.

Definition 2.3.2. Let $(G, \mathcal{G})$ be a paved set. The paving $\mathcal{G}$ is said to be *semi-compact* if every countable family of elements of $\mathcal{G}$, which has the finite intersection property, has a nonempty intersection.

It is easy to see that $(S_0, \mathcal{S}_0)$ is a semi-compact paved set. Moreover, $\mathcal{S}_0$ is closed under the complement, finite union, and finite intersection operations.

Definition 2.3.3. A subset A of S_0 is said to be $\mathcal{S}_0$-*analytic* if there exists an auxiliary set G with a semi-compact paving $\mathcal{G}$, and a subset $B \subset G \times S_0$ belonging to $(\mathcal{G} \times \mathcal{S}_0)_{\sigma\delta}$ such that A is the projection of B on S_0. We denote by $\mathcal{A}(\mathcal{S}_0)$ all the $\mathcal{S}_0$-analytic sets (we notice that $\mathcal{G} \times \mathcal{S}_0 = \{G_1 \times S_1 \mid G_1 \in \mathcal{G}$ and $S_1 \in \mathcal{S}_0\}$).

Lemma 2.3.4. *The σ-field $\sigma(\mathcal{S}_0)$ generated by $\mathcal{S}_0$ is contained in $\mathcal{A}(\mathcal{S}_0)$.*

Proof. For every $F \in \mathcal{S}_0$, $S_0 - F$ belongs to $\mathcal{S}_0$ also. By Meyer [282], Chap. III T12 Theorem, we know $\sigma(\mathcal{S}_0) \subset \mathcal{A}(\mathcal{S}_0)$. □

Definition 2.3.4. An extended real valued set function $I : 2^{S_0} \to [-\infty, +\infty]$, defined on all subsets 2^{S_0} of S_0, is called a *Choquet $\mathcal{S}_0$-capacity* if it satisfies the following properties:

(i) I is increasing, i.e., $A \subset B \Longrightarrow I(A) \leq I(B)$.
(ii) For every increasing sequence $\{A_n \mid n \in \mathbb{N}\}$ of subsets of S_0, we have

$$I\left(\bigcup_{n\in\mathbb{N}} A_n\right) = \sup_{n\in\mathbb{N}} I(A_n).$$

(iii) For every decreasing sequence $\{A_n \mid n \in \mathbb{N}\}$ of elements of $\mathcal{S}_0$, we have

$$I\left(\bigcap_{n\in\mathbb{N}} A_n\right) = \inf_{n\in\mathbb{N}} I(A_n).$$

We have reached one of our main results.

Theorem 2.3.1. *For each $\delta \in T$ and $\alpha \geq 0, \alpha \in {}^*\mathbb{R}$, we have the following results:*

(i) ${}^\circ\mathrm{Cap}_\alpha^{(\delta)}(\cdot)$ *is a Choquet $\mathcal{S}_0$-capacity.*
(ii) Every $\mathcal{S}_0$-analytic set is capacitable with respect to capacity ${}^\circ\mathrm{Cap}_\alpha^{(\delta)}(\cdot)$*. That is, for every $A \in \mathcal{A}(\mathcal{S}_0)$, we have*

$${}^\circ\mathrm{Cap}_\alpha^{(\delta)}(A) = \sup\left\{ {}^\circ\mathrm{Cap}_\alpha^{(\delta)}(B) \,\middle|\, B = \bigcap_{m\in\mathbb{N}} B_m, B_m \in \mathcal{S}_0 \text{ and } B \subset A \right\}.$$

(iii) Every subset A of S_0 belonging to $\sigma(\mathcal{S}_0)$ is capacitable with respect to the capacity ${}^\circ\mathrm{Cap}_\alpha^{(\delta)}(\cdot)$ *whenever* $0 < \mathrm{st}(\alpha) < \infty$.

Proof. By Lemma 2.3.1 and Proposition 2.3.1, we know that $°\mathrm{Cap}_\alpha^{(\delta)}(\cdot)$ is a Choquet $\mathcal{S}_0$-capacity. Therefore, (ii) holds by Meyer [282], Chap. III T19 Theorem. (iii) is the consequence of (ii) and Lemma 2.3.4. □

Definition 2.3.5. (i) A subset B of S_0 is said to be of *δ-zero capacity*, if $\mathrm{Cap}_1^{(\delta)}(B) \approx 0$.

(ii) A subset B of S_0 is said to be of *zero capacity* if $\mathrm{Cap}_1^{(\delta)}(B) \approx 0$ for some infinitesimal $\delta \in T$.

Remark 2.3.1. For any $B \in \mathcal{S}_0$ and $\delta \in T$, we have $m(B) \leq \mathrm{Cap}_1^{(\delta)}(B)$. Therefore, for any zero capacity subset B of S_0, we have $L(m)(B) = 0$.

2.4 Relation of Exceptionality and Capacity Theory

In regular Dirichlet space theory, we know that the concepts of exceptional sets and zero capacity sets are equivalent, see Fukushima [175], Theorem 4.3.1. As the fourth section of this chapter, we will discuss the corresponding problem in our hyperfinite Dirichlet space theory.

We shall continue the discussion of Sect. 2.3. Hence, we assume that all conditions in Sect. 2.2 are satisfied in this section as well, i.e., X and $\hat{X}$ are dual hyperfinite Markov chains, and $\mathcal{E}(\cdot,\cdot)$ and $\hat{\mathcal{E}}(\cdot,\cdot)$ are the hyperfinite quadratic form and co-form associated with X and $\hat{X}$, respectively. Let H be the hyperfinite dimensional space with an inner product $\langle\cdot,\cdot\rangle$ defined by (2.2.3) in Sect. 2.2 or (1.5.15) in Sect. 1.5, Chap. 1.

Lemma 2.4.1. *Let $\{B_n \mid n \in \mathbb{N}\}$ be a sequence of internal subsets of S_0. If $lim_{n\to\infty} °\mathrm{Cap}_1^{(\delta)}\left(\bigcap_{m=1}^n B_m\right) = 0$, then $\bigcap_{n=1}^\infty B_n$ is a δ-exceptional set, where $\delta \in T, \delta \approx 0$.*

Proof. Since $\mathcal{S}_0$ is closed under finite intersection, we may assume that $\{B_n \mid n \in \mathbb{N}\}$ is a decreasing sequence. Define a stopping time for each $n \in \mathbb{N}$,

$$\sigma_{B_n}^{(\delta)}(\omega) = \min\{t \in T_\delta \mid X(\omega,t) \in B_n\}.$$

We have

$$\begin{aligned} &L(P)\left\{\omega \middle| \exists t \in T_\delta^1\left(X(\omega,t) \in B_n\right)\right\} \\ &= °P\left\{\omega \middle| \exists t \in T_\delta^1\left(X(\omega,t) \in B_n\right)\right\} \\ &= °\!\!\int_{S_0} E_i\left(\omega \middle| \exists t \in T_\delta^1\left(X(\omega,t) \in B_n\right)\right) dm(i) \end{aligned}$$

$$
\begin{aligned}
&= {}^{\circ}\!\!\int_{S_0} E_i 1_{(\sigma_{B_n}^{(\delta)} \le 1)}\, dm(i) \\
&= {}^{\circ}\!\!\int_{S_0} E_i \left\{ \omega \,\middle|\, (1+\delta)^{-\sigma_{B_n}^{(\delta)}/\delta} \ge (1+\delta)^{-\frac{1}{\delta}} \right\} dm(i) \\
&\le {}^{\circ}\!\!\int_{S_0} E_i \left\{ \frac{(1+\delta)^{-\sigma_{B_n}^{(\delta)}/\delta}}{(1+\delta)^{-\frac{1}{\delta}}} \right\} dm(i) \\
&= e \cdot {}^{\circ}\!\!\int_{S_0} E_i\, (1+\delta)^{-\sigma_{B_n}^{(\delta)}/\delta}\, dm(i) \\
&= e \cdot {}^{\circ}\!\!\int_{S_0} e_1^{(\delta)}(B_n)(i)\, dm(i) \\
&\le e \cdot {}^{\circ}\mathcal{E}_1 \left(e_1^{(\delta)}(B_n), e_1^{(\delta)}(B_n) \right) \\
&= e \cdot {}^{\circ}\mathrm{Cap}_1^{(\delta)}(B_n) \longrightarrow 0,
\end{aligned}
\tag{2.4.1}
$$

where the last inequality comes from (2.3.1) in Sect. 2.3. Then, we have

$$
L(P) \left\{ \omega \,\middle|\, \exists t \in T_\delta^1 \left(X(\omega, t) \in \bigcap_{n=1}^{\infty} B_n \right) \right\} = 0. \tag{2.4.2}
$$

By the dual property of the Markov process $X(t)$ and (2.4.2), we also have

$$
L(P) \left\{ \omega \,\middle|\, \exists t \in T_\delta^{\text{fin}} \left(X(\omega, t) \in \bigcap_{n=1}^{\infty} B_n \right) \right\} = 0.
$$

Therefore, the set $\bigcap_{n=1}^{\infty} B_n$ is δ-exceptional. □

Theorem 2.4.1. *If a subset A of S_0 is of δ-zero capacity, it is δ-exceptional.*

Proof. Since $\mathrm{Cap}_1^{(\delta)}(A) \approx 0$, we can take a sequence of internal subsets $\{B_n \mid n \in \mathbb{N}\}$ satisfying

$$
A \subset \bigcap_{n=1}^{\infty} B_n, \quad \lim_{n \to \infty} {}^{\circ}\mathrm{Cap}_1^{(\delta)}\left(\bigcap_{m=1}^{n} B_m\right) = 0.
$$

Using Lemma 2.4.1, we know that $\bigcap_{n=1}^{\infty} B_n$ is δ-exceptional. Hence, A is δ-exceptional also. This completes the proof of Theorem 2.4.1. □

Lemma 2.4.2. *Let $\delta_1 \in T, \delta_1 \approx 0$. Assume that for all $f \in H$, if ${}^{\circ}(\mathcal{E}_1^{(\delta_1)}(f,f)) < \infty$ and $f(s) \approx 0$ for all $s \notin B$, where B is a δ_1-exceptional set, then we have $\mathcal{E}_1^{(\delta_1)}(f,f) \approx 0$. Let A be a subset of S_0. If A is δ_1-exceptional and there exists an internal subset B of S_0 such that*

$$A \subset B, \quad {}^\circ Cap_1^{(\delta_1)}(B) < \infty, \tag{2.4.3}$$

then A is of δ_1-zero capacity.

Proof. By using Proposition 2.1.1, there exists a properly δ_1-exceptional set

$$\bigcup_{m\in\mathbb{N}} \bigcap_{n\in\mathbb{N}} B_{m,n} \supset A.$$

For simplicity, we assume that $B_{m,n} \subset B$ for all $n, m \in \mathbb{N}$, and for each m, the sequence $\{B_{m,n} \mid n \in \mathbb{N}\}$ is decreasing with respect to n. In order to show that A has zero capacity, we first prove that $\bigcap_{n\in N} B_{m,n}$ has zero capacity for every m. From now on, we fix an $m \in \mathbb{N}$.

By the assumption (2.4.3), we know that ${}^\circ\mathrm{Cap}_1^{(\delta_1)}(B_{m,n}) < \infty$ for every n. Moreover, $\mathrm{Cap}_1^{(\delta_1)}(B_{m,n})$ is decreasing with n. Let $\{B_{m,n} \mid n \in {}^*\mathbb{N}\}$ be a decreasing extension of $\{B_{m,n} \mid n \in \mathbb{N}\}$. By saturation, there exists an infinite element $n_m \in {}^*\mathbb{N} - \mathbb{N}$ such that

$$\begin{aligned}&\lim_{n\to\infty} {}^\circ[\mathcal{E}_1^{(\delta_1)}(e_1^{(\delta_1)}(B_{m,n}), e_1^{(\delta_1)}(B_{m,n}))] \\ &\quad = {}^\circ[\mathcal{E}_1^{(\delta_1)}(e_1^{(\delta_1)}(B_{m,n_m}), e_1^{(\delta_1)}(B_{m,n_m}))].\end{aligned}$$

Therefore, we have

$$\begin{aligned}{}^\circ\!\left[\mathcal{E}_1^{(\delta_1)}\left(e_1^{(\delta_1)}(B_{m,n_m}), e_1^{(\delta_1)}(B_{m,n_m})\right)\right] &= {}^\circ\!\left[\mathrm{Cap}_1^{(\delta_1)}(B_{m,n_m})\right] \\ &< \infty.\end{aligned}$$

Besides, for every $i \in S_0$, it is easy to see that $\left\{e_1^{(\delta_1)}(B_{m,n})(i) \mid n \in \mathbb{N}\right\}$ is decreasing with respect to n. Since $\bigcup_{m\in\mathbb{N}}\bigcap_{n\in\mathbb{N}} B_{m,n}$ is properly δ_1-exceptional, we can show

$$e_1^{(\delta_1)}(B_{m,n_m})(i) \approx 0 \quad \text{for every} \quad i \notin \bigcup_{m\in\mathbb{N}}\bigcap_{n\in\mathbb{N}} B_{m,n}. \tag{2.4.4}$$

In fact, for every $M_0 \in [0, \infty)$, we have

$$\begin{aligned}e_1^{(\delta_1)}(B_{m,n})(i) &= E_i\left[(1+\delta_1)^{-\sigma_{m,n}^{(\delta_1)}/\delta_1}\right] \\ &= E_i\left[(1+\delta_1)^{-\sigma_{m,n}^{(\delta_1)}/\delta_1}\left(1_{(\sigma_{B_{m,n}} \geq M_0)} + 1_{(\sigma_{m,n} < M_0)}\right)\right] \\ &\leq (1+\delta_1)^{-M_0/\delta_1} + E_i 1_{(\sigma_{m,n} < M_0)}.\end{aligned}$$

By letting M_0 be sufficiently large, we know that $(1+\delta_1)^{-M_0/\delta_1}$ will be very small. Taking n sufficiently large, we see that the approximation (2.4.4) holds.

The assumption in the Lemma implies that

$$\mathcal{E}_1^{(\delta_1)}\left(e_1^{(\delta_1)}(B_{m,n_m}), e_1^{(\delta_1)}(B_{m,n_m})\right) \approx 0.$$

Therefore, we get

$$^\circ\left[\mathcal{E}_1^{(\delta_1)}\left(e_1^{(\delta_1)}(B_{m,n}), e_1^{(\delta_1)}(B_{m,n})\right)\right] \longrightarrow 0.$$

This says that

$$\mathrm{Cap}_1^{(\delta_1)}\left(\bigcap_{n\in\mathbb{N}} B_{m,n}\right) \approx 0. \tag{2.4.5}$$

By Proposition 2.3.1 (i) and (ii) and the approximation (2.4.5), we obtain

$$\begin{aligned} {}^\circ\mathrm{Cap}_1^{(\delta_1)}(A) &\leq {}^\circ\mathrm{Cap}_1^{(\delta_1)}\left(\bigcup_{m\in\mathbb{N}}\bigcap_{n\in\mathbb{N}} B_{m,n}\right) \\ &\leq \sum_{m\in\mathbb{N}} {}^\circ\mathrm{Cap}_1^{(\delta_1)}\left(\bigcap_{n\in\mathbb{N}} B_{m,n}\right) \\ &= 0. \end{aligned}$$

Thus, the set A has a δ_1-zero capacity. □

Theorem 2.4.2. *Let $\delta_1 \in T, \delta_1 \approx 0$. Assume for all $f \in H$, if $^\circ(\mathcal{E}_1^{(\delta_1)}(f,f)) < \infty$ and $f(s) \approx 0$ for all $s \notin B$, where B is a δ_1-exceptional set, then we have $\mathcal{E}_1^{(\delta_1)}(f,f) \approx 0$. Let A be a subset of S_0. If A is δ_1-exceptional and there exists a sequence of internal subsets $\{B_n \mid n \in \mathbb{N}\}$ of S_0 such that*

$$A \subset \bigcup_{n\in\mathbb{N}} B_n \quad \text{and} \quad {}^\circ\mathrm{Cap}_1^{(\delta_1)}(B_n) < \infty, \quad \forall n \in \mathbb{N},$$

then A is of δ_1-zero capacity.

Proof. The proof follows easily from Lemma 2.4.2 and Proposition 2.3.1 (ii). □

In Lemma 2.4.2 and Theorem 2.4.2, we talk about the hyperfinite quadratic form and co-form and make one assumption: for all $f \in H$, if $^\circ(\mathcal{E}_1^{(\delta_1)}(f,f)) < \infty$ and $f(s) \approx 0$ for all $s \notin B$, where B is a δ_1-exceptional set, then we have $\mathcal{E}_1^{(\delta_1)}(f,f) \approx 0$. This assumption is somewhat equivalent to say that for all f, if $^\circ(\mathcal{E}_1^{(\delta_1)}(f,f)) < \infty$, then $f \in \mathcal{D}(\overline{\mathcal{E}}^{(\delta)})$. The assumption, however, is not always easy to verify. In the following, we will give results for hyperfinite weak coercive quadratic forms, for which we do not make the assumption. The results were first proved for hyperfinite symmetric Dirichlet forms in Fan [166]. The following results extend those of Fan [166].

Lemma 2.4.3. *Assume that $\mathcal{E}^{(\delta)}(\cdot,\cdot)$ is a hyperfinite weak coercive quadratic form on the space H for all infinitesimal $\delta \in T$. Let $\delta_1 \in T, \delta_1 \approx 0$. Let A be a subset of S_0. If A is δ_1-exceptional and there exists an internal subset B of S_0 which satisfies condition (2.4.3), then there is an infinitesimal $\delta_0 \in T$ which is larger than δ_1 such that A is of δ-zero capacity for all $\delta \geq \delta_0, \delta \approx 0$.*

Proof. In the proof of Lemma 2.4.2, we have

$$^{\circ}\left[\mathcal{E}_1^{(\delta_1)}\left(e_1^{(\delta_1)}(B_{m,n_m}), e_1^{(\delta_1)}(B_{m,n_m})\right)\right] = {}^{\circ}\left[\mathrm{Cap}_1^{(\delta_1)}(B_{m,n_m})\right] < \infty.$$

By Corollary 1.2.4 and Theorem 1.4.2, there is a $\delta_m \approx 0$ such that $e_1^{(\delta_1)}(B_{m,n_m}) \in \mathcal{D}(\mathcal{E}^{(\delta)})$ for all infinitesimal $\delta \geq \delta_m$.

Moreover, we know from the proof of Lemma 2.4.2 that

$$e_1^{(\delta_1)}(B_{m,n_m})(i) \approx 0 \quad \text{for every} \quad i \notin \bigcup_{m\in\mathbb{N}} \bigcap_{n\in\mathbb{N}} B_{m,n}.$$

Since $e_1^{(\delta_1)}(B_{m,n_m}) \in \mathcal{D}(\mathcal{E}^{(\delta)})$ for all $\delta \geq \delta_m, \delta \approx 0, e_1^{(\delta_1)}(B_{m,n_m})$ is S^2-integrable in the sense of Albeverio et al. [25], page 77, Chap. 3. Hence, we have $e_1^{(\delta_1)}(B_{m,n_m}) \approx 0$ in the hyperfinite dimensional space H because

$$L(m)\Big(\bigcup_{m\in\mathbb{N}} \bigcap_{m\in\mathbb{N}} B_{m,n}\Big) = 0.$$

Therefore, we have $\mathcal{E}_1^{(\delta)}(e_1^{(\delta_1)}(B_{m,n_m}), e_1^{(\delta_1)}(B_{m,n_m})) \approx 0$ for all $\delta \geq \delta_m, \delta \approx 0$. This implies that

$$^{\circ}\left[\mathcal{E}_1^{(\delta)}\left(e_1^{(\delta)}(B_{m,n}), e_1^{(\delta)}(B_{m,n})\right)\right] \downarrow 0, n \longrightarrow \infty, \quad \text{for all} \quad \delta \geq \delta_m, \delta \approx 0.$$

Hence, we have

$$\mathrm{Cap}_1^{(\delta)}\left(\bigcap_{n\in\mathbb{N}} B_{m,n}\right) \approx 0 \quad \text{for all} \quad \delta \geq \delta_m, \delta \approx 0. \tag{2.4.6}$$

By saturation, there is a $\delta_0 \approx 0$ larger than all $\delta_m, m \in \mathbb{N}$. Therefore, it follows from the approximation (2.4.6) that for $\delta \geq \delta_0, \delta \approx 0$,

$$\mathrm{Cap}_1^{(\delta)}\left(\bigcap_{n\in\mathbb{N}} B_{m,n}\right) \approx 0 \quad \text{for all} \quad m \in \mathbb{N}. \tag{2.4.7}$$

By Proposition 2.3.1 (i) and (ii) and the approximation (2.4.7), we obtain

$$\begin{aligned}
{}^\circ\mathrm{Cap}_1^{(\delta)}(A) &\leq {}^\circ\mathrm{Cap}_1^{(\delta)}\left(\bigcup_{m\in\mathbb{N}}\bigcap_{n\in\mathbb{N}} B_{m,n}\right) \\
&\leq \sum_{m=1}^{\infty} {}^\circ\mathrm{Cap}_1^{(\delta)}\left(\bigcap_{n\in\mathbb{N}} B_{m,n}\right) \\
&= 0.
\end{aligned}$$

Therefore, the set A has δ-zero capacity. □

Theorem 2.4.3. *Let $\mathcal{E}^{(\delta)}(\cdot,\cdot)$ be a hyperfinite weak coercive quadratic form on the space H for all infinitesimal $\delta \in T$. Let $\delta_1 \in T, \delta_1 \approx 0$. Let A be a subset of S_0. If A is δ_1-exceptional and there exists a sequence of internal subsets $\{B_n \mid n \in \mathbb{N}\}$ of S_0 such that*

$$A \subset \bigcup_{n\in\mathbb{N}} B_n \quad \text{and} \quad {}^\circ\mathrm{Cap}_1^{(\delta_1)}(B_n) < \infty, \quad \forall n \in \mathbb{N},$$

then there is an infinitesimal $\delta_0 \in T$ which is larger than δ_1 such that A is of δ-zero capacity for all $\delta \geq \delta_0, \delta \approx 0$.

Proof. The proof follows easily from Lemma 2.4.3, Proposition 2.3.1 (ii), and saturation. □

2.5 Measures of Hyperfinite Energy Integrals

We have defined an inner product $\langle\cdot,\cdot\rangle$ on the hyperfinite dimensional space H by (2.2.3) in Sect. 2.2 or (1.5.15) in Sect. 1.5, Chap. 1. Let $||\cdot||$ be the norm generated by $\langle\cdot,\cdot\rangle$. Denote by $\mathrm{Fin}(H)$ the set of all elements in H with finite norm $||\cdot||$. By defining $u \approx v$ if $||u-v|| \approx 0$, we recall that the space

$${}^\circ H = \mathrm{Fin}(H)/\approx$$

is a Hilbert space with the inner product $({}^\circ u, {}^\circ v) = \mathrm{st}(\langle u, v\rangle)$, where ${}^\circ u$ denotes the equivalence class of u.

In this section, we shall continue our discussion of Sects. 2.2, 2.3, and 2.4. We assume that X and $\hat{X}$ are dual hyperfinite Markov chains. Let $\mathcal{E}(\cdot,\cdot)$ and $\hat{\mathcal{E}}(\cdot,\cdot)$ be the hyperfinite quadratic form and co-form associated with X and $\hat{X}$, respectively.

We know that for $\alpha \in {}^*\mathbb{R}$, $\alpha \geq 0$ and $\delta \in T$,

$$\mathcal{E}_\alpha^{(\delta)}(u,v) = \mathcal{E}^{(\delta)}(u,v) + \alpha\langle u,v\rangle.$$

Each of these forms generates a norm (possibly a semi-norm in the case $\alpha = 0$): $|u|_\alpha^{(\delta)} = [\mathcal{E}_\alpha^{(\delta)}(u,u)]^{\frac{1}{2}}$. Similarly, we denote by $\mathrm{Fin}_\alpha^{(\delta)}(H)$ the set of all elements in H with finite norm $|\cdot|_\alpha^{(\delta)}$. Define $u \approx_\alpha^{(\delta)} v$ if $|u-v|_\alpha^{(\delta)} \approx 0$. The space

$${}^\circ H_\alpha^{(\delta)} = \mathrm{Fin}_\alpha^{(\delta)}(H)/\approx_\alpha^{(\delta)}$$

is a Hilbert space if ${}^\circ\alpha > 0$ with respect to the inner product

$$([u]_\alpha^{(\delta)}, [v]_\alpha^{(\delta)})_\alpha^{(\delta)} = {}^\circ\left[\mathcal{E}_\alpha^{(\delta)}(u,v)\right],$$

where $[u]_\alpha^{(\delta)}$ denotes the equivalence class of u under the norm $|\cdot|_\alpha^{(\delta)}$, and $(\cdot,\cdot)_\alpha^{(\delta)}$ denotes the related inner product.

Definition 2.5.1. Let μ be a hyperfinite positive measure on S_0. For $\delta \in T$, if there exists a constant $K \in {}^*[0,\infty) = {}^*\mathbb{R}_+$ such that

$$\begin{aligned}\int_{S_0} |u(s)|\ \mu(ds) &= \sum_{i=1}^{N} |u(s_i)|\mu(i) \\ &\leq K\left[\mathcal{E}_1^{(\delta)}(u,u)\right]^{\frac{1}{2}} \qquad (2.5.1)\end{aligned}$$

for every $u \in H$, we say that μ is of *δ-hyperfinite energy integral.* Moreover, if there exists $K \in {}^*\mathbb{R}_+$ satisfying (2.5.1) and ${}^\circ K < \infty$, μ is said to be of *δ-finite energy integral.*

Henceforth, we will identify a hyperfinite measure μ on S_0 with the measure $\tilde{\mu}$ on S defined by $\tilde{\mu}(s_0) = 0$, $\tilde{\mu}(s_i) = \mu(s_i)$ for all $s_i \in S_0$.

Theorem 2.5.1. *Let $\mathcal{E}^{(\delta)}(\cdot,\cdot)$ be a hyperfinite weak coercive quadratic form. Then*

(i) A positive hyperfinite measure μ on S_0 is of δ-hyperfinite energy integral if and only if for each $\alpha \in {}^*\mathbb{R}, 0 < \mathrm{st}(\alpha) < \infty$, there exists an element $U_\alpha^{(\delta)}\mu \in H$ such that for every $v \in H$,

$$\mathcal{E}_\alpha^{(\delta)}(U_\alpha^{(\delta)}\mu, v) = \int_{S_0} v(s)\ \mu(ds). \qquad (2.5.2)$$

Moreover, if μ is of δ-finite energy integral, we have $U_\alpha^{(\delta)}\mu \in Fin_\alpha^{(\delta)}(H)$.

(ii) *A positive hyperfinite measure μ on S_0 is of δ-hyperfinite energy integral if and only if for each $\alpha \in {}^*\mathbb{R}, 0 < \mathrm{st}(\alpha) < \infty$, there exists an element $\hat{U}_\alpha^{(\delta)}\mu \in H$ such that for every $v \in H$,*

$$\mathcal{E}_\alpha^{(\delta)}(v, \hat{U}_\alpha^{(\delta)}\mu) = \int_{S_0} v(s)\ \mu(ds).$$

Moreover, if μ is of δ-finite energy integral, we have $\hat{U}_\alpha^{(\delta)}\mu \in Fin_\alpha^{(\delta)}(H)$.

Proof. The theorem is an easy consequence of Riesz's representation theorem. □

Remark 2.5.1. We call $U_\alpha^{(\delta)}\mu$ and $\hat{U}_\alpha^{(\delta)}\mu$ in Theorem 2.5.1 *hyperfinite α-potential* and *hyperfinite α-co-potential* of μ associated with $\mathcal{E}^{(\delta)}(\cdot,\cdot)$, respectively. An internal element $u \in H$ is called a *hyperfinite α-potential* (or *hyperfinite α-co-potential*) if $u = U_\alpha^{(\delta)}\mu$ (or $u = \hat{U}_\alpha^{(\delta)}\mu$) for some positive measure μ of δ-hyperfinite energy integral.

Definition 2.5.2. Fix $\alpha \in {}^*\mathbb{R}, \alpha \geq 0$ and $\delta \in T$.

(i) An element $u \in H$ is called *hyperfinite pre-α-excessive* associated with $\mathcal{E}^{(\delta)}(\cdot,\cdot)$, if $u(i) \geq 0, Q^\delta u(i) \leq (1+\alpha\delta)u(i)$ for every $i \in S_0$ such that $m(i) \neq 0$.
(ii) An element $u \in H$ is called *hyperfinite pre-α-co-excessive* associated with $\hat{\mathcal{E}}^{(\delta)}(\cdot,\cdot)$, if $u(i) \geq 0, \hat{Q}^\delta u(i) \leq (1+\alpha\delta)u(i)$ for every $i \in S_0$ such that $m(i) \neq 0$.

In order to develop our theory, we shall denote by $\{G_\beta^{(\delta)} \mid \beta \in {}^*(-\infty, 0)\}$ and $\{\hat{G}_\beta^{(\delta)} \mid \beta \in {}^*(-\infty, 0)\}$ the resolvent and co-resolvent of $\mathcal{E}^{(\delta)}(\cdot,\cdot)$, respectively. That is, they are defined by

$$\begin{aligned} G_\beta^{(\delta)} &= (A^{(\delta)} - \beta)^{-1}, \\ \hat{G}_\beta^{(\delta)} &= (\hat{A}^{(\delta)} - \beta)^{-1}. \end{aligned}$$

Hence, we have for $\alpha \in {}^*\mathbb{R}_+$,

$$\begin{aligned} \left(1+\alpha\delta - Q^\delta\right) G_{-\alpha}^{(\delta)} &= \delta(\alpha + A^{(\delta)})G_{-\alpha}^{(\delta)} \\ &= \delta \end{aligned}$$

and

$$\begin{aligned} \left(1+\alpha\delta - \hat{Q}^\delta\right) \hat{G}_{-\alpha}^{(\delta)} &= \delta(\alpha + \hat{A}^{(\delta)})\hat{G}_{-\alpha}^{(\delta)} \\ &= \delta. \end{aligned} \tag{2.5.3}$$

Theorem 2.5.2. *For $\delta \in T, \alpha \in {}^*\mathbb{R}, \alpha \geq 0$, and $u \in H$, the following conditions are equivalent:*

(i) u is hyperfinite pre-α-excessive associated with $\mathcal{E}^{(\delta)}(\cdot,\cdot)$.
(ii) There exists a hyperfinite positive measure μ on S_0 such that

$$\mathcal{E}_\alpha^{(\delta)}(u,v) = \int_{S_0} v(s)\ \mu(ds) \quad \text{for all} \quad v \in H.$$

(iii) $\mathcal{E}_\alpha^{(\delta)}(u,v) \geq 0$ for all $v \in H, v \geq 0$.
If $\mathcal{E}^{(\delta)}(\cdot,\cdot)$ is a hyperfinite weak coercive quadratic form, the above statements are equivalent to the following:
(iv) u is a hyperfinite α-potential of $\mathcal{E}^{(\delta)}(\cdot,\cdot)$.

Proof. (i) $\Longrightarrow$ (ii). Assume that u is hyperfinite pre-α-excessive associated with $\mathcal{E}^{(\delta)}(\cdot,\cdot)$. Define a hyperfinite positive measure μ on S_0 by

$$\begin{aligned}\mu(s_0) &= 0,\\ \mu(s_i) &= \frac{1}{\delta}\left((1+\alpha\delta)u(i) - Q^\delta u(i)\right) m(i) \quad \text{for} \quad i \in S_0.\end{aligned}$$

Since

$$\begin{aligned}\mu(s_i) &= \frac{1}{\delta}\left((1+\alpha\delta)u(i) - Q^\delta u(i)\right) m(i)\\ &= \left(A^{(\delta)}u(i) + \alpha u(i)\right) m(i),\end{aligned}$$

we see that for every $v \in H$,

$$\begin{aligned}\mathcal{E}_\alpha^{(\delta)}(u,v) &= \sum_{i=1}^{N}\left(A^{(\delta)}u(i) + \alpha u(i)\right) v(i)m(i)\\ &= \int_{S_0} v(s)\ \mu(ds).\end{aligned}$$

(ii) $\Longrightarrow$ *(iii)*. This is easy, and thus we omit it!

(iii) $\Longrightarrow$ *(i)*. Such as in Sect. 1.5 of Chap. 1, let $u^+ = u \vee 0$. Since $u^+ - u \geq 0$, we have $\mathcal{E}_\alpha^{(\delta)}(u, u^+ - u) \geq 0$. Taking into account Corollary 1.5.2, we have

$$\begin{aligned}\mathcal{E}_\alpha^{(\delta)}(u^+ - u, u^+ - u) &= \mathcal{E}_\alpha^{(\delta)}(u^+, u^+ - u) - \mathcal{E}_\alpha^{(\delta)}(u, u^+ - u)\\ &\leq \mathcal{E}_\alpha^{(\delta)}(u^+, u^+ - u)\\ &= -\mathcal{E}^{(\delta)}((-u)\wedge 0, -u - (-u)\wedge 0)\\ &\leq 0.\end{aligned}$$

This implies that

$$\mathcal{E}_{\alpha}^{(\delta)}(u^{+}-u, u^{+}-u) = 0.$$

Therefore, $u(i) = u(i) \vee 0 \geq 0$ for every $i \in S_0$ such that $m(i) \neq 0$.

Furthermore, it follows from (2.5.3) that for any $v \in H$,

$$\begin{aligned} \langle (1+\alpha\delta - Q^{\delta})u, v\rangle &= \langle u, (1+\alpha\delta - \hat{Q}^{\delta})v\rangle \\ &= \mathcal{E}_{\alpha}^{(\delta)}\left(u, (1+\alpha\delta - \hat{Q}^{\delta})\hat{G}_{-\alpha}^{(\delta)}v\right) \\ &= \mathcal{E}_{\alpha}^{(\delta)}(u, \delta v). \end{aligned} \tag{2.5.4}$$

Fix $i \in S_0$. Let $v \in H$ be an internal function defined by $v(l) = \delta_{il}, l \in S$. Then, we have from (2.5.4) that

$$\begin{aligned} \left(1+\alpha\delta - Q^{\delta}\right) u(i)m(i) &= \mathcal{E}_{\alpha}^{(\delta)}(u, \delta v) \\ &\geq 0. \end{aligned}$$

Therefore, the internal function u is hyperfinite pre-α-excessive associated with $\mathcal{E}^{(\delta)}(\cdot,\cdot)$.

If $\mathcal{E}^{(\delta)}(\cdot,\cdot)$ is a hyperfinite weak coercive quadratic form, (ii) is equivalent to (iv) by Theorem 2.5.1 (i). □

Similarly, we have

Theorem 2.5.3. *For $\delta \in T$, $\alpha \in {}^*\mathbb{R}, \alpha \geq 0,$ and $u \in H$, the following conditions are equivalent:*

(i) u is hyperfinite pre-α-co-excessive associated with $\mathcal{E}^{(\delta)}(\cdot,\cdot)$.
(ii) There exists a hyperfinite positive measure μ on S_0 such that

$$\mathcal{E}_{\alpha}^{(\delta)}(v,u) = \int_{S_0} v(s)\ \mu(ds) \quad \textit{for all} \quad v \in H.$$

(iii) $\mathcal{E}_{\alpha}^{(\delta)}(v,u) \geq 0$ for all $v \in H, v \geq 0$.
If $\mathcal{E}^{(\delta)}(\cdot,\cdot)$ is a hyperfinite weak coercive quadratic form, the above statements are equivalent to the following:
(iv) u is a hyperfinite α-co-potential of $\mathcal{E}^{(\delta)}(\cdot,\cdot)$.

We denote by $\tau_0(\delta)$ the family of all internal positive measures on S_0 of δ-hyperfinite energy integrals. Denote $\tau_0 = \cup\{\tau_0(\delta) \mid \delta \text{ is infinitesimal}, \delta \in T\}$.

Proposition 2.5.1. *Let $\mathcal{E}^{(\delta)}(\cdot,\cdot)$ be a hyperfinite weak coercive quadratic form. For $\delta \in T$, a hyperfinite positive measure μ on S_0 is of δ-hyperfinite* *energy integral if and only if for any* $\alpha \in {}^*\mathbb{R}_+$, *there exists a hyperfinite pre-*α*-excessive function* u *associated with* $\mathcal{E}^{(\delta)}(\cdot,\cdot)$ *such that*

$$\mu(i) = \frac{1}{\delta}\left((1+\alpha\delta)u(i) - Q^\delta u(i)\right) m(i) \quad \textit{for all} \quad i \in S_0. \tag{2.5.5}$$

Moreover, if $\bar{u} \in H$ *satisfies above equation also, then* $u(i) = \bar{u}(i)$ *for all* $i \in S_0$ *with* $m(i) \neq 0$.

Proof. $\Longleftarrow$ Let u satisfy the condition (2.5.5). Then, we have

$$\mathcal{E}_\alpha^{(\delta)}(u,v) = \int_{S_0} v(s)\, \mu(ds) \quad \text{for all} \quad v \in H.$$

$\Longrightarrow$ Assume that μ is of δ-hyperfinite energy integral. Let $u = U_1^{(\delta)}\mu \in H$ satisfy the condition (2.5.2). Then, we have

$$\begin{aligned}\mu(i) &= \left(G_{-1}^{(\delta)}\right)^{-1} u(i)m(i)\\ &= \frac{1}{\delta}\Big((1+\delta)u(i) - Q^\delta u(i)\Big)m(i).\end{aligned}$$

Therefore, we have for any $\alpha \in {}^*\mathbb{R}_+, v \in H$,

$$\begin{aligned}\int_S v(s)\, d\mu(s) &= \int_S v(i) \left(G_{-1}^{(\delta)}\right)^{-1} u(i)\, dm(i)\\ &= \mathcal{E}_\alpha^{(\delta)}\left(G_{-\alpha}^{(\delta)}((G_{-1}^{(\delta)})^{-1}u), v\right).\end{aligned}$$

Hence, $w = G_{-\alpha}^{(\delta)}((G_{-1}^{(\delta)})^{-1}u)$ is hyperfinite pre-α-excessive associated with $\mathcal{E}^{(\delta)}(\cdot,\cdot)$ by Theorem 2.5.2. Furthermore, we have

$$\mu(i) = \frac{1}{\delta}\left((1+\alpha\delta)w(i) - Q^\delta w(i)\right) m(i) \quad \text{for all} \quad i \in S_0.$$

□

Proposition 2.5.2. *Let* $\mathcal{E}^{(\delta)}(\cdot,\cdot)$ *be a hyperfinite weak coercive quadratic form. For* $\delta \in T$, *let* ν *be a hyperfinite positive measure on* S_0. *Define a measure* μ *on* S_0 *by*

$$\mu(s) = \nu(s)1_{(m(s)\neq 0)} \quad \textit{for} \quad s \in S_0.$$

Then μ *is of* δ*-hyperfinite energy integral.*

Proof. Define

$$f(s) = \frac{\nu(s)}{m(s)} 1_{(m(s)\neq 0)} \quad \text{for} \quad s \in S_0, \quad f(s_0) = 0.$$

For any $u \in H$, we have

$$\begin{aligned}\int_S u(s)\, d\mu(s) &= \int_S u(s) 1_{(m(s)\neq 0)}\, d\nu(s) \\ &= \int_S u(s) f(s)\, dm(s) \\ &= \mathcal{E}_1^{(\delta)}(G_{-1}^{(\delta)} f, u).\end{aligned}$$

Hence, μ is of δ-hyperfinite energy integral by Theorem 2.5.2. □

We recall that in Sect. 1.4 of Chap. 1, we have introduced the standard part $(E^{(\delta)}(\cdot,\cdot), D(E^{(\delta)}))$ on $^\circ H$ for a hyperfinite weak coercive quadratic form $\mathcal{E}^{(\delta)}(\cdot,\cdot)$.

Theorem 2.5.4. *Let $\mathcal{E}^{(\delta)}(\cdot,\cdot)$ be a hyperfinite weak coercive quadratic form. For $\alpha \in {}^*\mathbb{R}$, $0 < \mathrm{st}(\alpha) < \infty$ and $\delta \in T$, let μ be a hyperfinite positive measure of δ-finite energy integral, and let u be an α-potential of μ associated with $\mathcal{E}^{(\delta)}(\cdot,\cdot)$. Define*

$$g_n(i) = n\left(u(i) - nG_{-n-\alpha}^{(\delta)} u(i)\right), n \in \mathbb{N}.$$

Then for every $v \in Fin_\alpha^{(\delta)}(H)$, we have

$$(i) \quad {}^\circ\mathcal{E}_\alpha^{(\delta)}(G_{-\alpha}^{(\delta)} g_n, v) \longrightarrow E_\alpha^{(\delta)}({}^\circ u, {}^\circ v) \quad as \quad n \to \infty. \tag{2.5.6}$$

(ii) *If $u \in \mathcal{D}(\mathcal{E}^{(\delta)})$, let $\tilde{v} \approx_\alpha^\delta v$ and $\tilde{v} \in \mathcal{D}(\mathcal{E}^{(\delta)})$,*

$$lim_{n\to\infty} {}^\circ\!\!\int_{S_0} g_n(i) v(i)\, dm(i) = {}^\circ\!\!\int_{S_0} \tilde{v}(i)\, d\mu(i).$$

Proof. (i) First of all, we have

$$\begin{aligned}{}^\circ\mathcal{E}_\alpha^{(\delta)}(G_{-\alpha}^{(\delta)} g_n, v) &= {}^\circ(\langle g_n, v\rangle) \\ &= {}^\circ(\langle nu - n^2 G_{-n-\alpha}^{(\delta)} u, v\rangle) \\ &= {}^\circ\!\left(\langle (n+\alpha)u - (n+\alpha)^2 G_{-n-\alpha}^{(\delta)} u, v\rangle\right) \qquad (2.5.7) \\ &\quad + {}^\circ\!\left(\langle -\alpha u + \alpha^2 G_{-n-\alpha}^{(\delta)} u + 2n\alpha G_{-n-\alpha}^{(\delta)} u, v\rangle\right).\end{aligned}$$

It follows from Theorem 1.4.1 that

$$\overset{\circ}{\left[\langle (n+\alpha)u - (n+\alpha)^2 G^{(\delta)}_{-n-\alpha} u, v\rangle + \alpha\langle u, v\rangle\right]}$$
$$\longrightarrow E^{(\delta)}_{\alpha}({}^{\circ}u, {}^{\circ}v) \text{ as } n \to \infty. \tag{2.5.8}$$

By Lemma 1.4.4, we have

$$\overset{\circ}{\left(\langle \alpha^2 G^{(\delta)}_{-n-\alpha} u, v\rangle\right)} \longrightarrow 0 \text{ as } n \longrightarrow \infty. \tag{2.5.9}$$

Besides, it follows from Theorem 1.4.1

$$\begin{aligned}&\overset{\circ}{\left|\langle -2\alpha u + 2n\alpha G^{(\delta)}_{-n-\alpha} u, v\rangle\right|}\\ &\leq \overset{\circ}{\Big[}\frac{2\alpha n}{(n+\alpha)^2}\Big(\langle -\frac{(n+\alpha)^2}{n} u + (n+\alpha)u, v\rangle\\ &\quad + \langle -(n+\alpha)u + (n+\alpha)^2 G^{(\delta)}_{-n-\alpha} u, v\rangle\Big)\Big]\\ &\longrightarrow 0 \quad \text{as} \quad n \to \infty.\end{aligned} \tag{2.5.10}$$

From the relations (2.5.7), (2.5.8), (2.5.9), and (2.5.10), we know that the approximation (2.5.6) holds.

(ii) From the proof of (i), we get

$$\begin{aligned}\lim_{n\to\infty} \overset{\circ}{\int_{S_0}} g_n(s) v(s)\, m(ds) &= {}^{\circ}\mathcal{E}^{(\delta)}_{\alpha}(u, \tilde{v})\\ &= \overset{\circ}{\int_{S_0}} \tilde{v}(s)\, \mu(ds).\end{aligned}$$

□

Proposition 2.5.3. *Let $\mathcal{E}^{(\delta)}(\cdot,\cdot)$ be a hyperfinite weak coercive quadratic form with continuity constant C. For $\delta \in T$, let μ be a positive measure of δ-hyperfinite energy integral. Then for every $L(\mu)$ measurable subset A of S_0, we have*

$$L(\mu)(A) \leq \overset{\circ}{\left[C\sqrt{\mathcal{E}^{(\delta)}_1(U^{(\delta)}_1\mu, U^{(\delta)}_1\mu)}\right]} \sqrt{{}^{\circ}\mathrm{Cap}^{(\delta)}_1(A)}. \tag{2.5.11}$$

Proof. For simplicity, we assume that ${}^{\circ}\mathrm{Cap}^{(\delta)}_1(A) < \infty$. If A is internal, we have

$$\begin{aligned}{}^{\circ}\mu(A) &= \overset{\circ}{\int_{S_0}} 1_A(s)\, \mu(ds)\\ &\leq \overset{\circ}{\int_{S_0}} e^{(\delta)}_1(A)(s)\, \mu(ds)\end{aligned}$$

$$= {}^\circ\mathcal{E}_1^{(\delta)}(U_1^{(\delta)}\mu, e_1^{(\delta)}(A))$$
$$\le {}^\circ\left[C\sqrt{\mathcal{E}_1^{(\delta)}(U_1^{(\delta)}\mu, U_1^{(\delta)}\mu)}\sqrt{\mathcal{E}_1^{(\delta)}(e_1^{(\delta)}(A), e_1^{(\delta)}(A))}\right]$$
$$= {}^\circ\left[C\sqrt{\mathcal{E}_1^{(\delta)}(U_1^{(\delta)}\mu, U_1^{(\delta)}\mu)}\sqrt{\mathrm{Cap}_1^{(\delta)}(A)}\right].$$

Now it is easy to see that the inequality (2.5.11) holds for all $L(\mu)$ measurable subsets A. □

Corollary 2.5.1. *Let $\mathcal{E}^{(\delta)}(\cdot,\cdot)$ be a hyperfinite weak coercive quadratic form. Let μ be a positive measure of δ-finite energy integral on S_0. Then $L(\mu)$ charges no set of δ-zero capacity.*

Proof. The proof follows easily from Proposition 2.5.3. □

We shall now state and prove the following characterization theorem.

Theorem 2.5.5. *Let $A \subset S_0$ be an $\mathcal{A}(\mathcal{S}_0)$-measurable set (in particular, $A \in \sigma(\mathcal{S}_0)$) and let $\delta \in T$. We have*

(i) For any $\mu \in \tau_0(\delta), L(\mu)(A) = 0 \Longrightarrow {}^\circ\mathrm{Cap}_1^{(\delta)}(A) = 0$.
(ii) For any $\mu \in \tau_{00}(\delta), L(\mu)(A) = 0 \Longrightarrow {}^\circ\mathrm{Cap}_1^{(\delta)}(A) = 0$, where

$$\tau_{00}(\delta) = \left\{\mu \in \tau_0(\delta)\Big|\mu(S_0) = 1, {}^\circ\|U_1^{(\delta)}\mu\|_\infty < \infty\right\}$$

and

$$\|U_1^{(\delta)}\mu\|_\infty = \max\left\{|U_1^{(\delta)}\mu(s)|\Big|s \in S_0\right\}.$$

(iii) For any $\mu \in \hat{\tau}_{00}(\delta), L(\mu)(A) = 0 \Longrightarrow {}^\circ\mathrm{Cap}_1^{(\delta)}(A) = 0$, where

$$\hat{\tau}_{00}(\delta) = \left\{\mu \in \tau_0(\delta)\Big|\mu(S_0) = 1, {}^\circ\|\hat{U}_1^{(\delta)}\mu\|_\infty < \infty\right\}$$

and

$$\| \hat{U}_1^{(\delta)}\mu \|_\infty = \max\left\{|\hat{U}_1^{(\delta)}\mu(s)|\Big|s \in S_0\right\}.$$

Proof. (*i*) Assume that $\infty \ge {}^\circ\mathrm{Cap}_1^{(\delta)}(A) = \alpha > 0$. Since $A \in \mathcal{A}(\mathcal{S}_0)$, it follows from Theorem 2.3.1 (ii) that A is capacitable with respect to the capacity ${}^\circ\mathrm{Cap}_1^{(\delta)}(\cdot)$. That is

$${}^\circ\mathrm{Cap}_1^{(\delta)}(A) = \sup\left\{{}^\circ\mathrm{Cap}_1^{(\delta)}(B)\Big|B = \bigcap_{n\in\mathbb{N}} B_n, B_n \in \mathcal{S}_0 \quad \text{and} \quad B \subset A\right\}.$$

Therefore, there exists a sequence $\{B_n \mid n \in \mathbb{N}\}$ of decreasing internal subsets of S_0 such that

$$\bigcap_{n=1}^{\infty} B_n \subset A, \quad \infty \geq {}^\circ\mathrm{Cap}_1^{(\delta)}\left(\bigcap_{n=1}^{\infty} B_n\right) \geq \frac{\alpha}{2} > 0.$$

Let $\{B_n \mid n \in {}^*\mathbb{N}\}$ be an internal decreasing extension of $\{B_n \mid n \in \mathbb{N}\}$. Then, there exists an infinite $\gamma \in {}^*\mathbb{N} - \mathbb{N}$ such that

$$B_\gamma \subset A, \quad 0 < \frac{\alpha}{2} \leq {}^\circ\mathrm{Cap}_1^{(\delta)}\left(\bigcap_{n=1}^{\infty} B_n\right) = {}^\circ\mathrm{Cap}_1^{(\delta)}(B_\gamma) \leq \infty.$$

Set $B = B_\gamma$. Consider the internal function $e_1^{(\delta)}(B)$. Notice that if $i \notin B$, then

$$\begin{aligned}(1+\delta)e_1^{(\delta)}(B)(i) &= (1+\delta)E_i\left[(1+\delta)^{-\sigma_B^{(\delta)}/\delta}\right] \\ &= \sum_{j=1}^{N} e_1^{(\delta)}(B)(j)q_{ij}^{(\delta)} \\ &= Q^\delta e_1^{(\delta)}(B)(i), \end{aligned} \tag{2.5.12}$$

where $\sigma_B^{(\delta)} = \min\{t \in T_\delta \mid X(\omega, t) \in B\}$. If $i \in B$, then we have

$$\begin{aligned}(1+\delta)e_1^{(\delta)}(B)(i) &= (1+\delta) \\ &\geq \sum_{j=1}^{N} e_1^{(\delta)}(B)(j)q_{ij}^{(\delta)} \\ &= Q^\delta e_1^{(\delta)}(B)(i). \end{aligned} \tag{2.5.13}$$

It follows from the relations (2.5.12) and (2.5.13) that the function $e_1^{(\delta)}(B)$ is hyperfinite 1-excessive associated with $\mathcal{E}^{(\delta)}(\cdot,\cdot)$. Define a hyperfinite positive measure μ on S by

$$\begin{aligned}&\mu(s_0) = 0, \\ &\mu(s_i) = \frac{1}{\delta}\left((1+\delta)e_1^{(\delta)}(B)(i) - Q^\delta e_1^{(\delta)}(B)(i)\right) m(i), i \in S_0. \end{aligned}$$

Then, we have

$$\begin{aligned}\mu(B) &= \int_{S_0} \mu(ds) \\ &= \mathcal{E}_1^{(\delta)}(e_1^{(\delta)}(B), 1) \\ &= \mathrm{Cap}_1^{(\delta)}(B). \end{aligned}$$

This contradicts the assumption ${}^{\circ}L(\mu)(A) = 0$. Hence, we must have ${}^{\circ}\mathrm{Cap}_1^{(\delta)}(A) = 0$.

(*ii*) Assume that $0 < {}^{\circ}\mathrm{Cap}_1^{(\delta)}(A) = \alpha \leq \infty$. By the proof (*i*), we see that there exists an element $\nu \in \tau_0(\delta)$ such that (replacing μ in the proof of (i) by ν)

$$0 < {}^{\circ}\mathrm{Cap}_1^{(\delta)}(B) \leq \infty, \quad \mathrm{Cap}_1^{(\delta)}(B) = \nu(B) = \nu(S_0)$$
$$\text{and } U_1^{(\delta)}\nu(i) = e_1^{(\delta)}(B)(i) \leq 1$$

for any $i \in S_0$, where B is an internal set contained in A. Define $\mu(\cdot)$ by

$$\mu(\cdot) = \frac{\nu(\cdot)}{\mathrm{Cap}_1^{(\delta)}(B)}.$$

Then, we have

$$\begin{aligned} \mu(S_0) = 1, \mu \in \tau_0(\delta) \quad &\text{and} \\ {}^{\circ}\left(\max_{i\in S_0} |U_1^{(\delta)}\mu(i)|\right) &= {}^{\circ}\left(\max_{i\in S_0} \frac{|U_1^{(\delta)}\nu(i)|}{\mathrm{Cap}_1^{(\delta)}(B)}\right) \\ &\leq {}^{\circ}\left(\frac{1}{\mathrm{Cap}_1^{(\delta)}(B)}\right) \\ &< \infty. \end{aligned}$$

But

$$1 = {}^{\circ}\mu(B) \leq {}^{\circ}\mu(A) \leq 1, \textit{ i.e., } L(\mu)(A) = 1.$$

This contradicts our assumption of $L(\mu)(A) = 0$. We have proved that

$${}^{\circ}\mathrm{Cap}_1^{(\delta)}(A) = 0.$$

(*iii*) Recalling Corollary 2.2.1, we can show this result in the same way as that of above (ii). □

Theorem 2.5.6. *For $\delta \in T$, let $\mathcal{E}^{(\delta)}(\cdot,\cdot)$ be a hyperfinite weak coercive quadratic form, which has the following property:*

$$\begin{aligned} \forall A \subset S_0, \ {}^{\circ}\mathrm{Cap}_1^{(\delta)}(A) = \infty& \\ \Longrightarrow \exists B \subset A \quad \textit{such that} \quad 0 < {}^{\circ}\mathrm{Cap}_1^{(\delta)}(B) < \infty.& \quad (2.5.14) \end{aligned}$$

Let $\tau_{0f}(\delta)$ be the family of all positive measures of δ-finite energy integrals. Then the following statements are equivalent for $A \in \mathcal{A}(\mathcal{S}_0)$ (in particular, $A \in \sigma(\mathcal{S}_0)$):

(i) A is of zero δ-capacity, i.e., ${}^\circ\mathrm{Cap}_1^{(\delta)}(A) = 0$.
(ii) For any $\mu \in \tau_{0f}(\delta), L(\mu)(A) = 0$.
(iii) For any $\mu \in \tau_{00f}(\delta), L(\mu)(A) = 0$, where

$$\tau_{00f}(\delta) = \left\{\mu \in \tau_{0f}(\delta) \middle| \mu(S_0) = 1, {}^\circ||U_1^{(\delta)}\mu||_\infty < \infty\right\}.$$

(iv) For any $\mu \in \hat{\tau}_{00f}(\delta), L(\mu)(A) = 0$, where

$$\hat{\tau}_{00f}(\delta) = \left\{\mu \in \tau_{0f}(\delta) \middle| \mu(S_0) = 1, {}^\circ||\hat{U}_1^{(\delta)}\mu||_\infty < \infty\right\}.$$

Proof. $(i) \Longrightarrow (ii) \Longrightarrow (iii)$ and $(ii) \Longrightarrow (iv)$ are clear by Corollary 2.5.1. We can show $(ii) \Longrightarrow (i)$ and $(iii) \Longrightarrow (i)$ and $(iv) \Longrightarrow (i)$ in the same way as the proof of Theorem 2.5.5. □

2.6 Internal Additive Functionals and Associated Measures

Let X and $\hat{X}$ be dual hyperfinite Markov chains, and let $\mathcal{E}(\cdot,\cdot)$ and $\hat{\mathcal{E}}(\cdot,\cdot)$ be the hyperfinite quadratic form and co-form of X and $\hat{X}$, respectively. Again, let H be the hyperfinite dimensional space with an inner product $\langle\cdot,\cdot\rangle$ defined by (2.2.3) in Sect. 2.2 or (1.5.15) in Sect. 1.5, Chap. 1.

We recall that in Sect. 1.5 of Chap. 1, we have introduced the dual hyperfinite Markov chains $(\Omega, X^{(\delta)}, \{\mathcal{F}_t^{(\delta)} \mid t \in T_\delta\}, \{P_i \mid i \in S\})$ and $(\hat{\Omega}, \hat{X}^{(\delta)}, \{\hat{\mathcal{F}}_t^{(\delta)} \mid t \in T_\delta\}, \{\hat{P}_i \mid i \in S\})$ for $\delta \in T$. For simplicity, we assume that $\Omega = \hat{\Omega}$, $X^{(\delta)} = \hat{X}^{(\delta)}$, $\mathcal{F}_t^{(\delta)} = \hat{\mathcal{F}}_t^{(\delta)}, t \in T_\delta$. The hyperfinite quadratic form associated with $(X, \{P_i \mid i \in S\})$ (or $(\hat{X}, \{\hat{P}_i \mid i \in S\})$) are given by the expression (1.5.19) (or (1.5.22)) in Sect. 1.5. As in the study of standard Markov processes, we define a family of translation operators $\{\theta_t \mid t \in T\}$ of Ω. That is, for each $t \in T, \theta_t$ is a map from Ω to Ω defined by

$$\omega \in \Omega \Longrightarrow \theta_t\omega \in \Omega \quad \text{and for any} \quad s \in T, \theta_t\omega(s) = \omega(s+t).$$

Hence for each $\delta \in T$, we have a family of translation operators $\{\theta_t^{(\delta)} \mid t \in T_\delta\}$ induced by $\{\theta_t \mid t \in T\}$, $\theta_t^{(\delta)} = \theta_t$ for any $t \in T_\delta$. In other words, for each $t \in T_\delta$, $\theta_t^{(\delta)}$ is a map from Ω to Ω given by

$$\omega \in \Omega \Longrightarrow \theta_t^{(\delta)}\omega \in \Omega \quad \text{and for any} \quad s \in T_\delta, \theta_t^{(\delta)}\omega(s) = \omega(s+t).$$

Definition 2.6.1. For any $\delta \in T$, we call an internal $^*\mathbb{R}$-valued function $A(\omega,t)$ or $A_t(\omega), t \in T_\delta, \omega \in \Omega$, *$\delta$-internal additive functional* (abbreviated by δ-IAF) if it satisfies the following two conditions:

(1) For each $t \in T_\delta$, $A_t(\omega)$ is *non-anticipating* with respect to the filtration $(\Omega, \{\mathcal{F}_t^{(\delta)} \big| t \in T_\delta\})$, i.e., $A_t(\cdot)$ is $\mathcal{F}_t^{(\delta)}$-measurable, $\forall t \in T_\delta$.
(2) For each $\omega \in \Omega$, we have

$$\begin{aligned} A(\omega,0) &= 0, \\ A(\omega,t+s) &= A(\omega,s) + A(\theta_s^{(\delta)}\omega,t) \quad \text{for any} \quad t,s \in T_\delta. \end{aligned}$$

Proposition 2.6.1. *If $A(\omega,t)$ is a δ-internal additive functional, then there exists a hyperfinite measure $\mu_{<A>}$ on S_0 (not necessarily positive) such that $\mu_{<A>}(i) = 0$ whenever $m(i) = 0$ and for all $n \in {}^*\mathbb{N}, f, h \in H$,*

$$\begin{aligned} &\int_{S_0} h(i)E_i \sum_{k=0}^{n} f(X^{(\delta)}(k\delta))\Big(A(\omega,(k+1)\delta) - A(\omega,k\delta)\Big)\, dm(i) \\ &= \sum_{k=0}^{n} \int_{S_0} f(i)\hat{E}_i h(X^{(\delta)}(k\delta))\, d\mu(i)\delta. \end{aligned} \tag{2.6.1}$$

Proof. Define

$$\begin{aligned} \mu_{<A>}(0) &= 0, \\ \mu_{<A>}(i) &= \frac{1}{\delta} E_i A(\omega,\delta) m(i), 1 \le i \le N. \end{aligned} \tag{2.6.2}$$

Then, we have

$$\begin{aligned} &\int_{S_0} h(i)E_i \sum_{k=0}^{n} f(X^{(\delta)}(k\delta))\Big(A(\omega,(k+1)\delta) - A(\omega,k\delta)\Big)\, dm(i) \\ &= \sum_{k=0}^{n} \int_{S_0} h(i)E_i f(X^{(\delta)}(k\delta,\omega))A(\theta_{k\delta}^{(\delta)}\omega,\delta)\, dm(i) \\ &= \sum_{k=0}^{n} \int_{S_0} h(i)E_i f(X(k\delta))E_{X(k\delta)}A(\omega,\delta)\, dm(i) \\ &= \sum_{k=0}^{n} \int_{S_0} f(i)E_i A(\omega,\delta)\hat{E}_i h(X(k\delta))\, dm(i) \\ &= \sum_{k=0}^{n} \int_{S_0} f(i)\hat{E}_i h(X(k\delta))\, d\mu(i)\delta. \end{aligned}$$

□

Proposition 2.6.2. *Let μ be a hyperfinite measure on S_0 satisfying $\mu(i) = 0$ whenever $m(i) = 0$. Then for each $\delta \in T$, there exists a δ-IAF $A(\omega, t)$ such that (2.6.1) holds.*

Proof. First of all, let $f(s) = \frac{\mu(s)}{m(s)} 1_{(m(s) \neq 0)}$. For each $u \in H$, we have

$$\begin{aligned} \int_{S_0} u(s)\, d\mu(s) &= \int_{S_0} u(s) f(s)\, dm(s) \\ &= \mathcal{E}_1^{(\delta)}(u, \hat{G}_{-1}^{(\delta)} f). \end{aligned}$$

Define

$$\begin{aligned} A(\omega, 0) &= 0, \\ A(\omega, k\delta) &= \delta \sum_{l=1}^{k} f(X^{(\delta)}(\omega, (l-1)\delta)) \quad \text{for} \quad k \in {}^*\mathbb{N}, k \geq 1. \end{aligned}$$

It is easy to verify that $A(\omega, t)$ is a δ-IAF. Moreover, we have for each $i \in S_0$,

$$\begin{aligned} \frac{1}{\delta} E_i A(\omega, \delta) m(i) &= \frac{1}{\delta} E_i f(X^{(\delta)}(\omega, 0)) \delta m(i) \\ &= f(i) m(i) \\ &= \mu(i). \end{aligned}$$

Therefore, it follows from the proof of Proposition 2.6.1 that (2.6.1) holds. □

For $\delta \in T$, let $A(\omega, t)$ be a δ-IAF. Define

$$\begin{aligned} e(A) &= \frac{1}{2\delta} E(A(\omega, \delta))^2 \\ &= \frac{1}{2\delta} \int_{S_0} E_i A_\delta^2\, dm(i). \end{aligned} \tag{2.6.3}$$

We call $e(A)$ the *energy* of A. Furthermore, we define the *mutual energy* $e(A, B)$ for δ-internal additive functionals A and B by

$$e(A, B) = \frac{1}{2\delta} E\Big(A(\omega, \delta) B(\omega, \delta)\Big).$$

Let $\Delta A(\omega, k\delta)$ be the forward increment of $A(\omega, t)$ at time $k\delta$, i.e.,

$$\Delta A(\omega, k\delta) = A(\omega, (k+1)\delta) - A(\omega, k\delta) \quad \text{for} \quad k \in {}^*\mathbb{N}.$$

We define the quadratic variation $[A] : \Omega \times T_\delta \longrightarrow {}^*\mathbb{R}$ by

$$[A](\omega, 0) = 0,$$

$$[A](\omega, n\delta) = \sum_{k=0}^{n-1} (\Delta A(\omega, k\delta))^2 \quad \text{for} \quad n \in {}^*\mathbb{N}, n > 0.$$

Because

$$\begin{aligned}
[A](\omega, (n+m)\delta) &= \sum_{k=0}^{n-1} (\Delta A(\omega, k\delta))^2 + \sum_{k=n}^{n+m-1} (\Delta A(\omega, k\delta))^2 \\
&= \sum_{k=0}^{n-1} (\Delta A(\omega, k\delta))^2 + \sum_{k=0}^{m-1} (\Delta A(\omega, (k+n)\delta))^2 \\
&= \sum_{k=0}^{n-1} (\Delta A(\omega, k\delta))^2 + \sum_{k=0}^{m-1} (\Delta A(\theta_{n\delta}\omega, k\delta))^2,
\end{aligned}$$

$[A]$ is a positive δ-IAF. By Proposition 2.6.1 and its proof, we know that $\mu_{<[A]>}(i) = \frac{1}{\delta} E_i (A(\omega, \delta))^2 m(i)$ is the hyperfinite positive measure associated with $[A]$ in the sense (2.6.1). We call $\mu_{<[A]>}$ the *energy measure* of A. It is obviously from the relations (2.6.2) and (2.6.3) that

$$e(A) = \frac{1}{2} \mu_{<[A]>}(S_0). \tag{2.6.4}$$

Let $u \in H$. For $\delta \in T$, define a δ-IAF $A^{[u]}(\omega, t)$ by

$$A^{[u]}(\omega, t) = u(X^{(\delta)}(\omega, t)) - u(X^{(\delta)}(\omega, 0)) \quad \text{for} \quad t \in T_\delta.$$

Then, we have

$$\begin{aligned}
e(A^{[u]}) &= \frac{1}{2\delta} E \left(A^{[u]}(\omega, \delta) \right)^2 \\
&= \frac{1}{2\delta} E \left[u(X^{(\delta)}(\omega, \delta)) - u(X^{(\delta)}(\omega, 0)) \right]^2 \\
&= \frac{1}{2\delta} \sum_{i=0}^{N} E_i \left(u(X^{(\delta)}(\omega, \delta)) - u(i) \right)^2 m(i) \\
&= \frac{1}{2\delta} \sum_{i=0}^{N} \sum_{j=0}^{N} (u(j) - u(i))^2 q_{ij}^{(\delta)} m(i) \\
&= \frac{1}{2\delta} \sum_{i=1}^{N} \sum_{j=1}^{N} (u(j) - u(i))^2 q_{ij}^{(\delta)} m(i) + \frac{1}{2\delta} \sum_{i=1}^{N} (u(i))^2 q_{i0}^{(\delta)} m(i)
\end{aligned}$$

$$= \frac{1}{2\delta}\sum_{i=1}^{N}\sum_{j=1}^{N}(u(i)-u(j))\,u(i)q_{ij}^{(\delta)}m(i) + \frac{1}{2\delta}\sum_{i=1}^{N}(u(i))^2 q_{i0}^{(\delta)}m(i)$$
$$+\frac{1}{2\delta}\sum_{i=1}^{N}\sum_{j=1}^{N}(u(j)-u(i))\,u(j)\hat{q}_{ji}^{(\delta)}m(j) + \frac{1}{2\delta}\sum_{i=1}^{N}(u(i))^2 \hat{q}_{i0}^{(\delta)}m(i)$$
$$-\frac{1}{2\delta}\sum_{i=1}^{N}(u(i))^2 \hat{q}_{i0}^{(\delta)}m(i)$$
$$= \mathcal{E}^{(\delta)}(u,u) - \frac{1}{2\delta}\sum_{i=1}^{N}(u(i))^2 \hat{q}_{i0}^{(\delta)}m(i), \tag{2.6.5}$$

where we have used Lemma 1.5.1 (i) and (ii) in the last equation of (2.6.5).

Theorem 2.6.1. *For $u \in H, f \in H$, let $\mu_{<u>}(\cdot)$ be the energy measure of $A^{[u]}$. We have*

$$(1)\quad \mu_{<u>}(i) = \frac{1}{\delta}E_i\left(A^{[u]}(\omega,\delta)\right)^2 m(i)$$
$$= \frac{1}{\delta}\sum_{j=0}^{N}(u(j)-u(i))^2\, q_{ij}^{(\delta)}m(i).$$
$$(2)\quad \int_{S_0} f(s)\,\mu_{<u>}(ds) = 2\mathcal{E}^{(\delta)}(u,uf) - \mathcal{E}^{(\delta)}(u^2,f). \tag{2.6.6}$$

Proof. (1) The proof is quite immediate, and hence we omit it!

(2) On the one hand, we have

$$\int_{S_0} f(s)\,\mu_{<u>}(ds) = \frac{1}{\delta}\sum_{i=1}^{N}\sum_{j=0}^{N} f(i)\Big(u(j)-u(i)\Big)^2 q_{ij}^{(\delta)}m(i). \tag{2.6.7}$$

On the other hand, we have

$$2\mathcal{E}^{(\delta)}(u,fu) - \mathcal{E}^{(\delta)}(u^2,f)$$
$$= \frac{2}{\delta}\sum_{i=1}^{N}\left[(u(i))^2 f(i)m(i) - \sum_{j=1}^{N}(uf)(i)u(j)q_{ij}^{(\delta)}m(i)\right]$$
$$-\frac{1}{\delta}\sum_{i=1}^{N}\left[(u(i))^2 f(i)m(i) - \sum_{j=1}^{N}(u(j))^2 f(i)q_{ij}^{(\delta)}m(i)\right]$$
$$= \frac{1}{\delta}\sum_{i=1}^{N}(u(i))^2 f(i)m(i) - \frac{1}{\delta}\sum_{i=1}^{N}\sum_{j=1}^{N}(2u(i)-u(j))u(j)f(i)q_{ij}^{(\delta)}m(i)$$

$$= \frac{1}{\delta} \sum_{i=1}^{N} f(i)m(i) \left((u(i))^2 - \sum_{j=0}^{N} (2u(i) - u(j))u(j)q_{ij}^{(\delta)} \right)$$

$$= \frac{1}{\delta} \sum_{i=1}^{N} \sum_{j=0}^{N} f(i)(u(j) - u(i))^2 q_{ij}^{(\delta)} m(i). \qquad (2.6.8)$$

By (2.6.7) and (2.6.8), we get (2.6.6). □

2.7 Fukushima's Decomposition Theorem

As before, we shall work under the conditions (1.5.1), (1.5.2), (1.5.4), (1.5.5), (1.5.8), (1.5.9), (1.5.10), (1.5.11), (1.5.13), and (1.5.14) of Sect. 1.5, Chap. 1, in this section.

2.7.1 Decomposition Under the Individual Probability Measures P_i

Lemma 2.7.1. *For $\delta \in T$, let $\mathcal{E}^{(\delta)}(\cdot,\cdot)$ be a hyperfinite weak coercive quadratic form with continuity constant C. Let ν be a positive measure on S_0 of δ-hyperfinite energy integral. For any $u \in H, t \in T_\delta$, and $\varepsilon > 0$, we have*

$$P_\nu \left(\omega \middle| \exists s \in T_\delta^t (|u(X^{(\delta)}(\omega, s))| \geq \varepsilon) \right)$$
$$\leq \frac{2C^2(1+\delta)^{t/\delta}}{\varepsilon} \left[\mathcal{E}_1^{(\delta)}(U_1^{(\delta)}\nu, U_1^{(\delta)}\nu) \mathcal{E}_1^{(\delta)}(u,u) \right]^{\frac{1}{2}},$$

where $P_\nu(\cdot) = \int_{S_0} P_i(\cdot) d\nu(i)$.

Proof. Let $A = \{i \in S_0 \mid u(i) \geq \varepsilon\}$. Define

$$\sigma_A^{(\delta)}(\omega) = \min\{t \in T_\delta \mid X^{(\delta)}(\omega, t) \in A\}.$$

Then, we have

$$P_\nu \left(\omega \middle| \exists s \in T_\delta, s \leq t \left(u(X^{(\delta)}(\omega, s)) \geq \varepsilon \right) \right)$$
$$= \int_{S_0} P_i \left\{ \omega \middle| (1+\delta)^{-\sigma_A^{(\delta)}/\delta} \geq (1+\delta)^{-t/\delta} \right\} d\nu(i)$$
$$\leq \int_{S_0} E_i \left[(1+\delta)^{-\sigma_A^{(\delta)}/\delta} (1+\delta)^{t/\delta} \right] d\nu(i)$$

$$
\begin{aligned}
&= (1+\delta)^{t/\delta} \int_{S_0} e_1^{(\delta)}(A)(i)\, d\nu(i) \\
&= (1+\delta)^{t/\delta}\, \mathcal{E}_1^{(\delta)} \left(U_1^{(\delta)} \nu, e_1^{(\delta)}(A) \right) \\
&\leq C\,(1+\delta)^{t/\delta} \left[\mathcal{E}_1^{(\delta)} \left(U_1^{(\delta)} \nu, U_1^{(\delta)} \nu \right) \mathcal{E}_1^{(\delta)} \left(e_1^{(\delta)}(A), e_1^{(\delta)}(A) \right) \right]^{\frac{1}{2}} \\
&\leq \frac{C^2\,(1+\delta)^{t/\delta}}{\varepsilon} \left[\mathcal{E}_1^{(\delta)} \left(U_1^{(\delta)} \nu, U_1^{(\delta)} \nu \right) \mathcal{E}_1^{(\delta)}(u,u) \right]^{\frac{1}{2}},
\end{aligned}
$$

where the reason for the last step holding is

$$
\begin{aligned}
\mathcal{E}_1^{(\delta)} \left(e_1^{(\delta)}(A), e_1^{(\delta)}(A) \right) &\leq C^2 \mathcal{E}_1^{(\delta)} \left(\frac{|\, u \,|}{\varepsilon}, \frac{|\, u \,|}{\varepsilon} \right) \\
&\leq \frac{C^2}{\varepsilon^2} \mathcal{E}_1^{(\delta)}(u,u),
\end{aligned}
$$

by Corollary 2.2.3. Hence, we can prove Lemma 2.7.1 by applying the same argument to $-u$. □

Proposition 2.7.1. *For $\delta \in T$, let $\mathcal{E}^{(\delta)}(\cdot,\cdot)$ be a hyperfinite weak coercive quadratic form with continuity constant C. Assume that $(\mathcal{E}^{(\delta)}(\cdot,\cdot), \mathcal{D}(\mathcal{E}^{(\delta)}))$ satisfies condition (2.5.14) of Theorem 2.5.6. Let u, $\{u_n \mid n \in \mathbb{N}\}$ be the elements in H and let $\delta \in T$. Suppose that*

$$
{}^\circ\mathcal{E}_1^{(\delta)}(u_n - u, u_n - u) \longrightarrow 0 \quad \text{as} \quad n \to \infty.
$$

Then there exist a subsequence $\{u_{n_k} \mid k \in \mathbb{N}\}$ and a δ-exceptional set B such that for all $i \in S_0 - B, t \in T_\delta^{\text{fin}}$,

$$
\begin{aligned}
L(P_i)\Big({}^\circ u_{n_k}(X^{(\delta)}(\omega,s)) \ &\textit{converges uniformly to} \quad {}^\circ u(X^{(\delta)}(\omega,s)) \\
&\textit{in } s \textit{ on} \quad T_\delta^t \quad \textit{as} \quad n \longrightarrow \infty \Big) = 1.
\end{aligned}
$$

Proof. Let $\{n_k \mid k \in \mathbb{N}\}$ be a subsequence satisfying

$$
{}^\circ\mathcal{E}_1^{(\delta)}(u_{n_k} - u, u_{n_k} - u) \leq 2^{-4k}.
$$

Set

$$
\Lambda_k(t) = \left\{ \omega \,\middle|\, \exists s \in T_\delta^t \left(|u_{n_k}(X^{(\delta)}(\omega,s)) - u(X^{(\delta)}(\omega,s))| \geq 2^{-k} \right) \right\}.
$$

For $\nu \in \tau_{00f}(\delta)$, we have from Lemma 2.7.1 that

$$^\circ P_\nu\left(\Lambda_k(t)\right) \leq {}^\circ\!\left(C^2(1+\delta)^{\frac{t}{\delta}} 2^{-k+1}\left[\mathcal{E}_1^{(\delta)}(U_1^{(\delta)}\nu, U_1^{(\delta)}\nu)\right]^{\frac{1}{2}}\right).$$

Hence, we have

$$\sum_{k=1}^{\infty} L(P_\nu)(\Lambda_k(t)) < \infty.$$

By the Borel-Cantelli lemma, we get

$$L(P_\nu)\left(\bigcap_{k=1}^{\infty}\bigcup_{l=k}^{\infty}\Lambda_l(t)\right) = 0. \tag{2.7.1}$$

Set $\Lambda(t) = \bigcap_{k=1}^{\infty}\bigcup_{l=k}^{\infty}\Lambda_l(t)$. From Theorem 2.5.6, Theorem 2.4.1, and (2.7.1), there exists a δ-exceptional set $B(t)$ such that

$$L(P_i)(\Lambda(t)) = 0 \quad \text{for any} \quad i \in S_0 - B(t).$$

Now let us select a countable subset $\{t_n \mid n \in \mathbb{N}\} \subset T_\delta$ such that $t_n \approx n$. We define $\Lambda = \bigcup_{n=1}^{\infty}\Lambda(t_n), B = \bigcup_{n=1}^{\infty} B(t_n)$. It is easy to see that Proposition 2.7.1 holds. □

Lemma 2.7.2. *For $\delta \in T$, let A be a δ-IAF and let $\mu_{<A>}(i)$ be the hyperfinite measure defined by (2.6.2) in Sect. 2.6. Then for any $v \in H$, we have*

$$\mathcal{E}^{(\delta)}(f_t, v) = \int_{S_0}\left[v(i) - \hat{E}_i v(X^{(\delta)}(t))\right] d\mu_{<A>}(i) \quad \textit{for all} \quad t \in T_\delta, \tag{2.7.2}$$

where $f_t(i) = E_i A(\omega, t)$.

Proof. If $k = 1$, then we have

$$\begin{aligned}\mathcal{E}^{(\delta)}(f_\delta, v) &= \frac{1}{\delta}\int_{S_0}\left(v(i) - \hat{Q}^\delta v(i)\right) f_\delta(i)\, dm(i) \\ &= \int_{S_0}\left(v(i) - \hat{E}_i v(X^{(\delta)}(\delta))\right) d\mu_{<A>}(i).\end{aligned}$$

Assume that (2.7.2) holds whenever $k \leq n$. Then, we have

$$\begin{aligned}\mathcal{E}^{(\delta)}(f_{(n+1)\delta}, v) &= \frac{1}{\delta}\int_{S_0}\left(v(i) - \hat{Q}^\delta v(i)\right) E_i A(\omega, (n+1)\delta)\, dm(i) \\ &= \frac{1}{\delta}\int_{S_0}\left(v(i) - \hat{Q}^\delta v(i)\right)\left(E_i A(\omega, n\delta) + E_i A(\theta_{n\delta}^{(\delta)}\omega, \delta)\right) dm(i)\end{aligned}$$

$$
\begin{aligned}
&= \int_{S_0} \left(v(i) - \hat{E}_i v(X^{(\delta)}(n\delta))\right) \, d\mu_{<A>}(i) \\
&\quad + \frac{1}{\delta} \int_{S_0} \left(\hat{E}_i v(X^{(\delta)}(n\delta)) - \hat{E}_i v(X^{(\delta)}((n+1)\delta))\right) \\
&\quad \times E_i A(\omega, \delta) \, dm(i) \\
&= \int_{S_0} \left[v(i) - \hat{E}_i v(X^{(\delta)}((n+1)\delta))\right] \, d\mu_{<A>}(i).
\end{aligned}
$$

□

Lemma 2.7.3. *For $\delta \in T$, let $\mathcal{E}^{(\delta)}(\cdot,\cdot)$ be a hyperfinite weak coercive quadratic form with continuity constant C. Let $A(\omega, t)$ be a positive δ-IAF. For all positive hyperfinite measures ν on S_0 of δ-hyperfinite energy integral, we have*

$$
E_\nu(A_t) \leq (1+t)||\hat{U}_1^{(\delta)}\nu||_\infty \mu_{<A>}(S_0) \quad \textit{for all} \quad t \in T_\delta. \tag{2.7.3}
$$

Proof. It follows from Theorem 2.5.1 and Lemma 2.7.2 that

$$
\begin{aligned}
E_\nu A(\omega, t) &= \int_{S_0} E_i A(\omega, t) \, d\nu(i) = \mathcal{E}_1^{(\delta)}(f_t, \hat{U}_1^{(\delta)}\nu) \\
&= \int_{S_0} \left[\hat{U}_1^{(\delta)}\nu(i) - \hat{E}_i(\hat{U}_1^{(\delta)}\nu(X^{(\delta)}(t)))\right] \, d\mu_{<A>}(i) \\
&\quad + \int_{S_0} f_t(i)\hat{U}_1^{(\delta)}\nu(i) \, dm(i) \\
&\leq ||\hat{U}_1^{(\delta)}\nu||_\infty \left[\mu_{<A>}(S_0) + \int_{S_0} f_t(i) \, dm(i)\right], \qquad (2.7.4)
\end{aligned}
$$

where we have used Theorem 2.5.3 in the latter inequality above. Notice that

$$
\begin{aligned}
\int_{S_0} f_{(k+1)\delta}(i) \, dm(i) &= \int_{S_0} E_i A(\omega, (k+1)\delta) \, dm(i) \\
&= \int_{S_0} E_i A(\omega, k\delta) \, dm(i) + \int_{S_0} E_i A(\theta_{k\delta}\omega, \delta) \, dm(i) \\
&= \int_{S_0} E_i A(\omega, k\delta) \, dm(i) + \int_{S_0} \hat{E}_i 1_{(X(k\delta) \in S_0)} \\
&\quad \times E_i A(\delta) \, dm(i) \\
&\leq \int_{S_0} E_i A(\omega, k\delta) \, dm(i) + \int_{S_0} E_i A(\omega, \delta) \, dm(i).
\end{aligned}
$$

We can show

$$\int_{S_0} f_t(i)\, dm(i) \leq t\mu_{<A>}(S_0). \tag{2.7.5}$$

From the relations (2.7.4) and (2.7.5), we obtain the inequality (2.7.3). □

Definition 2.7.1. We call an internal process $A : \Omega \times T_\delta \to {}^*\mathbb{R}$ a *martingale* with respect to $(\Omega, \mathcal{F}_t^{(\delta)}, P_i, i \in S_0)$ if $\omega \to A(\omega, t)$ is $\mathcal{F}_t^{(\delta)}$ measurable for all $t \in T_\delta$, and for all $s, t \in T_\delta, s < t$, and all $B \in \mathcal{F}_s^{(\delta)}$,

$$E_i \left(1_B(A_t - A_s)\right) = 0.$$

It is easy to see that if $[\omega]_t^{(\delta)}$ is the equivalence class of ω defined by (1.5.35) in Sect. 1.5 of Chap. 1, then a non-anticipating process $A(\omega, t)$ is a martingale if and only if

$$\sum_{\tilde{\omega} \in [\omega]_t^{(\delta)}} \Delta A(\tilde{\omega}, t) P_i \{\tilde{\omega}\} = 0. \tag{2.7.6}$$

In the following, we shall use "*δ-quasi-everywhere*" or "*δ-q.e.*" to mean "except for a δ-exceptional set". A statement depending on $i \in S_0$ is said to be "*δ-quasi-everywhere*" or "*δ-q.e.*" if there exists a set $B \subset S_0$ of δ-exceptional such that the statement is true for every $i \in S_0 - B$.

Let us introduce the following set of *internal martingale additive functionals* (abbreviated by δ-IMAF) by

$$\begin{aligned}\mathcal{M}^{(\delta)} = \{M \mid M \text{ is a } \delta\text{-IAF}, {}^\circ(E_i M^2(t)) < \infty, \delta\text{-q.e.} i \in S_0 \text{ for all} \\ \text{finite } t \in T_\delta \text{ and } E_i M(t) = 0, \forall t \in T_\delta, \forall i \in S_0\}.\end{aligned}$$

For $M \in \mathcal{M}^{(\delta)}$, we have for all $s, t \in T_\delta$

$$\begin{aligned}E_i(M_{s+t} \mid \mathcal{F}_t^{(\delta)}) &= E_i(M_t + M_s(\theta_t^{(\delta)}) \mid \mathcal{F}_t^{(\delta)}) \\ &= M_t + E_{X(t)} M_s \\ &= M_t,\end{aligned}$$

for all $i \in S_0$. Hence, $M \in \mathcal{M}^{(\delta)}$ is a martingale with respect to P_i for all $i \in S_0$. Meantime, $M \in \mathcal{M}^{(\delta)}$ is also square integrable with respect to P_i for δ-q.e. $i \in S_0$. In the following, let

$$\mathring{\mathcal{M}}^{(\delta)} = \{M \in \mathcal{M}^{(\delta)} \mid {}^\circ e(M) < \infty\}.$$

Any element in $\mathring{\mathcal{M}}^{(\delta)}$ is called a *δ-internal martingale additive functional of finite energy.*

Proposition 2.7.2. *For $\delta \in T$, let $\mathcal{E}^{(\delta)}(\cdot,\cdot)$ be a hyperfinite weak coercive quadratic form with continuity constant C. Let $\{A_n \mid n \in \mathbb{N}\}$ be δ-internal additive functionals. Assume that ${}^\circ e(A_n - A_m) \to 0$ as $n, m \to \infty$, and for each $i \in S_0$, $\{\Omega, \mathcal{F}_t^{(\delta)}, A_n(t), P_i\}$ is a martingale for all $n \geq 1$. Then there exists a unique $A \in \mathring{\mathcal{M}}^{(\delta)}$ with inner product ${}^\circ e(\cdot,\cdot)$ such that ${}^\circ e(A_n - A) \to 0$ as $n \to \infty$. Moreover, there exist a subsequence $\{A_{n_k}(\omega, t) \mid k \in \mathbb{N}\}$ and a δ-exceptional set B such that for all $i \in S_0 - B, t \in T_\delta^{\mathrm{fin}}$,*

$$L(P_i)\left(\omega \,\middle|\, {}^\circ A_{n_k}(\omega, s) \longrightarrow {}^\circ A(\omega, s) \quad \textit{uniformly on} \quad T_\delta^t\right) = 1.$$

Proof. By the equality (2.6.4) in Sect. 2.6, we know that

$$\begin{aligned} {}^\circ\mu_{<[A_n - A_m]>}(S_0) &= {}^\circ\Big(2e(A_n - A_m)\Big) \\ &\longrightarrow 0 \quad \text{as} \quad n, m \to \infty. \end{aligned}$$

Hence, there exists a subsequence $\{n_k \mid k \in \mathbb{N}\}$ such that ${}^\circ\mu_{<[A_{n_{k+1}} - A_{n_k}]>}(S_0) < 2^{-3k}$. Since $A_n, n \geq 1$ are martingales with respect to $P_i, i \in S_0$, we get from Lemma 2.7.3 that for all $\nu \in \hat{\tau}_{00}(\delta), t \in T_\delta^{\mathrm{fin}}$:

$$\begin{aligned} {}^\circ\Big[E_\nu\left(A_{n_{k+1}}(t) - A_{n_k}(t)\right)^2\Big] &= {}^\circ\Big[E_\nu([A_{n_{k+1}} - A_{n_k}](t))\Big] \\ &\leq {}^\circ\Big[(1+t)\|\hat{U}_1^{(\delta)}\nu\|_\infty 2^{-3k}\Big]. \end{aligned} \tag{2.7.7}$$

From the relation (2.7.7) and Doob's inequality (for which we refer to [25] 4.2.8), we get

$$\begin{aligned} &{}^\circ\left[P_\nu\left(\max_{s \leq t}|A_{n_{k+1}}(s) - A_{n_k}(s)| \geq 2^{-k}\right)\right] \\ &\leq {}^\circ\left[2^{2k}E_\nu\left(\max_{s \leq t}|A_{n_{k+1}}(s) - A_{n_k}(s)|\right)^2\right] \\ &\leq {}^\circ\Big[2^{2k+2}E_\nu\left(A_{n_{k+1}}(t) - A_{n_k}(t)\right)^2\Big] \\ &\leq {}^\circ\Big[4(1+t)\|\hat{U}_1^{(\delta)}\nu\|_\infty 2^{-k}\Big]. \end{aligned} \tag{2.7.8}$$

It is easy to see from the inequality (2.7.8), by using Borel-Cantelli lemma, that

$$L(P_\nu)(\Lambda) = 1, \ \text{ for all } \nu \in \hat{\tau}_{00},$$

where $\Lambda = \{\omega \mid {}^\circ A_{n_{k+1}}(\omega,\cdot)$ converges uniformly on each finite interval $T_\delta^t\}$. By Theorem 2.5.5 and Theorem 2.4.1, we have $L(P_i)(\Lambda) = 1$, δ-q.e.$i \in S_0$. Let $\{A_{n_k} \mid k \in {}^*\mathbb{N}\}$ be an internal extension of $\{A_{n_k} \mid k \in \mathbb{N}\}$. Define $A(\omega,t) = A_{n_K}(\omega,t)$ for some $K \in {}^*\mathbb{N} - \mathbb{N}$. Then, $A(\omega,t)$ is a P_i-martingale δ-IAF for all $i \in S_0$. Since $A_n(t)$ converges in $L^2(P_\nu)$, we have $({}^\circ E_\nu A^2(t)) < \infty, \forall t \in T_\delta^{\mathrm{fin}}$. Therefore, we have shown $A \in \mathcal{M}^{(\delta)}$. For $\varepsilon > 0$, choose N such that $e(A_n - A_m) < \varepsilon, n, m > N$. By Fatou's lemma, we have $e(A_n - A) \leq \varepsilon, n > N$. Thus, we have $A \in \mathring{\mathcal{M}}^{(\delta)}$ and ${}^\circ e(A_n - A) \longrightarrow 0$ as $n \longrightarrow \infty$. □

Definition 2.7.2. (1) We call an internal function $u \in H$ *S-bounded* if there is a positive constant $C \in [0,\infty)$ such that $|u(i)| \leq C$ for all $i \in S_0$.
(2) An internal function $f : T_\delta \longrightarrow {}^*\mathbb{R}$ is called *S-continuous* if $f(s) \approx f(t)$ whenever $s \approx t$ and s and t are nearstandard.

Lemma 2.7.4. *For $\delta \in T, u \in Fin(H)$, define a δ-IAF A^δ by*

$$A^\delta(\omega,0) = 0,$$
$$A^\delta(\omega,n\delta) = \delta \sum_{k=1}^{n} u(X(\omega,(k-1)\delta)), n \in \mathbb{N}, n \geq 1.$$

Then, there exists a properly δ-exceptional set $B \subset S_0$ such that for all $i \in S_0 - B$,

$$L(P_i)\left\{\omega \middle| A^\delta(\omega,\cdot) \text{ is } S\text{-continuous}\right\} = 1.$$

Proof. First of all, we assume that u is S-bounded. We have for any $\omega \in \Omega$,

$$|A^\delta(\omega,t) - A^\delta(\omega,s)| \leq |t-s| \left(\max_{i \in S_0} |u(i)| + 1\right).$$

This implies that $A^\delta(\omega,t)$ is S-continuous.

Next, we suppose that $u \in \mathrm{Fin}(H)$. For each n, set

$$B_n = \{s \in S_0 \mid |u(s)| \geq n\}.$$

In addition, we define

$$\sigma_{B_n}(\omega) = \min\{t \in T_\delta \mid X(t) \in B_n\}.$$

Then, we have

$$P\left(\omega \middle| \exists t \in T_\delta^1(X(\omega,t) \in B_n)\right)$$
$$= P\left(\omega \middle| \sigma_{B_n}(\omega) \leq 1\right)$$

$$\leq n^{-2} \int_{S_0} E_i[u(X(\sigma_{B_n}))]^2 dm(i)$$
$$\leq n^{-2} \int_{S_0} [u(i)]^2 \; dm(i).$$

Thus, we get

$$L(P) \left(\omega \middle| \exists t \in T_\delta^1 (X(\omega, t) \in \bigcap_{n=1}^{\infty} B_n) \right) = 0.$$

This implies that $\{s \in S_0 \mid {}^\circ|u(s)| = \infty\}$ is a δ-exceptional set. Let B be a properly δ-exceptional set containing $\{s \in S_0 \mid {}^\circ|u(s)| = \infty\}$ (Proposition 2.1.1). For each $n \in \mathbb{N}$, define

$$\sigma_n(\omega) = \min\left\{t \in T_\delta \mid |u(X(\omega, t))| \geq n\right\}.$$

Then for each $\omega \in \Omega, A^\delta(\omega, \cdot)$ is continuous in $[0, \sigma_n(\omega))$ for every n. Moreover, for each $i \in S_0 - B$, we have

$$L(P_i)\left\{\omega \mid {}^\circ\sigma_n(\omega) \uparrow \infty \quad \text{as} \quad n \to \infty\right\} = 1,$$

which proves Lemma 2.7.4. □

Set

$$\mathcal{N}_C^{(\delta)} = \{N \mid L(P_i)\{\omega \mid N(\omega, \cdot) \text{ is an } S\text{-continuous } \delta\text{-IAF}\} = 1, \\ \text{q.e.} i \in S_0 \text{ and } e(N) \approx 0\}.$$

For $N \in \mathcal{N}_C^{(\delta)}$, the variation vanishes in the following sense

$$E([N](t)) \approx 0 \text{ for all } t \in T_\delta^{\text{fin}}.$$

In fact, we have for all finite $t \in T_\delta$,

$$\begin{aligned} E([N](t)) &= \sum_{0<s<t} E(\Delta N(\omega, s))^2 \\ &= \sum_{0<s<t} E\Big(N(\omega, s+\delta) - N(\omega, s)\Big)^2 \\ &= \sum_{0<s<t} E\Big(N(\theta_s^{(\delta)}\omega, \delta)\Big)^2 \\ &\leq \sum_{0<s<t} E(\Delta N(\omega, \delta))^2 \\ &= te(N) \\ &\approx 0. \end{aligned}$$

Theorem 2.7.1. *For infinitesimal $\delta \in T$, let $\mathcal{E}^{(\delta)}(\cdot,\cdot)$ be a hyperfinite weak coercive quadratic form with continuity constant C. Assume that $(\mathcal{E}^{(\delta)}(\cdot,\cdot), \mathcal{D}(\mathcal{E}^{(\delta)}))$ satisfies the property (2.5.14) in Sect. 2.5. For any $u \in \mathcal{D}(\mathcal{E}^{(\delta)})$, there are two δ-internal additive functionals $M^{[u]}(\omega,t) \in \mathring{\mathcal{M}}^{(\delta)}$ and $N^{[u]}(\omega,t) \in \mathcal{N}_C^{(\delta)}$ such that*

$$A^{[u]}(\omega,t) = M^{[u]}(\omega,t) + N^{[u]}(\omega,t), \forall \omega \in \Omega, \forall t \in T_\delta.$$

Proof. Step 1. Define

$$\begin{aligned}
N^{[u]}(\omega,0) &= 0,\\
\Delta N^{[u]}(\omega,t) &= Q^\delta u(X^{(\delta)}(\omega,t)) - u(X^{(\delta)}(\omega,t))\\
&= -\delta A^{(\delta)} u(X^{(\delta)}(\omega,t)), t \in T_\delta,
\end{aligned}$$

and

$$M^{[u]}(\omega,t) = u(X^{(\delta)}(\omega,t)) - u(X^{(\delta)}(\omega,0)) - N^{[u]}(\omega,t).$$

For each $t \in T_\delta$, we have

$$M^{[u]}(\omega,t+\delta) - M^{[u]}(\omega,t) = u(X^{(\delta)}(\omega,t+\delta)) - Q^\delta u(X^{(\delta)}(\omega,t)). \tag{2.7.9}$$

It is easy to see from (2.7.6) and (2.7.9) that $(\Omega, \mathcal{F}_t^{(\delta)}, M^{[u]}(\omega,t), P_i)$ is a martingale for each $i \in S_0$. Furthermore, it is easy to see that

$$A^{[u]}(\omega,t) = M^{[u]}(\omega,t) + N^{[u]}(\omega,t).$$

Step 2. Assume that ${}^\circ\langle A^{(\delta)}u, A^{(\delta)}u\rangle < \infty$. We have from Lemma 2.7.4 that there exists a properly δ-exceptional set $B \subset S_0$ such that for all $i \in S_0 - B$,

$$L(P_i)\left\{\omega \middle| N^{[u]}(\omega,\cdot) \quad \text{is } S\text{-continuous}\right\} = 1.$$

Moreover, we have

$$\begin{aligned}
E\left(\Delta N^{[u]}(t)\right)^2 &= \delta^2 E\left(A^{(\delta)} u(X^{(\delta)}(\omega,t)\right)^2\\
&\leq \delta^2 \langle Au, Au\rangle.
\end{aligned} \tag{2.7.10}$$

From the relation (2.7.10), we obtain

$$E[N^{[u]}](t) \leq t\delta\langle Au, Au\rangle \approx 0 \quad \text{for each} \quad t \in T_\delta^{\text{fin}},$$

and

$$
\begin{aligned}
e(N^{[u]}) &= \frac{1}{2\delta} E\left(N^{[u]}(\omega,\delta)\right)^2 \\
&\le \frac{1}{2}\delta\langle Au, Au\rangle \\
&\approx 0.
\end{aligned}
\tag{2.7.11}
$$

Therefore, we deduce from (2.6.5) in Sect. 2.6 and the relation (2.7.11) that

$$
\begin{aligned}
e(M^{[u]}) &= e(A^{[u]} - N^{[u]}) \\
&\approx e(A^{[u]}) \\
&\le \mathcal{E}^{(\delta)}(u,u).
\end{aligned}
\tag{2.7.12}
$$

By using Lemma 2.7.3, (2.6.4) in Sect. 2.6, and the relation (2.7.12), we get for any $\nu \in \hat{\tau}_{00}(\delta), t \in T_\delta^{\text{fin}}$

$$
\begin{aligned}
{}^\circ\left(E_\nu(M^{[u]}(\omega,t))^2\right) &= {}^\circ\left(E_\nu[M^{[u]}](t)\right) \\
&\le {}^\circ\left[(1+t)\|\hat{U}_1^{(\delta)}\nu\|_\infty \mu_{<[M^{[u]}]>}(S_0)\right] \\
&= {}^\circ\left[(1+t)\|\hat{U}_1^{(\delta)}\nu\|_\infty 2e(M^{[u]})\right] \\
&\le {}^\circ\left[(1+t)\|\hat{U}_1^{(\delta)}\nu\|_\infty 2\mathcal{E}^{(\delta)}(u,u)\right] \\
&< \infty.
\end{aligned}
\tag{2.7.13}
$$

Therefore, it follows from Theorem 2.5.5 and the relation (2.7.13) that there exists a δ-zero capacity set $A_1 \subset S_0$ such that

$$
{}^\circ\left[E_i\left(M^{[u]}(\omega,t)\right)^2\right] < \infty \quad \text{for all} \quad t \in T_\delta^{\text{fin}}, i \in S_0 - A_1.
$$

Now it is easy to see that all results of Theorem 2.7.1 hold whenever

$$
{}^\circ\langle A^{(\delta)}u, A^{(\delta)}u\rangle < \infty.
$$

Step 3. For general $u \in \mathcal{D}(\mathcal{E}^{(\delta)})$, put $u_n = nG_{-n}u, n \in \mathbb{N}$. Then

$$
\begin{aligned}
\Delta N^{[u_n]}(\omega,t) &= Q^\delta u_n(X(\omega,t)) - u_n(X(\omega,t)) \\
&= -\delta A^{(\delta)} u_n(X(\omega,t)).
\end{aligned}
$$

First of all, we have

$$\begin{aligned}
{}^{\circ}\int_{S_0} \left(A^{(\delta)}u_n(i)\right)^2 dm(i) &= {}^{\circ}\int_{S_0} \left[(A^{(\delta)}+n)u_n(i) - nu_n(i)\right]^2 dm(i) \\
&= {}^{\circ}\int_{S_0} [nu(i) - nu_n(i)]^2 \, dm(i) \\
&\leq {}^{\circ}\left[4n^2 \int_{S_0} (u(i))^2 \, dm(i)\right] \\
&< \infty. \qquad (2.7.14)
\end{aligned}$$

Therefore, we know from the relation (2.7.14) and Lemma 2.7.4 that, for each $n \in \mathbb{N}$, there exists a properly δ-exceptional set $B_n \subset S_0$ such that for all $i \in S_0 - B_n$,

$$L(P_i)\left\{\omega \,\middle|\, N^{[u_n]}(\omega,\cdot) \text{ is S-continuous}\right\} = 1. \qquad (2.7.15)$$

It follows from the relation (2.7.12) that

$$\begin{aligned}
{}^{\circ}e\left(M^{[u_m]} - M^{[u_n]}\right) &\leq {}^{\circ}\mathcal{E}^{(\delta)}(u_m - u_n, u_m - u_n) \\
&\longrightarrow 0 \text{ as } n, m \to \infty. \qquad (2.7.16)
\end{aligned}$$

The relation (2.7.16), Propositions 2.7.1 and 2.7.2 imply that there exist a subsequence $\{n_k \mid k \in \mathbb{N}\}$ and a δ-exceptional set B and an element $M \in \mathring{\mathcal{M}}$ such that

$$L(P_i)(\Omega_0) = 1 \quad \text{for all} \quad i \in S_0 - B, \qquad (2.7.17)$$

where $\Omega_0 = \{\omega \in \Omega \mid {}^{\circ}u_{n_k}(X^{(\delta)}(\omega,s))$ and ${}^{\circ}M^{[u_{n_k}]}(\omega,s)$ converge uniformly to ${}^{\circ}u(X^{(\delta)}(\omega,s))$ and ${}^{\circ}M(\omega,s)$ on each S-bounded subset of T_δ, respectively$\}$. Moreover, we have ${}^{\circ}e(M^{[u_n]} - M) \longrightarrow 0, n \longrightarrow \infty$. Actually, we may take some $K \in {}^*\mathbb{N} - \mathbb{N}$. Then from the proof of Proposition 2.7.2, we may take $M = M^{[u_{n_K}]}$, where $u_{n_K} \in \{u_n \mid n \in {}^*\mathbb{N}\}$ is an internal extension of $\{u_n \mid n \in \mathbb{N}\}$. Since

$$e(M^{[u]} - M) = e(M^{[u]} - M^{[u_{n_K}]}) \approx 0$$

from (2.6.5) in Sect. 2.6, we may replace $M = M^{[u_{n_K}]}$ by $M = M^{[u]}$.

Set $A = B \cup (\cup_{n=1}^{\infty} B_n)$. Since

$$N^{[u_{n_k}]}(\omega,t) = u_{n_k}(X^{(\delta)}(\omega,t)) - u_{n_k}(X^{(\delta)}(\omega,0)) - M^{[u_{n_k}]}(\omega,t),$$

we know from (2.7.15) and (2.7.17) that

$$L(P_i)\left\{\omega\middle|N^{[u]}(\omega,\cdot)\text{ is S-continuous}\right\}=1,\forall i\in S_0-A.$$

On the other hand, we have

$$N^{[u]}(\omega,t)=A^{[u-u_k]}(\omega,t)-\left(M^{[u]}(\omega,t)-M^{[u_n]}(\omega,t)\right)+N^{[u_n]}(\omega,t).$$

Hence, we have

$${}^{\circ}e(N^{[u]})\leq{}^{\circ}\left[3e(A^{[u-u_n]})+3e(M^{[u]}-M^{[u_n]})\right].$$

By letting $n\longrightarrow\infty$, we know from (2.6.5) in Sect. 2.6 that $e(N^{[u]})\approx 0$. □

2.7.2 Decomposition Under the Whole Measure P

In the following, we shall consider a similar decomposition as in Theorem 2.7.1 under the whole measure P. Just as Lemma 2.7.1, we have

Lemma 2.7.5. *For $\delta\in T$, let $\mathcal{E}^{(\delta)}(\cdot,\cdot)$ be a hyperfinite weak coercive quadratic form with continuity constant C. For all $u\in H, t\in T_\delta$ and $\varepsilon>0$, we have*

$$P\left(\omega\middle|\exists s\in T_\delta^t(|u(X^{(\delta)}(\omega,s))|\geq\varepsilon)\right)\leq\frac{2C^2(1+\delta)^{t/\delta}}{\varepsilon^2}\mathcal{E}_1^{(\delta)}(u,u).$$

Proof. Let $A=\{i\in S_0\mid u(i)\geq\varepsilon\}$, and

$$\sigma_A^{(\delta)}(\omega)=\min\{t\in T_\delta\mid X^{(\delta)}(t)\in A\}.$$

Then, we have

$$\begin{aligned}
&P\left(\omega\middle|\exists s\in T_\delta, s\leq t\left(u(X^{(\delta)}(\omega,s))\geq\varepsilon\right)\right)\\
&=\int_{S_0}P_i\left\{\omega\middle|(1+\delta)^{-\sigma_A^{(\delta)}/\delta}\geq(1+\delta)^{-t/\delta}\right\}\,dm(i)\\
&\leq\int_{S_0}E_i\left[(1+\delta)^{-\sigma_A^{(\delta)}/\delta}(1+\delta)^{t/\delta}\right]\,dm(i)\\
&=(1+\delta)^{t/\delta}\int_{S_0}e_1^{(\delta)}(A)(i)\,dm(i)\\
&\leq(1+\delta)^{t/\delta}\,\mathcal{E}_1^{(\delta)}\left(e_1^{(\delta)}(A),e_1^{(\delta)}(A)\right),
\end{aligned}$$

where the reason for the last step holding is (2.3.1) in Sect. 2.3. Since $\mathcal{E}^{(\delta)}(\cdot,\cdot)$ has the Markov property, we get from Corollary 2.2.3 that

$$\mathcal{E}_1^{(\delta)}\left(e_1^{(\delta)}(A), e_1^{(\delta)}(A)\right) \le C^2 \mathcal{E}_1^{(\delta)}\left(\frac{|u|}{\varepsilon}, \frac{|u|}{\varepsilon}\right) \le \frac{C^2}{\varepsilon^2}\mathcal{E}_1^{(\delta)}(u,u).$$

Hence, we have

$$P\left(\omega \middle| \exists s \in T_\delta^t(u(X^{(\delta)}(\omega,s)) \ge \varepsilon)\right) \le \frac{C^2(1+\delta)^{t/\delta}}{\varepsilon^2}\mathcal{E}_1^{(\delta)}(u,u).$$

Applying the same argument to $-u$, the lemma follows. □

Corollary 2.7.1. *For $\delta \in T$, let $\mathcal{E}^{(\delta)}(\cdot,\cdot)$ be a hyperfinite weak coercive quadratic form with continuity constant C. Let $u, u_n, n \in \mathbb{N}$ be elements in H, and assume that*

$${}^\circ\mathcal{E}_1^{(\delta)}(u-u_n, u-u_n) \longrightarrow 0 \text{ as } n \longrightarrow \infty.$$

There is a subsequence $\{u_{n_k}\}$ such that for almost every ω, ${}^\circ u_{n_k}(X^{(\delta)}(\omega,\cdot))$ converges uniformly to ${}^\circ u(X^{(\delta)}(\omega,\cdot))$ on all S-bounded subsets of T_δ.

Proof. The result follows from Lemma 2.7.5 and basic measure theory. □

Definition 2.7.3. A martingale M with respect to $(\Omega, \left\{\mathcal{F}_t^{(\delta)}\right\}, P)$ is called a λ^2*-martingale* if ${}^\circ E(M_t^2) < \infty$ for all $t \in T_\delta^{\text{fin}}$.

Theorem 2.7.2. *For infinitesimal $\delta \in T$, let $\mathcal{E}^{(\delta)}(\cdot,\cdot)$ be a hyperfinite weak coercive quadratic form with continuity constant C. For any $u \in \mathcal{D}(\mathcal{E}^{(\delta)})$, there are two δ-internal additive functionals $M^{[u]}(\omega,t)$ and $N^{[u]}(\omega,t)$ such that*

(i) $A^{[u]}(\omega,t) = M^{[u]}(\omega,t) + N^{[u]}(\omega,t)$.
(ii) $M^{[u]}$ *is a* λ^2*-martingale with respect to* $(\Omega, \left\{\mathcal{F}_t^{(\delta)}\right\}, P)$.
(iii) $N^{[u]}(\omega,\cdot)$ *is S-continuous for almost all paths $\omega \in \Omega$ under $L(P)$, and* $E[N^{[u]}](t) \approx 0$ *for all* $t \in T_\delta^{\text{fin}}$.

Proof. The proof of this result is similar to the one of Theorem 2.7.1. We also refer the reader to the detailed proof of the corresponding result in the symmetric case given in Albeverio et al. [25]. □

2.8 Internal Multiplicative Functionals

2.8.1 Internal multiplicative functionals

Definition 2.8.1. For $\delta \in T$, an internal function $M(\omega, t)$, $t \in T_\delta$, $\omega \in \Omega$, is said to be a *δ-internal multiplicative functional* (abbreviated by δ-IMF) of $X^{(\delta)}(\omega, t)$ if and only if

(i) For each $t \in T_\delta$, $M_t(\cdot)$ is $\mathcal{F}_t^{(\delta)}$-measurable.
(ii) For each $t \in T_\delta$, $\omega \in \Omega$, we have $M(\omega, t) \in {}^*[0, 1]$.
(iii) For each $\omega \in \Omega$, we have

$$M(\omega, 0) = 1,$$
$$M(\omega, t+s) = M(\omega, s)M(\theta_s^{(\delta)}\omega, t), \forall t, s \in T_\delta.$$

Remark 2.8.1. Let $A(\omega, t)$ be a nonnegative δ-internal additive functional. We can define a δ-IMF $M(\omega, t)$ by

$$M(\omega, t) = \exp\left(-A(\omega, t)\right) \quad \text{for all} \quad \omega \in \Omega, t \in T_\delta.$$

If $M(\omega, t)$ is a δ-internal multiplicative functional, let us define a family of operators $\{P^t \mid t \in T_\delta\}$ on H by

$$P^t f(i) = E_i\left\{f(X^{(\delta)}(\omega, t))M(\omega, t)\right\}.$$

Then, we have for all $t, s \in T_\delta, i \in S$,

$$\begin{aligned} P^{t+s} f(i) &= E_i\left\{f(X^{(\delta)}(\omega, t+s))M(\omega, t+s)\right\} \\ &= E_i\left\{f(X^{(\delta)}(\theta_s^{(\delta)}\omega, t))M(\omega, s)M(\theta_s^{(\delta)}\omega, t)\right\} \\ &= E_i\left\{M(\omega, s)E_{X^{(\delta)}(\omega, s)}[f(X^{(\delta)}(t))M_t]\right\} \\ &= P^s P^t f(i). \end{aligned} \tag{2.8.1}$$

Hence, $\{P^t \mid t \in T_\delta\}$ is a semigroup. We call it the *semigroup* generated by $(X^{(\delta)}, M)$. Moreover, we have for all $f \in H$,

$$\begin{aligned} P^0 f(i) &= E_i[f(X^{(\delta)}(\omega, 0))M(\omega, 0)] \\ &= f(i). \end{aligned}$$

In particular, we have

$$P^0 1 = 1. \tag{2.8.2}$$

Since $M(\omega, \delta)$ is $\mathcal{F}_\delta^{(\delta)}$-measurable, we have

$$M(\omega, \delta) = \sum_{i,j=0}^{N} 1_{[\hat{\omega}](ij)}(\omega) M_{ij},$$

where $[\hat{\omega}](ij) = \{\omega \mid \omega(0) = i \text{ and } \omega(\delta) = j\}$ and $\{M_{ij} \mid i, j = 0, 1, 2, \ldots, N\}$ is a family of positive hyperreals. Moreover, $M_{ij} \in {}^*[0,1]$. Therefore, the transition matrix $\{p_{ij}^{(\delta)} \mid i, j = 0, 1, 2, \ldots, N\}$ of $\{P^t \mid t \in T_\delta\}$ is given by

$$p_{ij}^{(\delta)} = q_{ij}^{(\delta)} M_{ij}, \quad M_{ij} \in {}^*[0,1], i, j = 0, 1, 2, \ldots, N. \tag{2.8.3}$$

From the relation (2.8.3), we know that for all nonnegative internal functions $f \in H$,

$$P^t f(i) \leq Q^t f(i), \forall i \in S, t \in T_\delta. \tag{2.8.4}$$

2.8.2 Subordinate Semigroups

Definition 2.8.2. A semigroup $\{P^t \mid t \in T_\delta\}$ of positive linear operators from H to H is said to be *subordinate* to $\{Q^t \mid t \in T_\delta\}$ if and only if $P^t f(i) \leq Q^t f(i)$ for all $t \in T_\delta$, $f \in H, f(i) \geq 0, i \in S$ and $P^0 = I$.

Theorem 2.8.1. *Let $\{P^t \mid t \in T_\delta\}$ be a semigroup on H. Then the following two conditions are equivalent:*

(1) $\{P^t \mid t \in T_\delta\}$ is subordinate to $\{Q^t \mid t \in T_\delta\}$.
(2) There exists a δ-IMF $M(\omega, t)$ of $X^{(\delta)}(\omega, t)$ generating $\{P^t \mid t \in T_\delta\}$.

Proof. (2) $\Longrightarrow$ (1). It follows from the relations (2.8.2) and (2.8.4).

(1) $\Longrightarrow$ (2). Let $\{p_{ij}^{(\delta)} \mid i, j = 0, 1, 2, \ldots, N\}$ be the transition matrix of the semigroup $\{P^t \mid t \in T_\delta\}$. Then for each $f \in H$, we have

$$Q^\delta f(i) = \sum_{j=1}^{N} q_{ij}^{(\delta)} f(j),$$

$$P^\delta f(i) = \sum_{j=1}^{N} p_{ij}^{(\delta)} f(j).$$

Since $\{P^t \mid t \in T_\delta\}$ is subordinate to $\{Q^t \mid t \in T_\delta\}$, we see that

$$p_{ij}^{(\delta)} \leq q_{ij}^{(\delta)}, \forall i, j = 0, 1, 2, \ldots, N.$$

Define

$$\hat{M}(i,j) = \frac{p_{ij}^{(\delta)}}{q_{ij}^{(\delta)}} 1_{(q_{ij}^{(\delta)} \neq 0)},$$

where we set $\frac{a}{0} = 0, a \in {}^*[0, \infty)$. Put

$$\begin{aligned} M(\omega, 0) &= 1, \\ M(\omega, \delta) &= \hat{M}\left(X^{(\delta)}(\omega, 0), X^{(\delta)}(\omega, \delta)\right). \end{aligned} \tag{2.8.5}$$

For all $i \in S$, we have for any $f \in H$,

$$\begin{aligned} E_i[f(X^{(\delta)}(\omega, \delta))M(\omega, \delta)] &= \sum_{j=1}^{N} f(j)\hat{M}(i,j)q_{ij}^{(\delta)} \\ &= P^\delta f(i). \end{aligned}$$

By using mathematical induction, we define

$$M(\omega, (k+1)\delta) = M(\omega, k\delta)M(\theta_{k\delta}^{(\delta)}\omega, \delta) \quad \text{for all} \quad k \in {}^*\mathbb{N}.$$

It is then easy to show that $M(\omega, t)$ is a δ-IMF generating $\{P^t \mid t \in T_\delta\}$. □

2.8.3 Subprocesses

In Sect. 1.5 of Chap. 1, we defined a hyperfinite Markov chain $X^{(\delta)}(\omega, t)$ associated with the hyperfinite Dirichlet form $\mathcal{E}^{(\delta)}(\cdot, \cdot)$. Let us denote $\left\{\underline{\Omega}, \underline{\mathcal{F}}_t^{(\delta)}, Y^{(\delta)}(\underline{\omega}, t), \underline{\theta}_t^{(\delta)}, \underline{P}_i\right\}_{t \in T_\delta}$ a hyperfinite Markov chain with state space $(S, \mathcal{S})$. We call $Y^{(\delta)}(\underline{\omega}, t)$ a *subprocess* of $X^{(\delta)}(\omega, t)$ if and only if the semigroup $\{\underline{P}^t \mid t \in T_\delta\}$ of $Y^{(\delta)}(\underline{\omega}, t)$ is subordinate to $\{Q^t \mid t \in T_\delta\}$.

Let $Y^{(\delta)}(\underline{\omega}, t)$ be a subprocess of $X^{(\delta)}(\omega, t)$. From Theorem 2.8.1, there exists a δ-IMF $M(\omega, t)$ of $X^{(\delta)}(\omega, t)$ such that

$$E_{\underline{P}_i} f(Y^{(\delta)}(\underline{\omega}, t)) = E_i[f(X^{(\delta)}(\omega, t))M(\omega, t)] \quad \text{for all } t \in T_\delta, \tag{2.8.6}$$

where $E_{\underline{P}_i}(\cdot)$ is the expectation operator corresponding to $Y^{(\delta)}(\underline{\omega}, t)$.

We are interested in the following question. Given a δ-IMF $M(\omega, t)$, could we construct a subprocess $Y^{(\delta)}(\underline{\omega}, t)$ such that the relation (2.8.6) holds? The answer is yes! In fact, let $\underline{\Omega} = \Omega$, $\underline{\mathcal{F}}_t^{(\delta)} = \mathcal{F}_t^{(\delta)}$, $Y^{(\delta)}(\omega, t) = X^{(\delta)}(\omega, t)$,

$\underline{\theta}_t^{(\delta)} = \theta_t^{(\delta)}$. Furthermore, let $\{\underline{P}^t \mid t \in T_\delta\}$ be the semigroup given by the relation (2.8.1), and let $\{\underline{p}_{ij}^{(\delta)} \mid i,j = 0,1,2,\ldots\}$ be the transition matrix of $\{\underline{P}^t \mid t \in T_\delta\}$. Define for $\omega \in \underline{\Omega} = \Omega, k \in {}^*\mathbb{N}$,

$$\underline{P}_i\left([\omega]_{k\delta}\right) = \delta_{i\omega(0)} \prod_{n=0}^{k-1} \underline{p}\Big(\omega(n\delta), \omega((n+1)\delta)\Big).$$

It is obvious that the relation (2.8.6) holds with respect to $Y^{(\delta)}(\omega,t)$. We call $Y^{(\delta)}(\omega,t)$ the *canonical subprocess* associated with $(X^{(\delta)}, M)$. We notice that $\{\underline{p}_{ij}^{(\delta)} \mid i,j = 0,1,2,\ldots,N\}$ need not to have the regularities (1.5.1) and (1.5.2) in Sect. 1.5 of Chap. 1.

2.8.4 Feynman-Kac Formulae

Let $\{q_i^{(\delta)} \mid i = 1,2,\ldots,N\}$ be a $1 \times N$ matrix satisfying

$$0 \le q_i^{(\delta)} \le q_{ii}^{(\delta)}, i = 1,2,\ldots,N.$$

Define a transition matrix $P^{(\delta)} = \{p_{ij}^{(\delta)} \mid i,j = 0,1,2,\ldots,N\}$ by

$$\begin{aligned} p_{ij}^{(\delta)} &= q_{ij}^{(\delta)} - q_i^{(\delta)}\delta_{ij} \quad \text{for} \quad i,j = 1,2,\ldots,N, \\ p_{i0}^{(\delta)} &= q_{i0}^{(\delta)} + q_i^{(\delta)} \quad \text{for} \quad i = 1,2,\ldots,N, \text{ and} \\ p_{00}^{(\delta)} &= 1, p_{i0}^{(\delta)} = 0 \quad \text{for} \quad i = 1,2,\ldots,N. \end{aligned}$$

Similarly, define the transition matrix $\hat{P}^{(\delta)} = \{\hat{p}_{ij}^{(\delta)} \mid i,j = 0,1,2,\ldots,N\}$ by

$$\begin{aligned} \hat{p}_{ij}^{(\delta)} &= \hat{q}_{ij}^{(\delta)} - q_i^{(\delta)}\delta_{ij} \quad \text{for} \quad i,j = 1,2,\ldots,N, \\ \hat{p}_{i0}^{(\delta)} &= \hat{q}_{i0}^{(\delta)} + q_i^{(\delta)} \quad \text{for} \quad i = 1,2,\ldots,N \text{ and} \\ \hat{p}_{00}^{(\delta)} &= 1, \hat{p}_{i0}^{(\delta)} = 0 \quad \text{for} \quad i = 1,2,\ldots,N. \end{aligned}$$

Let $\{P^t \mid t \in T_\delta\}$ be the semigroup with $P^{(\delta)} = \{p_{ij}^{(\delta)} \mid i,j = 0,1,2,\ldots,N\}$ as its transition matrix, and let $\{\hat{P}^t \mid t \in T_\delta\}$ be the semigroup with $\hat{P}^{(\delta)} = \{\hat{p}_{ij}^{(\delta)} \mid i,j = 0,1,2,\ldots,N\}$ as its transition matrix. Then, we have

Theorem 2.8.2. *We have*

(i) The semigroups $\{P^t \mid t \in T_\delta\}$ *and* $\{\hat{P}^t \mid t \in T_\delta\}$ *are dual with respect to* m.

(ii) The Dirichlet form associated with $P^{(\delta)}$ *and* m *is given by*

$$\underline{\mathcal{E}}^{(\delta)}(u,v) = \mathcal{E}^{(\delta)}(u,v) + \sum_{i=1}^{N} u(i)v(i)q_i^{(\delta)}m(i).$$

(iii) There exists a δ*-IMF* $M(\omega,t)$ *such that for any* $f \in H$, $i \in S$, $t \in T_\delta$,

$$P^t f(i) = E_i[f(X^{(\delta)}(\omega,t))M(\omega,t)].$$

Proof. (i) and (ii) are obvious. (iii) follows easily from Theorem 2.8.1. □

2.9 Alternative Expression of Hyperfinite Dirichlet Forms

Suppose that $\mu(\cdot)$ is a positive internal measure on S. In this section, we shall find conditions on μ and μ-dual transition matrices $P = \{p_{ij} \mid i,j = 0,1,\ldots,N\}$ and $\hat{P} = \{\hat{p}_{ij} \mid i,j = 0,1,2,\ldots,N\}$ with the regularities (1.5.1) and (1.5.2) in Sect. 1.5 such that $\mathcal{E}(\cdot,\cdot)$ is the hyperfinite Dirichlet form associated with m and Q.

First we observe that if μ and P and $\hat{P}$ have been found, then for all $u,v \in H$, we have

$$\mathcal{E}(u,v) = \frac{1}{\Delta t}\sum_{i=1}^{N}\left[u(i)v(i)\mu(i) - \sum_{j=1}^{N} u(j)v(i)p_{ij}\mu(i)\right] \tag{2.9.1}$$

and

$$\mathcal{E}(u,v) = \frac{1}{\Delta t}\sum_{i=1}^{N}\left[u(i)v(i)\mu(i) - \sum_{j=1}^{N} u(j)v(i)\hat{p}_{ji}\mu(j)\right]. \tag{2.9.2}$$

Therefore, we have from the expression (1.5.19) in Sect. 1.5 and the expression (2.9.1) that

$$m(i)[1-q_{ii}] = \mu(i)[1-p_{ii}] \quad \text{for all} \quad i = 1,2,\ldots,N, \tag{2.9.3}$$
$$m(i)q_{ij} = \mu(i)p_{ij} \quad \text{for all} \quad i,j \in \{1,2,\ldots,N\}, i \neq j. \tag{2.9.4}$$

Similarly, we have

$$m(i)[1 - \hat{q}_{ii}] = \mu(i)[1 - \hat{p}_{ii}] \quad \text{for all} \quad i = 1, 2, \ldots, N, \tag{2.9.5}$$

$$m(i)\hat{q}_{ij} = \mu(i)\hat{p}_{ij} \quad \text{for all} \quad i, j \in \{1, 2, \ldots, N\}, i \neq j. \tag{2.9.6}$$

On the other hand, if the pairs of (P, μ) and (Q, m) satisfy the conditions (2.9.3) and (2.9.4) (or the pairs of $(\hat{P}, \mu)$ and $(\hat{Q}, m)$ satisfy the conditions (2.9.5) and (2.9.6)), then the expression (2.9.1) (or (2.9.2)) holds also. Hence, we get

Theorem 2.9.1. *Let $\mu(\cdot)$ be a positive internal measure on S, and let P and $\hat{P}$ be μ-dual transition matrices. Then the conditions (2.9.3), (2.9.4), (2.9.5), and (2.9.6) are sufficient and necessary conditions such that the expression (1.5.19) in Sect. 1.5 and the expressions (2.9.1) and (2.9.2) hold.*

2.10 Transformations of Symmetric Dirichlet Forms

In this section, we assume that

$$\overline{q}_{ii} = 0 \text{ for all } i = 1, 2, \ldots, N. \tag{2.10.1}$$

In fact, this assumption will not affect our theory. The reason comes from the proof of Proposition 1.5.1. Actually, for the general hyperfinite quadratic form $\mathcal{E}(\cdot, \cdot)$ of the expression (1.5.19) in Sect. 1.5, we define

$$\tilde{m}(i) = (1 - q_{ii})m(i),$$
$$\tilde{q}_{ii} = 0 \quad \text{for all} \quad i = 1, 2, \ldots, N.$$

Moreover, if $i = 1, 2, \ldots, N$ and $q_{ii} < 1$, define

$$\tilde{q}_{ij} = \frac{q_{ij}}{1 - q_{ii}} \quad \text{for} \quad j \neq i, j \in \{0, 1, 2, \ldots, N\};$$

if $i = 1, 2, \ldots, N$ and $q_{ii} = 1$, define

$$\tilde{q}_{ij} = 0 \quad \text{for} \quad j \neq 0, i \text{ and } \tilde{q}_{i0} = 1.$$

Besides, let $\tilde{q}_{00} = 1, \tilde{q}_{0j} = 0, j = 1, 2, \ldots, N$. It is very easy to verify from the Beurling–Deny formulae, Lemma 1.5.1 that the hyperfinite quadratic form associated with $\tilde{m}$ and $\tilde{Q} = \{\tilde{q}_{ij} \mid i, j = 0, 1, 2, \ldots, N\}$ is $\mathcal{E}(\cdot, \cdot)$. Since

$$q_{ii}m(i) = \hat{q}_{ii}m(i), \forall i = 1, 2, \ldots, N,$$

we may suppose $\hat{q}_{ii} = 0, \forall i = 1, 2, \ldots, N$. Hence, we may assume

$$\begin{aligned}\overline{q}_{ii} &= \frac{1}{2}(q_{ii} + \hat{q}_{ii}) \\ &= 0, i = 1, 2, \ldots, N.\end{aligned}$$

Let Φ be an internal nonnegative function in H. We define the following quadratic form

$$\overline{\mathcal{E}}^{\Phi}(u, v) = \frac{1}{\Delta t} \sum_{1 \le i < j \le N} \Big(u(i) - u(j)\Big)\Big(v(i) - v(j)\Big)\Phi(i)\Phi(j)\overline{q}_{ij}m(i). \tag{2.10.2}$$

It is easy to see that $\overline{\mathcal{E}}^{\Phi}(\cdot,\cdot)$ satisfies Proposition 1.5.1 (iii). Moreover, $\overline{\mathcal{E}}^{\Phi}(\cdot,\cdot)$ is symmetric. Thus by Proposition 1.5.1, there exists a transition matrix $\underline{P} = \{\underline{p}_{ij} \mid 0 \le i, j \le N\}$ and a symmetric measure $\underline{m}(\cdot)$ such that

$$\overline{\mathcal{E}}^{\Phi}(u, v) = \int_{S_0} u(i)\Big(v(i) - \underline{P}^{\Delta t}v(i)\Big)\frac{1}{\Delta t}\, d\underline{m}(i). \tag{2.10.3}$$

In the following, we will find the $\underline{P}$ and $\underline{m}$. From the expression (2.10.3) and the Beurling–Deny formulae (Lemma 1.5.1), we have

$$\begin{aligned}\overline{\mathcal{E}}^{\Phi}(u, v) = \frac{1}{\Delta t}\Big[&\sum_{1 \le i < j \le N} \Big(u(i) - u(j)\Big)\Big(v(i) - v(j)\Big)\underline{p}_{ij}\underline{m}(i) \\ &+ \sum_{i=1}^{N} u(i)v(i)\underline{p}_{i0}\underline{m}(i)\Big].\end{aligned} \tag{2.10.4}$$

Comparing the expressions (2.10.2) and (2.10.4), we must have

$$\begin{aligned}&\Phi(i)\Phi(j)\overline{q}_{ij}m(i) = \underline{p}_{ij}\underline{m}(i) \quad \text{for all} \quad 1 \le i, j \le N, \\ &\underline{p}_{i0}\underline{m}(i) = 0 \quad \text{for all} \quad i = 1, 2, \ldots, N.\end{aligned} \tag{2.10.5}$$

Therefore, we get

$$\begin{aligned}\underline{m}(i) &= \sum_{j=0}^{N} \underline{p}_{ij}\underline{m}(i) \\ &= \sum_{j \ne i} \Phi(j)\overline{q}_{ij}\Phi(i)m(i) + \underline{p}_{ii}\underline{m}(i) \\ &= \overline{E}_i[\Phi(X(\Delta t))]\Phi(i)m(i) + \underline{p}_{ii}\underline{m}(i).\end{aligned} \tag{2.10.6}$$

Define

$$\begin{aligned} \underline{p}_{00} &= 1, \\ \underline{p}_{0i} &= 0 \quad \text{for} \quad i = 1, 2, \ldots, N, \\ \underline{p}_{ii} &= 0 \quad \text{for} \quad i = 1, 2, \ldots, N. \end{aligned}$$

Then from (2.10.6), we obtain

$$\underline{m}(i) = \overline{E}_i[\Phi(X(\Delta t))]\Phi(i)m(i).$$

For $i = 1, 2, \ldots, N$, if $\overline{E}_i\Phi(X(\Delta t)) \neq 0$, we see from the relation (2.10.5) that for all $j = 1, 2, \ldots, N, j \neq i$,

$$\begin{aligned} \underline{p}_{ij} &= \frac{\overline{q}_{ij}\Phi(j)}{\overline{E}_i[\Phi(X(\Delta t))]} \\ &= \frac{\overline{q}_{ij}\Phi(j)}{\sum_{l \neq i} \overline{q}_{il}\Phi(l)}. \end{aligned} \tag{2.10.7}$$

For $i = 1, 2, \ldots, N$, if $\overline{E}_i[\Phi(X(\Delta t))] = 0$, we can define $\underline{p}_{ij}, 1 \leq j \leq N, j \neq i$ arbitrarily such that

$$\sum_{j=1}^{N} \underline{p}_{ij} = 1. \tag{2.10.8}$$

From (2.10.7) and (2.10.8), we get $\underline{p}_{i0} = 0$ for all $i \neq 0$. In the following discussion, we suppose that

$$\overline{E}_i[\Phi(X(\Delta t))] > 0 \quad \text{for all} \quad i = 1, 2, \ldots, N. \tag{2.10.9}$$

Hence, we have from (2.10.7) that for all $f \in H$,

$$\underline{P}f(i) = \frac{\overline{E}_i[(f\Phi)(X(\Delta t))]}{\overline{E}_i[\Phi(X(\Delta t))]}.$$

Let $\{\underline{P}_t \mid t \in T\}$ be the semigroup generated by $\underline{P}$. Then it is easy to see that $\{\underline{P}^t \mid t \in T\}$ is subordinate to $\{Q^t \mid t \in T\}$. By using Theorem 2.8.1, there exists a Δt-IMF $M(\omega, t)$ of $X(\omega, t)$ generating $\{\underline{P}^t \mid t \in T\}$. From the relation (2.8.5) in Sect. 2.8, we know that

$$M(\omega, \Delta t) = \frac{\Phi(X(\omega, \Delta t))}{\overline{E}_{X(\omega,0)}[\Phi(X((\Delta t))]}.$$

Therefore, we have for all $k \in {}^*\mathbb{N}$,

$$M(\omega, (k+1)\Delta t) = \prod_{l=0}^{k} \frac{\Phi(\omega, (l+1)\Delta t)}{\overline{E}_{X(\omega, l\Delta t)}[\Phi(X(\Delta t))]}. \tag{2.10.10}$$

On the other hand, let Φ be an internal function in H satisfying the hypothesis (2.10.9). We define $M(\omega, t)$ directly by the relation (2.10.10). Then, we have the following:

Theorem 2.10.1. *Assume that the hypotheses (2.10.1) and (2.10.9) hold. Then*

(1) The semigroup $\{(\overline{Q}^{\Phi})^t \mid t \in T\}$ generated by $M(\omega, t)$ is symmetric with respect to the measure $\overline{E}_i[\Phi(X(\Delta t))]\Phi(i)m(i) = \underline{m}(i)$.
(2) $\overline{\mathcal{E}}^{\Phi}(\cdot, \cdot)$ is the hyperfinite Dirichlet form associated with $\{(\overline{Q}^{\Phi})^t \mid t \in T\}$ and $\underline{m}(\cdot)$.

Proof. The proof is straightforward. □

Chapter 3
Standard Representation Theory

The purpose of this chapter is to study the standard projection of hyperfinite Markov chains, and the standard projection of hyperfinite Markov chains associated with hyperfinite quadratic forms. At first, we shall introduce in Sect. 3.1 a concept of irregularity; and then we shall prove that if a hyperfinite Markov chain $X(\cdot, t)$ has a set of irregularities, its standard part $x(\cdot, t)$ is a strong Markov process. In Sect. 3.2, we shall find conditions on hyperfinite quadratic forms which guarantee that the modified standard parts of associated hyperfinite Markov chains are strong Markov processes.

The old version of the materials of this chapter can be found in Chap. 5, Albeverio et al. [25]. However, Chap. 5 of Albeverio et al. [25] only handles symmetric hyperfinite Dirichlet form. Moreover, many restrictions in Albeverio et al. [25] are not necessary. For instance, we simplify the definition of exceptional sets in Chap. 2 and remove the unnecessary assumptions. Using the updated version of exceptional sets, we will define a concept of exceptional irregularities in Sect. 3.1. We will show that a hyperfinite Markov chain with exceptional irregularities has a standard part, which is a strong Markov process.

In Sect. 3.2, we shall impose conditions directly on hyperfinite quadratic forms to achieve our goal. Again, many assumptions of Albeverio et al. [25] will be simplified. For instance, the separation of compacts of Albeverio et al. [25] will be replaced by separation of points; all concepts involving exceptional sets will be updated using our definition in Chap. 2. It is our hope that the materials of this chapter update and significantly improve the related materials of Chap. 5, Albeverio et al. [25].

One may want to notice that we will impose conditions directly on a standard Dirichlet form in Chap. 4 to get related strong Markov processes. The materials of Chaps. 1, 2, and 3 can be thought as the theory of hyperfinite Dirichlet forms and hyperfinite Markov chains. With the solid work of these three chapters, we will be able to attack the important construction of strong Markov processes associated with standard Dirichlet forms in Chap. 4.

S. Albeverio et al., *Hyperfinite Dirichlet Forms and Stochastic Processes*, Lecture Notes of the Unione Matematica Italiana 10,
DOI 10.1007/978-3-642-19659-1_3, © Springer-Verlag Berlin Heidelberg 2011

3.1 Standard Parts of Hyperfinite Markov Chains

In this section, we shall study standard parts of hyperfinite Markov chains. In order to take standard parts we need a topology. We shall assume that with the exception of the trap s_0, the state space S is embedded in the nonstandard version ${}^*Y$ of some Hausdorff space Y. If X is a hyperfinite Markov chain taking values in S, we want to find conditions that guarantee that the standard part of X exists and is a Y-valued Markov process. It turns out that there are two difficulties we shall have to overcome. The first is that the paths of X may be so irregular that no natural standard part process exists. The second is that even when a standard part does exist, there is no reason why it should automatically be a Markov process – taking standard parts we may lump together states that should be kept apart. By using the theory of right standard parts developed by Albeverio et al. [25], Chap. 4, we can solve the first of these problems. Thus, most of our work will be directed to the second problem under discussion. Before we delve into the technicalities, we shall discuss the problem informally in some more details.

Assume that $x : \Omega \times \mathbb{R}_+ \longrightarrow Y$ is the standard part of X. We would like to prove that x is a Markov process with respect to the filtration it generates. Given that $x(t) = y$, in general there will be several states $s \in S_0$ such that $y = \mathrm{st}(s)$, and X may be in any one of them. From the nonstandard point of view, these states are totally unrelated, and hence the past and the future of the process may differ widely from one state to the next. Observation of the whole past may indicate which states are more likely to occur, and thus influence our prediction of the future. This explains why in general x is not a Markov process.

Note, however, that if the process started at s_i and the process started at s_j have the "same" future whenever $s_i \approx s_j$, the above argument breaks down, and it is reasonable to expect that x is Markov. One way of formulating this condition is to demand that

$$L(P_i)\{\omega \mid x(\omega, t) \in B\} = L(P_j)\{\omega \mid x(\omega, t) \in B\} \tag{3.1.1}$$

for all $t \in \mathbb{R}_+$ and all Borel sets B. But the above relation turns out to be too strict for the applications. Instead of demanding that it holds for all infinitely close s_i, s_j, we shall only demand that it holds for all such s_i, s_j outside an exceptional set (i.e., a set which the process hits with probability zero). It may at first seem that little is achieved by allowing a condition to fail on an exceptional set, but in fact the extra freedom and flexibility we gain will turn out to be very useful.

The present section falls into two halves. In the first part, we develop the necessary theory for the inner standard part of sets and a general theory of exceptional sets. In the second part, we shall show that the standard part of hyperfinite Markov chain is a strong Markov process under some conditions.

3.1.1 Inner Standard Part of Sets

By a *hyperfinite Markov chain* X in this section we shall understand a stationary Markov chain as described in Sect. 1.5, (1.5.1)–(1.5.7). If X has a dual hyperfinite Markov chain $\hat{X}$ as introduced in Sect. 1.5, we have for all $k \in {}^*\mathbb{N}$

$$\begin{aligned} P\{\omega \mid X(\omega, k\Delta t) = s_i\} &= \sum_{j=0}^{N} P\{\omega \mid X(\omega, 0) = s_j, X(\omega, k\Delta t) = s_i\} \\ &= \sum_{j=0}^{N} q_{ji}^{(k\Delta t)} m(j) \\ &= \sum_{j=0}^{N} \hat{q}_{ij}^{(k\Delta t)} m(i) \\ &= m(i). \end{aligned}$$

However, we shall NOT assume the duality of X in the following discussion, i.e., we do not need condition (1.5.13) in Sect. 1.5. We shall instead assume that for each $i \in S_0$, the function

$$t \mapsto P\{\omega \mid X(\omega, t) = s_i\}$$

is decreasing.

We shall further assume that $S_0 \subset {}^*Y$ for some topological space Y, but that the trap s_0 is not an element of ${}^*Y$. Similarly as in Sect. 2.1, we define the exceptional sets and properly exceptional sets of X using Definition 2.1.1 and Definition 2.1.2. Furthermore, Lemma 2.1.1 and Proposition 2.1.1 hold for X.

Most of the exceptional sets we encounter in the theory of Markov processes are sets we would like to avoid. From the standard point of view, this means that a point y in Y should be avoided if all its nonstandard representations $s_i \in \mathrm{st}^{-1}(y) \cap S_0$ are in the exceptional set. To study this relationship closely, we define the *inner standard part* A° of a subset A of S by

$$A^\circ = \left\{ y \in Y \mid \mathrm{st}^{-1}(y) \cap S_0 \subset A \right\}.$$

The *standard part of a set* A is defined by

$$\begin{aligned} {}^\circ A &= \mathrm{st}(A) \\ &= \{y \in Y \mid \exists s \in A(\mathrm{st}(s) = y)\}. \end{aligned}$$

The inner standard part can also be defined in terms of the standard part operation:

$$A^\circ = \mathbf{C}(\mathrm{st}(\mathbf{C}A)), \tag{3.1.2}$$

where the outer complement is with respect to Y, and the inner complement is with respect to S_0.

It is trivial to check that standard parts commute with arbitrary unions. Using the operation (3.1.2), we get that inner standard parts commute with arbitrary intersections. The other way around is less nice, the standard parts and intersections do not commute, and neither do inner standard parts and unions. All we can say is the following:

Lemma 3.1.1. *(1) If* $\{A_n \mid n \in \mathbb{N}\}$ *is a decreasing family of internal sets, then we have*

$$^\circ\left(\bigcap_{n\in\mathbb{N}} A_n\right) = \bigcap_{n\in\mathbb{N}} {}^\circ A_n. \tag{3.1.3}$$

(2) If $\{B_n \mid n \in \mathbb{N}\}$ *is an increasing family of internal sets, then we have*

$$\left(\bigcup_{n\in\mathbb{N}} B_n\right)^\circ = \bigcup_{n\in\mathbb{N}} B_n^\circ. \tag{3.1.4}$$

Proof. Obviously, the left hand of the property (3.1.3) is contained in the set on the right. To prove the converse, let

$$x \in \bigcap_{n\in\mathbb{N}} {}^\circ A_n.$$

For each $n \in \mathbb{N}$, pick $y_n \in A_n$ such that $x = \mathrm{st}(y_n)$. Extend $\{A_n \mid n \in \mathbb{N}\}$ to be a decreasing internal family $\{A_n \mid n \in {}^*\mathbb{N}\}$ and $\{y_n \mid n \in \mathbb{N}\}$ to be an internal sequence $\{y_n \mid n \in {}^*\mathbb{N}\}$ such that $y_n \in A_n$ for all $n \in {}^*\mathbb{N}$.

For each neighborhood O of x, consider the set

$$N_O = \{n \in {}^*\mathbb{N} \mid y_n \in {}^*O\}.$$

All these sets are internal and contain $\mathbb{N}$. Hence, we can find an infinite integer η that is in all of them. But then, we must have

$$x = \mathrm{st}(y_\eta).$$

Since the family $\{A_n \mid n \in {}^*\mathbb{N}\}$ is decreasing, we have $y_\eta \in A_\eta \subset \cap_{n\in\mathbb{N}} A_n$. This proves the property (3.1.3).

We now get the property (3.1.4) from the property (3.1.3) by using the operation (3.1.2):

$$\begin{aligned}\left(\bigcup_{n\in\mathbb{N}} B_n\right)^\circ &= \mathbf{C}\left(\mathrm{st}\left(\mathbf{C}\left(\bigcup_{n\in\mathbb{N}} B_n\right)\right)\right)\\ &= \mathbf{C}\left(\mathrm{st}\left(\bigcap_{n\in\mathbb{N}} (\mathbf{C}B_n)\right)\right)\\ &= \mathbf{C}\left(\bigcap_{n\in\mathbb{N}} (\mathrm{st}\,(\mathbf{C}B_n))\right)\\ &= \bigcup_{n\in\mathbb{N}} \mathbf{C}\,(\mathrm{st}(\mathbf{C}B_n))\\ &= \bigcup_{n\in\mathbb{N}} B_n^\circ.\end{aligned}$$

□

Since in general

$$\left(\bigcup_{m\in\mathbb{N}}\bigcap_{n\in\mathbb{N}} B_{m,n}\right)^\circ \neq \bigcup_{m\in\mathbb{N}}\bigcap_{n\in\mathbb{N}} B_{m,n}^\circ,$$

there is no obvious reason to believe that the inner standard part of a properly exceptional set is always Borel. However, we shall now prove that it must at least be universally measurable.

Lemma 3.1.2. *Assume* $A = \cup_{m\in\mathbb{N}} \cap_{n\in\mathbb{N}} B_{m,n}$ *for a family* $\{B\}_{m,n\in\mathbb{N}}$ *of internal sets. For any completed Borel probability measure* μ *on* Y*, the inner standard part* A° *is* μ *measurable. Moreover, there exists a family* $\{D_{m,n} \mid m, n \in \mathbb{N}\}$ *of internal sets such that*

$$A \subset \bigcap_{n\in\mathbb{N}}\bigcup_{m\in\mathbb{N}} D_{m,n}$$

and

$$\mu\left(\bigcap_{n\in\mathbb{N}}\bigcup_{m\in\mathbb{N}} D^\circ_{m,n} - A^\circ\right) = 0.$$

Proof. We may assume that the family $\{B_{m,n}\}$ is increasing in m and decreasing in n. Note that

$$\begin{aligned} A &= \bigcup_{m\in\mathbb{N}}\bigcap_{n\in\mathbb{N}} B_{m,n} \\ &= \bigcap_{f\in\mathbb{N}^\mathbb{N}}\bigcup_{m\in\mathbb{N}} B_{m,f(m)}. \end{aligned}$$

Let $\bar{f}(m)$ be the sequence $\langle f(0), f(1), \cdots, f(m)\rangle$, and define

$$C_{\bar{f}(m)} = \bigcup_{k\le m} B_{k,f(k)}.$$

For fixed f, the sequence $\{C_{\bar{f}(m)}\}_{m\in\mathbb{N}}$ is increasing. Hence by Lemma 3.1.1 and the fact that inner standard parts and arbitrary intersections commute, we have

$$A^\circ = \bigcap_{f\in\mathbb{N}^\mathbb{N}}\bigcup_{m\in\mathbb{N}} C^\circ_{\bar{f}(m)}. \tag{3.1.5}$$

Since $\mathbf{C}C^\circ_{\bar{f}(m)} = \mathrm{st}(\mathbf{C}C_{\bar{f}(m)})$ is closed, the complement

$$\mathbf{C}A^\circ = \bigcup_{f\in\mathbb{N}^\mathbb{N}}\bigcap_{m\in\mathbb{N}} \mathbf{C}C^\circ_{\bar{f}(m)}$$

of A can be derived from the closed sets by the Souslin operation, and hence A° is measurable with respect to any completed Borel measure (see [321], page 50, for an easy proof).

It only remains to find the family $\{D_{m,n}\}$. From (3.1.5) it follows that for each $\varepsilon \in \mathbb{R}_+$, there is a function $f_\varepsilon : \mathbb{N} \longrightarrow \mathbb{N}$ such that

$$\mu\left(\bigcap_{g\le f_\varepsilon}\bigcup_{m\in\mathbb{N}} C^\circ_{\bar{g}(m)} - A^\circ\right) < \varepsilon,$$

where $g \le f_\varepsilon$ means that $g(n) \le f_\varepsilon(n)$ for all $n \in \mathbb{N}$. Since our original sequence $\{B_{m,n} \mid n \in \mathbb{N}\}$ is decreasing, we have

$$\bigcap_{g\leq f_\varepsilon}\bigcup_{m\in\mathbb{N}} C^\circ_{\bar{g}(m)} = \bigcup_{m\in\mathbb{N}} C^\circ_{\bar{f}_\varepsilon(m)}.$$

Putting $D_{m,n} = C_{\bar{f}_{1/n}(m)}$, the lemma follows. □

Remark 3.1.1. The argument above shows that a subset of a Hausdorff space can be derived from the closed sets by using the Souslin operation if and only if it is the standard part of a set derived from the internal sets by the same operation. This result and the proof we have given are due to Henson [201].

We have now reached the last lemma we shall need before we can return to our Markov processes. It will be used to pick hyperfinite representations of measures avoiding properly exceptional sets.

Lemma 3.1.3. *Let D be a subset of S_0 of the form $D = \bigcap_{n\in\mathbb{N}}\bigcup_{m\in\mathbb{N}} D_{m,n}$ for a family $\{D_{m,n}\}$ of internal sets. If μ is a Radon probability measure on Y with $\mu(D^\circ) = 0$, there exists an internal probability measure ν on S_0 such that $\mu = L(\nu)\circ \mathrm{st}^{-1}$ and $L(\nu)(D) = 0$.*

Proof. We may obviously assume that the family $\{D_{m,n}\}$ is increasing in m and decreasing in n. Define a new measure $\mu_{m,n}$ by $\mu_{m,n}(B) = \mu(B - D^\circ_{m,n})$. Then, we have

$$\mu(B) = \sup_{n\in\mathbb{N}}\inf_{m\in\mathbb{N}} \mu_{m,n}(B). \tag{3.1.6}$$

The set $S_0 - D_{m,n}$ is S-dense in $Y - D^\circ_{m,n}$. Hence, there is an internal measure $\nu_{m,n}$ concentrated on $S_0 - D_{m,n}$ such that

$$\mu_{m,n} = L(\nu_{m,n})\circ \mathrm{st}^{-1}. \tag{3.1.7}$$

We may choose these measures such that the family $\{\nu_{m,n}\}$ is decreasing in m and increasing in n. Extending to an internal sequence $\{\nu_{m,n} \mid m,n \in {}^*\mathbb{N}\}$, we first pick a $\gamma \in {}^*\mathbb{N} - \mathbb{N}$ such that

$${}^\circ\nu_{\gamma,n}(S_0) = \lim_{m\to\infty} {}^\circ\nu_{m,n}(S_0)$$

for all $n \in \mathbb{N}$, and then an $\eta \in {}^*\mathbb{N} - \mathbb{N}$ such that

$${}^\circ\nu_{\gamma,\eta}(S_0) = \lim_{n\to\infty} {}^\circ\nu_{\gamma,n}(S_0).$$

By using equations (3.1.6) and (3.1.7), and the definitions of γ and η, we see that $L(\nu_{\gamma,\eta})(D) = 0$ and $\mu = L(\nu_{\gamma,\eta})\circ \mathrm{st}^{-1}$. The measure $\nu_{\gamma,\eta}$ has all the properties of the desired measure, except that we still have to show that it is a probability measure. However, this is clear since we have $\nu_{\gamma,\eta}(S_0) = 1+\varepsilon$ for some $\varepsilon \approx 0$. Putting $\nu = (1+\varepsilon)^{-1}\nu_{\gamma,\eta}$, the lemma is then proved. □

By combining Lemmas 3.1.2 and 3.1.3, we get:

Corollary 3.1.1. *Let $A \subset S_0$ be a properly exceptional set, and let μ be a completed Borel probability measure on Y such that $\mu(A^\circ) = 0$. Then there is an internal probability measure ν on S_0 such that $\mu = L(\nu) \circ \mathrm{st}^{-1}$ and $L(\nu)(A) = 0$.*

3.1.2 Strong Markov Processes and Modified Standard Parts

Having completed our study of exceptional sets and inner standard part of sets, we now turn to the real subject matter of this section, an investigation of standard parts of hyperfinite Markov chains. Let us first describe what kind of processes we would like to obtain as standard parts [152].

Let Y be a Hausdorff space and let Δ be an extra element. Set $Y_\Delta = Y \cup \{\Delta\}$. Consider the topological Borel field $\mathcal{B}(Y)$ on Y. Let Y_Δ have the σ-algebra $\mathcal{B}(Y_\Delta)$ generated by the Borel sets on Y and the singleton $\{\Delta\}$. Our standard processes will be Y_Δ valued, and the new element Δ will serve as a trap.

Assume that (Ω, Π) is a measurable space and $\{\Pi_t \mid t \in \mathbb{R}_+\}$ is a family of sub-σ-algebras of Π satisfying

$$\Pi_s \subset \Pi_t \quad \text{for all} \quad s \leq t \quad \text{and} \quad \Pi_t = \bigcap_{s>t} \Pi_s \quad \text{for all} \quad t \in \mathbb{R}_+. \qquad (3.1.8)$$

We call $\{\Pi_t \mid t \in \mathbb{R}_+\}$ a *right continuous filtration* on Ω if it satisfies the condition (3.1.8). Set $\Pi_\infty = \sigma\{\Pi_t \mid t \in \mathbb{R}_+\}$. In our situation, we can take $\Pi = \Pi_\infty$.

A map $\sigma : \Omega \to [0, \infty)$ is called a *stopping time* of $\{\Pi_t \mid t \in \mathbb{R}_+\}$ if for all t

$$\{\omega \mid \sigma(\omega) \leq t\} \in \Pi_t.$$

For each stopping time σ, we introduce a σ-algebra Π_σ by

$$\Pi_\sigma = \{A \in \Pi_\infty \mid \forall t (A \cap (\sigma \leq t) \in \Pi_t)\}.$$

Denote by $\mathcal{M}(Y)$ the set of all positive Radon measures on $(Y, \mathcal{B}(Y))$ with finite masses. Given an element $\nu \in \mathcal{M}(Y)$, the completion of the σ-field $\mathcal{B}(Y)$ with respect to ν is denoted by $\overline{\mathcal{B}(Y)^\nu}$. A set or a function is called *universally measurable* if it is measurable with respect to $\bigcap_{\nu \in \mathcal{M}(Y)} \overline{\mathcal{B}(Y)^\nu}$.

If for each $y \in Y_\Delta$, we are given a probability measure Θ_y on (Ω, Π). Then for each $\nu \in \mathcal{M}(Y)$, we have

$$\Theta_\nu(A) = \int_Y \Theta_y(A)\, d\nu(y), A \in \Pi,$$

provided, of course, that this makes sense, i.e., $y \mapsto \Theta_y(A)$ is μ measurable.

Finally, if $x(t)$ is a Y_Δ valued process, its *lifetime* ζ is defined by

$$\zeta(\omega) = \inf\{t \geq 0 \mid x(\omega, t) = \Delta\}.$$

Definition 3.1.1. We call $(\Omega, \Pi_\infty, \{\Pi_t \mid t \in \mathbb{R}_+\}, x(t), \Theta_y)$ a *strong Markov process*, if $\{\Pi_t \mid t \in \mathbb{R}_+\}$ is a right continuous filtration of Π, each Θ_y is a probability measure on Π_∞, and $x : \Omega \times [0, \infty] \longrightarrow Y_\Delta$ is a stochastic process on $(\Omega, \Pi_\infty, \Theta_y)$ for each $y \in Y_\Delta$. Moreover, the following conditions are satisfied:

(i) For all $t \geq 0$ and all measurable $E \subset Y$, the map $y \mapsto \Theta_y(x(t) \in E)$ is universally measurable in $y \in Y$.
(ii) $x(\omega, \infty) = \Delta$, $\forall \omega \in \Omega$. $x(\omega, t) = \Delta, \forall t \geq \zeta(\omega)$.
(iii) For each $y \in Y_\Delta$, the process x is adapted to $(\Omega, \{\Pi_t\}, \Theta_y), t \mapsto x_t(\omega)$ is right continuous from $[0, \infty)$ to Y_Δ, Θ_y-a.e.ω and $\lim_{s\uparrow t} x_s(\omega)$ exists in Y for all $t \in (0, \zeta(\omega)), \Theta_y$-a.e.$\omega$ (a.e. represents almost every).
(iv) $\Theta_\Delta(x(t) = \Delta) = 1, \forall t \geq 0$; $\Theta_y(x(0) = y) = 1, \forall y \in Y$.
(v) For all $\{\Pi_t\}$ stopping times σ, all measurable $E \subset Y_\Delta$, all $\mu \in \mathcal{M}(Y)$, and all $s \in \mathbb{R}_+$, we have

$$\Theta_\mu(x(\sigma + s) \in E \mid \Pi_\sigma) = \Theta_{x(\sigma)}(x(s) \in E), \Theta_\mu\text{-a.e.} \tag{3.1.9}$$

In order to prove that a class of hyperfinite Markov chains have standard parts that are strong Markov processes, we shall have to overcome two main difficulties: the construction of the family of measures $\{\Theta_y\}$, and the proof of the strong Markov property (3.1.9). But first we must introduce the necessary regularity conditions in our nonstandard processes.

Here, we would remind the reader that we use P_i to mean the internal probability measure on $(\Omega, \mathcal{F}_t), i \in S, t \in T$, and we use Θ_y to mean the standard probability measure on $(\Omega, \Pi_\infty), y \in Y$. We hope this will not cause confusion.

Definition 3.1.2. Let $f : T \longrightarrow {}^*\mathbb{R}$ be internal. We say that $r \in \mathbb{R}$ is the *S-right limit* of f at $t \in [0, \infty)$ if for any standard $\varepsilon > 0$, there is a standard $\delta > 0$ such that if $s \in T$ and $t < {}^\circ s < t + \delta$, then $|f(s) - r| < \varepsilon$, we write $r = S\text{-}\lim_{s\downarrow t} f(s)$. The *S-left limit*, $S\text{-}\lim_{s\uparrow t} f(s)$ is defined analogously.

Recall that $\overline{S}_0 = S_0 \cap Ns(^*Y)$ and that $X^{(\delta)} = X|T_\delta$. The *lifetime* ζ_δ of $X^{(\delta)}$ is defined by

$$\zeta_\delta(\omega) = \inf\left\{{}^\circ t \mid X^{(\delta)}(\omega, t) \notin \overline{S}_0\right\}.$$

We define the *right standard part* ${}^\circ X^{(\delta)+}$ as follows. If $t < \zeta_\delta(\omega)$, let

$${}^\circ X^{(\delta)+}(\omega, t) = S\text{-}\lim_{s \downarrow t} X^{(\delta)}(\omega, s)$$

if this limit exists and for all $t_1 < t$, the S-$lim_{s\downarrow t_1} X^{(\delta)}(\omega, s)$ exists also, and

$${}^\circ X^{(\delta)+}(\omega, t) = \Delta$$

else. If $t \geq \zeta_\delta(\omega)$, we always put ${}^\circ X^{(\delta)+}(\omega, t) = \Delta$.

Definition 3.1.3. A subset A of S_0 is called a *set of irregularities* of X if there is a positive infinitesimal $\delta_0 \in T$ satisfying:

(i) For all $s_i \in \overline{S}_0 - A$ and $L(P_i)$-a.e. ω, the path $X^{(\delta_0)}(\omega, \cdot)$ has S-right and S-left limits at all $t < \zeta_{\delta_0}(\omega)$.
(ii) For all $s_i \in \overline{S}_0 - A$, the set

$$\left\{\omega \middle| \exists t \in T^{\text{fin}}_{\delta_0}({}^\circ t > \zeta_{\delta_0}(\omega) \wedge X^{(\delta_0)}(\omega, t) \in \overline{S}_0)\right\}$$

has $L(P_i)$ measure zero.
(iii) For all infinitely close $s_i, s_j \in \overline{S}_0 - A$,

$$L(P_i)\{{}^\circ X^{(\delta_0)+}(\omega, t) \in B\} = L(P_j)\{{}^\circ X^{(\delta_0)+}(\omega, t) \in B\}$$

for all finite $t \in [0, \infty)$ and all Borel sets B.

The hyperfinite Markov chain X has δ_0-*exceptional irregularities* if the set A of irregularities of X in the Definition 3.1.3 can be taken as a δ_0-exceptional set. Hence, X has *exceptional irregularities* if it has δ_0-exceptional irregularities for some $\delta_0 \approx 0, \delta_0 \in T$.

The first condition of the above Definition 3.1.3 guarantees that the hyperfinite Markov chain$X$ has a reasonable standard part. The second says that "infinite" is a trap. The third one is a version of the condition (3.1.1).

Putting $\text{st}(s_i) = \Delta$ when $s_i \in S - \overline{S}_0$, we have the following definition of the modified standard part of $X(\omega, t)$.

Definition 3.1.4. Assume that X has exceptional irregularities, and let A be a properly δ_0-exceptional set of irregularities, where δ_0 is as in Definition 3.1.3. Let $x : \Omega \times \mathbb{R}_+ \longrightarrow Y_\Delta$ be defined by

(i) if $X(\omega,0) \notin A$, then $x(\omega,t) = {}^\circ X^{(\delta_0)+}(\omega,t)$.
(ii) if $X(\omega,0) \in A$, then $x(\omega,t) = \mathrm{st}(X(\omega,0))$ for all $t \in \mathbb{R}_+$.

We call x *modified standard part* of $X(\omega,t)$.

Our aim is to prove that if X has exceptional irregularities, then with an appropriate definition of the family $\{\Theta_y\}$ of measures, the modified standard parts of X are strong Markov processes. The first step toward the definition of $\{\Theta_y\}$ is the following version of condition (iii) in Definition 3.1.3.

If ν is an internal probability measure on S, let P_ν be the measure on Ω defined by

$$P_\nu(C) = \int P_i(C)\, d\nu(s_i). \tag{3.1.10}$$

Lemma 3.1.4. *For $\delta_0 \approx 0, \delta_0 \in T$, let A be a properly δ_0-exceptional set of irregularities of X, and let x be the corresponding modified standard part of X. Let ν_1, ν_2 be two internal probability measures on S_0 such that $L(\nu_1)(A) = L(\nu_2)(A) = 0$ and $L(\nu_1) \circ \mathrm{st}^{-1} = L(\nu_2) \circ \mathrm{st}^{-1}$, then for all $t \in \mathbb{R}_+$ and all Borel sets B*

$$L(P_{\nu_1})\{x(\omega,t) \in B\} = L(P_{\nu_2})\{x(\omega,t) \in B\}.$$

Proof. Let $\mu = L(\nu_1) \circ \mathrm{st}^{-1} = L(\nu_2) \circ \mathrm{st}^{-1}$. Choose $\tilde{t} \approx t$ so large that

$${}^\circ X^{(\delta_0)}(\omega,\tilde{t}) = x(\omega,t), L(P_{\nu_1}) \text{ and } L(P_{\nu_2})\text{-a.e.}\omega \in \Omega$$

and pick an internal set $\tilde{B}$ such that

$$L(P_{\nu_i})(\{X^{(\delta_0)}(\omega,\tilde{t}) \in \tilde{B}\} \Delta \{{}^\circ X^{(\delta_0)}(\omega,\tilde{t}) \in B\}) = 0$$

for $i = 1, 2$. Define a function $f : Y \longrightarrow \mathbb{R}$ as follows:

(1) If $y \notin A^\circ$, let $f(y) = L(P_i)\{x(\omega,t) \in B\}$ for some (then for all) $s_i \in \mathrm{st}^{-1}(y) \cap S_0 - A$.
(2) If $y \in A^\circ$, define $f(y)$ arbitrarily.

The function $s_i \mapsto P_i\{X^{(\delta_0)}(\omega,\tilde{t}) \in \tilde{B}\}$ is a lifting of f with respect to both ν_1 and ν_2. Therefore, we have

$$\begin{aligned} L(P_{\nu_1})\{x(\omega,t) \in B\} &= {}^\circ P_{\nu_1}\{X^{(\delta_0)}(\omega,\tilde{t}) \in \tilde{B}\} \\ &= {}^\circ\!\!\int P_i\{X^{(\delta_0)}(\omega,\tilde{t}) \in \tilde{B}\}\, d\nu_1(s_i) \\ &= \int f(y)\, d\mu(y) \end{aligned}$$

$$= \int^{\circ} P_i\{X^{(\delta_0)}(\omega, \tilde{t}) \in \tilde{B}\}\, d\nu_2(s_i)$$
$$= {}^{\circ}P_{\nu_2}\{X^{(\delta_0)}(\omega, \tilde{t}) \in \tilde{B}\}$$
$$= L(P_{\nu_2})\{x(\omega, t) \in B\},$$

and the lemma is proved. □

Lemma 3.1.5. *For $\delta_0 \approx 0, \delta_0 \in T$, assume that X has a properly δ_0-exceptional set A of irregularities, and let x be the modified standard part. For all infinitely close $s_i, s_j \in \overline{S}_0 - A$, all finite sequences $t_1 < t_2 < \cdots < t_n$ from $\mathbb{R}_+$ and all Borel sets $B_1, B_2, \cdots, B_n$, we have*

$$L(P_i)\left(\bigcap_{l=1}^{n}\{x(t_l) \in B_l\}\right) = L(P_j)\left(\bigcap_{l=1}^{n}\{x(t_l) \in B_l\}\right).$$

Proof. We shall prove this by induction on the length n of the sequences $t_1 < t_2 < \cdots < t_n, B_1, B_2, \cdots, B_n$. The case of $n = 1$ is part of Definition 3.1.3. Assume that the lemma holds for all sequences of length $n-1$, and pick $\tilde{t}_1, \tilde{t}, \cdots, \tilde{t}_n \in T_{\delta_0}$ such that $\tilde{t}_1 \approx t_1, \tilde{t}_2 \approx t_2, \cdots, \tilde{t}_n \approx t_n$ and

$${}^{\circ}X^{(\delta_0)}(\omega, \tilde{t}_l) = x(\omega, t_l), L(P_k)\text{-a.e.}$$

for $l = 1, 2, \cdots, n$ and $k = i, j$. Choose internal sets $\tilde{B}_1, \tilde{B}_2, \cdots, \tilde{B}_n$ such that

$$L(P_k)\Big(\{X^{(\delta_0)}(\omega, \tilde{t}_l) \in \tilde{B}_l\} \Delta \{{}^{\circ}X^{(\delta_0)}(\omega, \tilde{t}_l) \in B_l\}\Big) = 0$$

for $l = 1, 2, \cdots, n$ and $k = i, j$.

We define two measures ν_i, ν_j on S by putting

$$\nu_k(s) = P_k\{\omega \mid X^{(\delta_0)}(\omega, \tilde{t}_{n-1}) = s \text{ and for all } l < n-1, X^{(\delta_0)}(\omega, \tilde{t}_l) \in \tilde{B}_l\}$$

for all $s \in S$ and $k = i, j$. By the induction hypothesis

$$L(\nu_i) \circ \mathrm{st}^{-1} = L(\nu_j) \circ \mathrm{st}^{-1},$$

and since A is properly δ_0-exceptional, we have $L(\nu_i)(A) = L(\nu_j)(A) = 0$. Applying Lemma 3.1.4 with $t = t_n - t_{n-1}$, we get

$$L(P_{\nu_i})\{x(t_n - t_{n-1}) \in B_n\} = L(P_{\nu_j})\{x(t_n - t_{n-1}) \in B_n\}.$$

Since X is Markov and time homogeneous, the lemma follows. □

In the following, we will define $\{\Pi_t\}$ and $\{\Theta_y\}$ of our standard Markov processes. For $t \in \mathbb{R}_+$, let Π_t° be the σ-algebra generated by the sets

$$\left\{\omega \middle| x(\omega,t_1) \in B_1 \wedge \cdots \wedge x(\omega,t_n) \in B_n\right\},$$

where $0 \le t_1 < t_2 < \cdots < t_n \le t$ and $B_1, \cdots, B_n$ are Borel sets in Y. Define

$$\Pi_t = \bigcap_{s>t} \Pi_s^\circ .$$

The filtration $\{\Pi_t \mid t \in [0,\infty)\}$ is right continuous. Let us set

$$\Pi_\infty = \vee \{\Pi_t \mid t \in [0,\infty)\} .$$

It follows from Lemma 3.1.5 that for all $C \in \Pi_\infty$ and all infinitely close s_i, s_j in $\overline{S}_0 - A$

$$L(P_i)(C) = L(P_j)(C).$$

This observation makes it possible to give the following definition of a family $\{\Theta_y \mid y \in Y_\Delta\}$ of measures on Π_∞.

If $y \notin A^\circ$, let for all $C \in \Pi_\infty$

$$\Theta_y(C) = L(P_i)(C) \quad \text{for all} \quad s_i \in \mathrm{st}^{-1}(y) - A.$$

If $y \in A^\circ \cup \{\Delta\}, C \in \Pi_\infty$, let

$$\Theta_y(C) = \begin{cases} 1, & \text{if } C \text{ contains all constant paths } x(t) = y, \\ 0, & \text{else.} \end{cases}$$

Observe that since a set $C \in \Pi_\infty$ contains either all or none of the constant paths $x(\omega,t) = y$, the set function Θ_y is a measure.

We have reached our goal:

Theorem 3.1.1. *Assume that S_0 is a hyperfinite subset of $^*Y$ for some Hausdorff space. Let $X : \Omega \times T \longrightarrow S$ be a hyperfinite Markov chain with exceptional irregularities. Assume that for each $i \in S_0$, the function $t \mapsto P\{\omega \mid X(\omega,t) = s_i\}$ is decreasing. If x is a modified standard part of X, then $(\Omega, \{\Pi_t \mid t \in \mathbb{R}_+\}, \{\Theta_y \mid y \in Y_\Delta\}, x(t))$ is a strong Markov process.*

Proof. Since we already know that $\{\Pi_t\}$ is right continuous, all we have to do is to check conditions (i)-(v) of Definition 3.1.1. Obviously, the (ii), (iii) and (iv) of Definition 3.1.1 are satisfied by our construction. We shall prove (i) and (v).

Given a Radon probability measure μ on Y, we define two new measures μ_0 and μ_1 on Y by

$$\mu_0(B) = \mu(B - A^\circ),$$
$$\mu_1(B) = \mu(B \cap A^\circ),$$

where A is the properly δ_0-exceptional set of irregularities used in the construction of x. By Corollary 3.1.1, we can find an internal measure ν on S_0 satisfying

$$\mu_0 = L(\nu) \circ \mathrm{st}^{-1},$$
$$L(\nu)(A) = 0. \tag{3.1.11}$$

We first prove that (i) in Definition 3.1.1 is satisfied:

(i) Given $\alpha \in [0,1]$, a Borel set $E \subset Y$, and $t \in \mathbb{R}_+$, we must show that

$$\{y \in Y \mid \Theta_y(x(t) \in E) > \alpha\}$$

is μ-measurable for all finite Radon measures μ. Since

$$\{y \in A^\circ \mid \Theta_y\,\{x(t) \in E\} > \alpha\} = \begin{cases} A^\circ \cap E, & \alpha \in [0,1) \\ \emptyset, & \alpha = 1 \end{cases}$$

is universally measurable by Lemma 3.1.2, it suffices to show that

$$\{y \notin A^\circ \mid \Theta_y(x(t) \in E) > \alpha\} \tag{3.1.12}$$

is μ-measurable. In fact, we only have to show that the set defined by (3.1.12) is μ_0 measurable by the definition of μ_0.

Since $X^{(\delta_0)}$ has S-right limits a.e. with respect to the probability measure P_ν constructed in (3.1.11) by using the relation (3.1.10), we can choose $\hat{t} \approx t$ so large that

$$L(P_\nu)(\{x(t) \in E\}\Delta\{{}^\circ X^{(\delta_0)}(\tilde{t}) \in E\}) = 0.$$

There must be an internal set $\tilde{E}$ such that

$$L(P_\nu)\{X^{(\delta_0)}(\tilde{t}) \in (\mathrm{st}^{-1}(E)\Delta\tilde{E})\} = 0.$$

Hence, we have

$$L(P_i)\{x(t) \in E\} = {}^\circ P_i\{X^{(\delta_0)}(\tilde{t}) \in \tilde{E}\}$$

for $L(\nu)$-a.e.s_i. Combining this with the definition of Θ_y, we see that $y \mapsto \Theta_y\{x(t) \in E\}$ has $i \mapsto P_i\{X^{(\delta_0)}(\tilde{t}) \in \tilde{E}\}$ as a ν-lifting. Hence, it is a μ_0 measurable function. This proves that the set defined by (3.1.12) is μ_0 measurable, and Definition 3.1.1 (i) follows.

(v) we must show that for all $\{\Pi_t\}$ stopping times σ, all sets $B \in \Pi_\sigma$ and all $s \in [0, \infty)$, the equation

$$\Theta_\mu\{\omega \in B \mid x_{\sigma+s} \in E\} = \int_B \Theta_{x_\sigma}\{x_s \in E\}\, d\Theta_\mu \tag{3.1.13}$$

holds for all Radon probability measure μ on Y and all Borel sets E.

First notice that since the paths of x are constant Θ_{μ_1}-a.e., we have

$$\Theta_{\mu_1}\{\omega \in B \mid x_{\sigma+s} \in E\} = \int_B \Theta_{x_\sigma}\{x_s \in E\}\, d\Theta_{\mu_1}.$$

Equation (3.1.13) will hold if we may prove

$$\Theta_{\mu_0}\{\omega \in B \mid x_{\sigma+s} \in E\} = \int_B \Theta_{x_\sigma}\{x_s \in E\}\, d\Theta_{\mu_0}. \tag{3.1.14}$$

In the sequel, we will prove (3.1.14). Let ν be the nonstandard representation of μ_0 given in the construction (3.1.11). The hyperfinite counterpart of (3.1.14) is

$$P_\nu\{\omega \in C \mid X^{(\delta_0)}_{\tau+s} \in F\} = \int_C P_{X^{(\delta_0)}_\tau}\{X^{(\delta_0)}_s \in F\}\, dP_\nu, \tag{3.1.15}$$

where C and F are internal sets and C is measurable in the *−algebra generated by the internal stopping time τ. Since $X^{(\delta_0)}$ is a time homogeneous Markov chain, (3.1.15) holds. Our plan is to deduce (3.1.14) from (3.1.15).

We first pick an internal stopping time τ such that $^\circ\tau = \sigma, L(P_\nu)$-a.e., and such that there is a τ measurable set C satisfying

$$L(P_\nu)(B \Delta C) = 0. \tag{3.1.16}$$

Let P^τ be an internal measure on Ω given by

$$P^\tau(D) = \int P_{X^{(\delta_0)}_\tau}(D)\, dP_\nu.$$

Since A is properly exceptional, we observe that

$$L(P^\tau)\{X^{(\delta_0)}_t \in A\} = 0$$

for all $t \in T^{\mathrm{fin}}_{\delta_0}$ (where δ_0 is the infinitesimal used in the construction of x, see Definition 3.1.4). Hence, we can choose an $\tilde{s} \in T^{\mathrm{fin}}_{\delta_0}$, $\tilde{s} = s$, such that

$$^{\circ}X_{\tilde{s}}^{(\delta_0)} = x_s, L(P^{\tau})\text{-a.e.} \tag{3.1.17}$$

and

$$^{\circ}X_{\tau+\tilde{s}}^{(\delta_0)} = x_{\sigma+s}, L(P_{\nu})\text{-a.e.} \tag{3.1.18}$$

Finally, we pick an internal set F such that

$$L(P_{\nu})\{X_{\tau+\tilde{s}}^{(\delta_0)} \in \text{st}^{-1}(E)\Delta F\} = 0, \tag{3.1.19}$$
$$L(P^{\tau})\{X_{\tilde{s}}^{(\delta_0)} \in \text{st}^{-1}(E)\Delta F\} = 0. \tag{3.1.20}$$

We now have

$$\begin{aligned}
&\Theta_{\mu_0}\{\omega \in B \mid x_{\sigma+s} \in E\} \\
&= \int \Theta_y\{\omega \in B \mid x_{\sigma+s} \in E\}\, d\mu_0(y) && \text{(by def. of } \Theta_{\mu_0}\text{)} \\
&= \int L(P_i)\{\omega \in B \mid x_{\sigma+s} \in E\}\, dL(\nu)(s_i) && \text{(by def. of } \Theta_y\text{)} \\
&= L(P_\nu)\{\omega \in B \mid x_{\sigma+s} \in E\} && \text{(by def. of } P_\nu\text{)} \\
&= {}^{\circ}P_\nu\{\omega \in C \mid X_{\tau+\tilde{s}} \in F\} && \text{(by (3.1.16), (3.1.18), (3.1.19))} \\
&= {}^{\circ}\!\!\int_C P_{X_\tau}\{X_{\tilde{s}} \in F\}\, dP_\nu && \text{(by (3.1.15))} \\
&= \int_B L(P_{X_\tau})\{x_s \in E\}\, dL(P_\nu) && \text{(by (3.1.16), (3.1.17), (3.1.20))} \\
&= \int_B \Theta_{x_\sigma}\{x_s \in E\}\, d\Theta_{\mu_0} && \text{(by def. of } \nu, \tau \text{ and } \{\Theta_y\}\text{).}
\end{aligned}$$

This proves (3.1.14) and the theorem. □

Example 3.1.1. Let $\mathbb{Z} = \{\cdots, -2, -1, 0, 1, 2, \cdots\}$ (i.e., set of all the integers). Let $\eta \in {}^*\mathbb{N} - \mathbb{N}$, and set

$$S_0 = \left\{ \frac{k}{\sqrt{\eta}} \middle| k \in {}^*\mathbb{Z}, |k| \le \eta \right\}.$$

Define transition probabilities $q_{s,s'}$ by

$$q_{s,s'} = \begin{cases} \frac{1}{2} & \text{if } |s - s'| = \frac{1}{\sqrt{\eta}}, \\ 0 & \text{otherwise.} \end{cases}$$

If $s, s' \in S_0$, and if s_0 is the trap, let

$$q_{s,s'} = \begin{cases} \frac{1}{2} & \text{if } s = \pm\sqrt{\eta}, \\ 0 & \text{otherwise.} \end{cases}$$

Let m be an internal measure on S_0 given by

$$m(s) = \frac{1}{\sqrt{\eta}}$$

for all $s \in S_0$, and let a time-line be

$$T = \left\{ \frac{k}{\eta} \middle| k \in {}^*\mathbb{N}_0 \right\}.$$

The process X is Anderson's random walk with a uniform initial distribution corresponding to the Lebesgue measure. To prove that the standard part of X is a strong Markov process, we show that the empty set is a set of irregularities of X. Since X is S-continuous $L(P_i)$-a.e. for all nearstandard $s_i \in S_0$, the two first conditions in Definition 3.1.3 are obviously satisfied. If $s_i \approx s_j$, the paths starting at s_i look exactly like the paths starting at s_j except for an infinitesimal translation. Hence, they induce the same standard paths and the property (iii) in Definition 3.1.3 follows.

3.2 Hyperfinite Dirichlet Forms and Markov Processes

In this section, we shall obtain conditions on hyperfinite quadratic forms which guarantee that the modified standard parts of the associated hyperfinite Markov chains are strong Markov processes. The method we shall apply is simple, we just use the relationship between forms and processes established in Sect. 1.5 to translate the conditions of Theorem 3.1.1 into the language of hyperfinite quadratic forms. One by one we shall reformulate the conditions of Definition 3.1.3 in terms of hyperfinite quadratic forms.

In this section, we need a slightly stronger assumption on the state space. Let Y be a *regular Hausdorff space* (or a T_3 *space*), i.e., for all closed sets F and $x \notin F$, there are disjoint open sets O_1, O_2 such that $x \in O_1$, $F \subset O_1$. Suppose that $S_0 = \{s_1, s_2, \cdots, s_N\}$ is a hyperfinite subset of ${}^*Y$ and m is a hyperfinite positive measure on S_0, $N \in {}^*\mathbb{N} - \mathbb{N}$. Let H be the linear space of all internal functions $u : S_0 \longrightarrow {}^*\mathbb{R}$ with the inner product

$$\langle u, v \rangle = \sum_{i=1}^{N} u(s_i)v(s_i)m(s_i).$$

Given an infinitesimal $\Delta t \approx 0$, denote by $T = \{k\Delta t \mid k \in {}^*\mathbb{N}_0\}$ the hyperfinite time line. Let X and $\hat{X}$ be the dual hyperfinite Markov chains introduced in Sect. 1.5 with m as dual measure. That is, we assume that conditions (1.5.1), (1.5.2), (1.5.4), (1.5.5), (1.5.8), (1.5.9), (1.5.10), (1.5.11), (1.5.13), and (1.5.14) in Sect. 1.5 are fulfilled. Let $\mathcal{E}(\cdot,\cdot)$ be the hyperfinite quadratic form on H associated with the transition matrix Q and the dual measure m, and $\hat{\mathcal{E}}(\cdot,\cdot)$ be the co-form of $\mathcal{E}(\cdot,\cdot)$.

Given $x \in Y$, we recall that the *monad*[1] of x is the set of ${}^*Y$ defined by

$$\mu(x) = \bigcap \{{}^*O \big| x \in O \text{ and } O \text{ is open}\}.$$

3.2.1 Separation of Points

The following assumption will take care of Definition 3.1.3 (i). In Albeverio et al. [25], 5.5.1. Definition, a definition of separation of compacts was given for symmetric hyperfinite Dirichlet forms. The following definition simplifies the condition since a single point can be viewed as a compact set.

Definition 3.2.1. Let Z be a subset of Y. A hyperfinite quadratic form $\mathcal{E}(\cdot,\cdot)$ *separates points* of Z if there exists a countable family $\pi = \{u_n \mid n \in \mathbb{N}\}$ of internal functions such that for all $x, y \in Z$, $x \neq y$, there is an element $u_n \in \pi$ satisfying

$$^\circ u_n(s_1) \neq {}^\circ u_n(s_2), s_1 \in \mu(x), s_2 \in \mu(y), s_i \in S_0, i = 1, 2. \tag{3.2.1}$$

Moreover, we have

$$^\circ\mathcal{E}_1(u_n, u_n) < \infty, \forall u_n \in \pi. \tag{3.2.2}$$

We call $\pi = \{u_n \mid n \in \mathbb{N}\}$ a *separating family* of Z by $\mathcal{E}(\cdot,\cdot)$.

We recall from Sect. 2.1 that if $\delta \in T$, the sub-line T_δ is defined by

$$T_\delta = \{k\delta \mid k \in {}^*\mathbb{N}_0\},$$

$X^{(\delta)}$ is the restriction of $X|T_\delta$. Let $\zeta = \zeta_\delta$ be the lifetime of $X^{(\delta)}$, i.e.,

$$\zeta_\delta(\omega) = \inf\{^\circ t \mid X^{(\delta)}(\omega, t) \in \overline{S}_0\},$$

where $\overline{S}_0 = S_0 \cap Ns({}^*Y)$. For any subset $A \subset S_0$, let

$$\tau_A^{(\delta)}(\omega) = \inf\{^\circ t \big| X(t) \in \text{st}^{-1}(A^\circ) \text{ and } t \in T_\delta\}.$$

---

[1] For the general concept of monad, we refer to Albeverio et al. [25].

Lemma 3.2.1. *Let Y be a regular Hausdorff space, and let A be a subset of S_0. Suppose that $\mathcal{E}(\cdot,\cdot)$ separates the points of $Y - A^\circ$ and π is the separating family. For $\omega \in \Omega$, if the path $X(\cdot,\omega)$ fails to have an S-left or S-right limit at $t < \zeta_{\Delta t}(\omega) \wedge \tau_A^{(\Delta t)}(\omega)$, then so does $u(X(\omega,\cdot))$ for some $u \in \pi$.*

Proof. Let ω be a fixed point in Ω. Fix $t < \zeta_{\Delta t}(\omega) \wedge \tau_A^{(\Delta t)}(\omega), t \in [0,\infty)$. Given a sequence $\{t_n \mid n \in \mathbb{N}\}$ from T such that the standard part $^\circ t_n$ increases strictly to t, we shall prove that the sequence

$$\left\{ {}^\circ X(\omega,t_n) \middle| n \in \mathbb{N} \right\}$$

has a cluster point in $Y - A^\circ$.

Suppose for a contradiction that

$$\left\{ {}^\circ X(\omega,t_n) \middle| n \in \mathbb{N} \right\}$$

has no cluster point in Y. Then for each $y \in Y$, there is a neighborhood O_y and an integer $n_y \in \mathbb{N}$ such that

$${}^\circ X(\omega,t_n) \notin O_y$$

when $n \geq n_y$. Since Y is regular, we can find a neighborhood G_y of y such that the closure $\overline{G}_y$ is contained in O_y. Hence, we have

$$X(\omega,t_n) \notin {}^*G_y$$

when $n \geq n_y$. Extend $\{t_n \mid n \in \mathbb{N}\}$ to be an internal sequence $\{t_n \mid n \in {}^*\mathbb{N}\}$ of elements of T less than t. Consider the set

$$A_y = \{n \in {}^*\mathbb{N} \mid n \leq n_y \quad \text{or} \quad X(\omega,t_n) \notin {}^*G_n\}.$$

Since A_y is internal and contains $\mathbb{N}$, there is an $\eta_y \in {}^*\mathbb{N} - \mathbb{N}$ such that all $\eta \leq \eta_y$ are elements of A_y. By saturation, there is an infinite η less than all η_y. But then, we have

$$X(\omega,t_\eta) \notin {}^*G_y$$

for all y. This implies that $X(\omega,t_\eta)$ is not nearstandard, contradicting our assumption that $t < \zeta_{\Delta t}(\omega)$.

Suppose that $y \in Y$ is a cluster point of $\{{}^\circ X(\omega,t_n)\}_{n\in\mathbb{N}}$. Since $t_n \uparrow t$, $t < \tau_A^{(\Delta t)}(\omega)$, it is easy to see that $y \notin A^\circ$.

Let $x \in Y - A^\circ \hat{=} Z$ be a cluster point of $\{{}^\circ X(\omega,t)\}$. If x is not the S-left limit of $X(\omega,\cdot)$ at t, there must be another sequence $\{s_n \mid n \in \mathbb{N}\}$ increasing to t such that x is not a cluster point of $\{{}^\circ X(\omega, s_n) \mid n \in \mathbb{N}\}$. Repeating the argument above, we see that $\{{}^\circ X(\omega, s_n) \mid n \in \mathbb{N}\}$ must have a cluster point $y \in Z$. Let $u \in \pi$ be the function satisfying the conditions of (3.2.1) and (3.2.2). Obviously, $u(X(\omega, \cdot))$ does not have an S-left limit at time t. This proves the S-left case of the lemma. The S-right limit case can be treated in a similar way. □

We shall use Lemma 3.2.1 and Fukushima's decomposition Theorem 2.7.2 to show that if $\mathcal{E}(\cdot, \cdot)$ separates points, then the associated Markov chain X satisfies Definition 3.1.3 (i). But we must get some preparation at first.

It is easy to see that

$$\{\omega \mid \exists t \in T_\delta^r(X(\omega, t) \in \bigcup_{n\in\mathbb{N}} A_n)\} = \bigcup_{n\in\mathbb{N}} \{\omega \mid \exists t \in T_\delta^r(X(\omega, t) \in A_n)\}, \tag{3.2.3}$$

for all sequences $\{A_n \mid n \in \mathbb{N}\}$ of subsets of S. However, the corresponding formula for intersections is false in general. In particular, it does hold if the sequence is decreasing and consists of internal sets. This is the observation behind the next lemma.

Lemma 3.2.2. *Let $A \subset S_0$, and assume that there is a family $\{B_{m,n} \mid m, n \in \mathbb{N}\}$ of internal sets such that*

$$A = \bigcup_{m\in\mathbb{N}} \bigcap_{n\in\mathbb{N}} B_{m,n}$$

and for each m, the sequence $\{B_{m,n} \mid n \in \mathbb{N}\}$ is decreasing. Then

$$\{\omega \mid \exists t \in T_\delta^r(X(\omega, t) \in A)\} = \bigcup_{m\in\mathbb{N}} \bigcap_{n\in\mathbb{N}} \left\{\omega \mid \exists t \in T_\delta^r\Big(X(\omega, t) \in B_{m,n}\Big)\right\}. \tag{3.2.4}$$

Proof. It suffices to prove that

$$\left\{\omega \middle| \exists t \in T_\delta^r \left(X(\omega, t) \in \bigcap_{n\in\mathbb{N}} B_{m,n}\right)\right\} = \bigcap_{n\in\mathbb{N}} \left\{\omega \middle| \exists t \in T_\delta^r\Big(X(\omega, t) \in B_{m,n}\Big)\right\} \tag{3.2.5}$$

for all m, since (3.2.4) then follows from (3.2.3). Also, it is immediately clear that the left hand of (3.2.5) is included in the right hand side. To prove the opposite inclusion, choose

$$\omega_0 \in \bigcap_{n\in\mathbb{N}} \{\omega \mid \exists t \in T_\delta^r(X(\omega,t) \in B_{m,n})\}.$$

Consider the set

$$\{n \in {}^*\mathbb{N} \mid \exists t \in T_\delta^r(X(\omega_0,t) \in \tilde{B}_{m,n})\},$$

where $\{\tilde{B}_{m,n} \mid n\in{}^*\mathbb{N}\}$ is some internal, decreasing extension of $\{B_{m,n} \mid n\in\mathbb{N}\}$. By the choice of ω_0, this set contains $\mathbb{N}$. Since it is internal, it must have an infinite member η. Thus, we have

$$\omega_0 \in \{\omega \mid \exists t \in T_\delta^r(X(\omega,t) \in \tilde{B}_{m,\eta})\} \subset \{\omega \mid \exists t \in T_\delta^r(X(\omega,t) \in \bigcap_{n\in\mathbb{N}} B_{m,n})\}.$$

The lemma is proved. □

We are now in the position to prove the following:

Proposition 3.2.1. *Let Y be a regular Hausdorff space, and let $\mathcal{E}^{(\delta)}(\cdot,\cdot)$ be a hyperfinite weak coercive quadratic form for every infinitesimal $\delta \in T$. Assume that A is an exceptional set and $\mathcal{E}(\cdot,\cdot)$ separates the points of $Y - A^\circ$. There exists an infinitesimal $\delta_0 \in T$ such that for all infinitesimal $\delta \in T_{\delta_0}$, there is a δ-exceptional set $A_0(\delta)$ satisfying that for all $s_i \in S_0 - A_0(\delta)$, the hyperfinite Markov chain $X^{(\delta)}(t)$ has S-left and S-right limits at all $t < \zeta_\delta, L(P_i)$-a.e.*

Proof. To find δ_0, we look at the property of the separating family $\pi = \{u_n \mid n \in \mathbb{N}\}$. From Theorem 1.4.2, we know that $\mathcal{D}(\mathcal{E}^{(\delta)}) = \mathcal{D}(\overline{\mathcal{E}}^{(\delta)})$ for all $\delta \in T, \delta \approx 0$. For each $u \in \pi$, we have ${}^\circ\mathcal{E}_1(u,u) < \infty$. Hence, we can find a $\delta_u \approx 0$ such that $u \in \mathcal{D}(\overline{\mathcal{E}}^{(\delta)}) = \mathcal{D}(\mathcal{E}^{(\delta)})$ for all infinitesimal $\delta \geq \delta_u$ by Corollary 1.2.4. By saturation, there is a $\delta_0 \approx 0$ larger than all δ_u and A is δ_0-exceptional.

Turning to $A_0(\delta)$ for infinitesimal $\delta \in T_{\delta_0}$, we first observe that by Lemma 3.2.1 and the countability of π, it suffices to show that for each $u_n \in \pi = \{u_n \mid n \in \mathbb{N}\}$, there is a δ-exceptional set $A_n(\delta)$ such that for all $s_i \in S_0 - A_n(\delta)$, the process $u_n(X^{(\delta)}(t))$ has S-left and S-right limits at all $t < \infty, L(P_i)$-a.e. Actually, we define

$$A_0(\delta) = A_0 \bigcup \left(\bigcup_{n=1}^{\infty} A_n(\delta) \right),$$

where A_0 is a properly δ-exceptional set containing A. Then, $A_0(\delta)$ is δ-exceptional by Lemma 2.1.1 and for all $s_i \in S_i - A_0(\delta), u_n \in \pi$, $u_n(X^{(\delta)}(t))$ has S-right and S-left limits for all $t < \infty, L(P_i)$-a.e. Therefore, the hyperfinite Markov chain $X^{(\delta)}(t)$ has S-left and S-right limits at all $t < \zeta_\delta(\omega)$,

$L(P_i)$-a.e. by Lemma 3.2.1. From this observation, we only need to find the exceptional set $A_n(\delta)$ for every $u_n \in \pi$. For the simplicity of notation, we write $u = u_n$ in the following.

Given two standard rationals $p, q, -\infty < p < q < \infty$, we define a sequence $\{\tau^n_{(p,q)}\}$ of stopping times as follows:

$$\begin{aligned}
\tau^0_{(p,q)} &= \min\{t \in T_\delta \mid u(X^{(\delta)})(\omega, t)) \leq p\}, \\
\tau^{2n}_{(p,q)} &= \min\{t \in T_\delta \mid t > \tau^{2n-1}_{(p,q)}(\omega) \wedge u(X^{(\delta)})(\omega, t)) \leq p\}, \\
\tau^{2n+1}_{(p,q)} &= \min\{t \in T_\delta \mid t > \tau^{2n}_{(p,q)}(\omega) \wedge u(X^{(\delta)})(\omega, t)) \geq q\}.
\end{aligned}$$

Let $B \subset S_0$ be defined by

$$B = \bigcup_{(p,q)} \bigcup_{m \in \mathbb{N}} \bigcup_{k \in \mathbb{N}} \bigcap_{n \in \mathbb{N}} \{i \mid P_i\{\tau^n_{(p,q)} \leq m\} \geq \frac{1}{k}\}.$$

If $X^{(\delta)}$ fails to have S-left or S-right limits with positive $L(P_i)$ probability, then we have $s_i \in B$. We must show that B is δ-exceptional.

By Lemma 3.2.2, we have for all $r \in [0, \infty)$

$$\begin{aligned}
&\{\omega \mid \exists t \in T^r_\delta (X(\omega, t) \in B)\} \\
&\quad = \bigcup_{(p,q)} \bigcup_{m \in \mathbb{N}} \bigcup_{k \in \mathbb{N}} \bigcap_{n \in \mathbb{N}} \left\{\omega \middle| \exists t \in T^r_\delta \left(P_{X^{(\delta)}(\omega,t)}\{\tau^n_{(p,q)} \leq m\} \geq \frac{1}{k}\right)\right\}.
\end{aligned}$$

If B is not δ-exceptional, there must be a pair (p, q) of rationals, as well as integers $m, k \in \mathbb{N}$, $r \in [0, \infty)$ and an infinite number $\eta \in {}^*\mathbb{N}$ such that

$$L(P) \left\{\omega \middle| \exists t \in T^r_\delta \left(P_{X^{(\delta)}(\omega,t)}\{\tau^\eta_{(p,q)} \leq m\} \geq \frac{1}{k}\right)\right\} > 0.$$

This implies that with positive probability, $u(X^{(\delta)})$ jumps back and forth between p and q more than η times before time $m + r$.

Since $u \in \mathcal{D}(\mathcal{E}^{(\delta)})$, Fukushima's decomposition Theorem 2.7.2 tells us that

$$u(X^{(\delta)}) = N^{[u]} + M^{[u]},$$

where $N^{[u]}$ is S-continuous $L(P)$-a.e. and $M^{[u]}$ is a λ^2-martingale. If $u(X^{(\delta)})$ jumps η times between p and q before $t = m+r$, there must be an infinitesimal interval where it jumps back and forth infinitely many times. Since N is S-continuous – and hence almost constant on infinitesimal intervals – most of this jumping is done by M. Hence the quadratic variation of M is infinite on a set of positive measure, contradicting the fact that it is a λ^2-martingale.

We then conclude that B must be δ-exceptional, and the proposition is thus proved. □

3.2.2 Nearstandardly Concentrated Forms

Now we are in a position to find a condition on $\mathcal{E}(\cdot,\cdot)$ that implies a similar property like (ii) in Definition 3.1.3. For any internal subset C of S_0, we set

$$H_C = \{u \in H \mid u(s_i) = 0 \text{ for all } s_i \in S_0 - C\}.$$

For each $u \in H$ and any internal set $C \subset S_0, \delta \in T$, we define

$$u_C^{(\delta)}(i) = u(i) - E_i\left[(1+\delta)^{-\sigma_{S_0-C}^{(\delta)}/\delta} u\Big(X^{(\delta)}(\sigma_{S_0-C}^{(\delta)})\Big)\right], \tag{3.2.6}$$

where

$$\sigma_{S_0-C}^{(\delta)}(\omega) = \min\Big\{t \in T_\delta \mid X^{(\delta)}(\omega,t) \in S_0 - C\Big\}.$$

It is easy to see that

$$u_C^{(\delta)} \in H_C.$$

Definition 3.2.2. (1) For $\delta \in T$, a hyperfinite quadratic form $\mathcal{E}(\cdot,\cdot)$ is said to be *δ-nearstandardly concentrated* if there exists an increasing sequence $\{B_n \mid n \in \mathbb{N}\}$ of internal subsets of S_0 such that:

(i) The set $\cup_{n\in\mathbb{N}} B_n - \overline{S}_0$ is δ-exceptional.
(ii) For each $F \in H$, if ${}^\circ\mathcal{E}_1^{(\delta)}(F,F) < \infty$, then we have

$${}^\circ\mathcal{E}_1^{(\delta)}(F - F_{B_m}^{(\delta)}, F - F_{B_m}^{(\delta)}) \longrightarrow 0, m \longrightarrow \infty.$$

(2) A hyperfinite quadratic form $\mathcal{E}(\cdot,\cdot)$ is said to be *nearstandardly concentrated* if it is δ-nearstandardly concentrated for some infinitesimal $\delta \in T$.

Notice that above definition of nearstandardly concentrated is different from that of Albeverio et al. [25], 5.5.4 Definition. Our definition is motivated by both Albeverio et al. [25], 5.5.4 Definition and Condition (I) of Theorem 4.1.1 in Chap. 4.

Lemma 3.2.3. *For $\delta \in T$, let $\mathcal{E}^{(\delta)}(\cdot,\cdot)$ be a hyperfinite weak coercive quadratic form with continuity constant C. Assume that $\mathcal{E}(\cdot,\cdot)$ is δ-nearstandardly concentrated. Let X be the related hyperfinite Markov chain. Then for each $F \in H$ with ${}^\circ\mathcal{E}_1^{(\delta)}(F,F) < \infty$, the set*

$$A(F) = \left\{ s_i \in S_0 - \bigcup_{m\in\mathbb{N}} B_n \middle| L(P_i)\left\{\omega \middle| \exists t \in T_\delta^{\text{fin}}\left(F(X^{(\delta)}(t)) \not\approx 0\right)\right\} > 0\right\} \tag{3.2.7}$$

is δ-exceptional.

Proof. Step 1. For $\varepsilon > 0$, set

$$\begin{aligned} D &= \left\{ s_i \middle| |F(s_i)| > \varepsilon \right\}, \\ \sigma_D^{(\delta)}(\omega) &= \min\left\{ t \in T_\delta \mid X^{(\delta)}(\omega,t) \in D \right\}, \\ e_1^{(\delta)}(D) &= E_i\left[(1+\delta)^{-\sigma_D^{(\delta)}(\omega)/\delta}\right]. \end{aligned}$$

From Corollary 2.2.3, we have

$$\mathcal{E}_1^{(\delta)}\left(e_1^{(\delta)}(D), e_1^{(\delta)}(D)\right) \le \frac{C^2}{\varepsilon^2}\mathcal{E}_1^{(\delta)}(F,F).$$

Hence, we have

$${}^\circ\mathcal{E}_1^{(\delta)}\left(e_1^{(\delta)}(D), e_1^{(\delta)}(D)\right) < \infty.$$

This implies that

$${}^\circ\mathcal{E}_1^{(\delta)}\left(e_1^{(\delta)}(D) - (e_1^{(\delta)}(D))_{B_m}^{(\delta)}, e_1^{(\delta)}(D) - (e_1^{(\delta)}(D))_{B_m}^{(\delta)}\right) \longrightarrow 0, m \longrightarrow 0, \tag{3.2.8}$$

where $(e_1^{(\delta)}(D))_{B_m}^{(\delta)}$ is defined in the same way as the definition (3.2.6).

Step 2. From the approximation (3.2.8), there exists an increasing sequence $\{m_k\}_{k\in\mathbb{N}}$ satisfying

$$\begin{aligned} \delta_k &= \sqrt{\mathcal{E}_1^{(\delta)}\left(e_1^{(\delta)}(D) - (e_1^{(\delta)}(D))_{B_{m_k}}^{(\delta)}, e_1^{(\delta)}(D) - (e_1^{(\delta)}(D))_{B_{m_k}}^{(\delta)}\right)} \\ &< 2^{-k}. \end{aligned}$$

Set

$$A_k(F,\varepsilon) = \left\{ s_i \in S_0 - B_{m_k} \middle| |e_1^{(\delta)}(D)(s_i)| \geq \sqrt{\delta_k} \right\}, k \in \mathbb{N}.$$

Note that if $s_i \in A_k(F,\varepsilon)$, we have

$$|e_1^{(\delta)}(D)(s_i) - (e_1^{(\delta)}(D))_{B_{m_k}}^{(\delta)}(s_i)| \geq \sqrt{\delta_k}.$$

Hence, we have from Lemma 2.7.5 the following

$$\begin{aligned}
&P\left(\omega \middle| \exists t \in T_\delta^1 \left(X^{(\delta)}(\omega,t) \in A_k(F,\varepsilon)\right)\right) \\
&\leq P\left[\omega \middle| \exists t \in T_\delta^1 \left(|e_1^{(\delta)}(D)(X^{(\delta)}(\omega,t)) - (e_1^{(\delta)}(D))_{B_{m_k}}^{(\delta)}(X^{(\delta)}(\omega,t))| \geq \sqrt{\delta_k}\right)\right] \\
&\leq \frac{2C^2(1+\delta)^{\frac{1}{\delta}}}{\delta_k} \mathcal{E}_1^{(\delta)}\left(e_1^{(\delta)}(D) - (e_1^{(\delta)}(D))_{B_{m_k}}^{(\delta)}, e_1^{(\delta)}(D) - (e_1^{(\delta)}(D))_{B_{m_k}}^{(\delta)}\right) \\
&= 2C^2(1+\delta)^{\frac{1}{\delta}}\delta_k \\
&\leq C^2(1+\delta)^{\frac{1}{\delta}} 2^{-k+1}. \qquad (3.2.9)
\end{aligned}$$

For each $K \in \mathbb{N}$, we have

$$\begin{aligned}
A(F,\varepsilon) &\hat{=} \left\{ s_i \in S_0 - \bigcup_{m\in\mathbb{N}} B_n \middle| L(P_i)\left\{\exists t \in T_\delta^{\text{fin}}\left(|F(X^{(\delta)}(t))| > \varepsilon\right)\right\} > 0 \right\} \\
&\subset \bigcup_{k=K}^{\infty} A_k(F,\varepsilon).
\end{aligned}$$

Since $\delta_k < 2^{-k}$, it follows from the relation (3.2.9) that

$$\begin{aligned}
P\{\omega \mid \exists t \in T_\delta^1(X(\omega,t) \in A(F,\varepsilon))\} &\leq 2C^2(1+\delta)^{\frac{1}{\delta}} \sum_{k=K}^{\infty} \delta_k \\
&\leq C^2(1+\delta)^{\frac{1}{\delta}} 2^{-K+2}.
\end{aligned}$$

Hence, $A(F,\varepsilon)$ is δ-exceptional.

Step 3. Now it is easy to see that $A(F)$ is δ-exceptional, since $A(F) = \bigcup_{m\in\mathbb{N}} A(F,\frac{1}{m})$. □

Proposition 3.2.2. *For $\delta \in T$, let $\mathcal{E}^{(\delta)}(\cdot,\cdot)$ be a hyperfinite weak coercive quadratic form with continuity constant C. Assume that $\mathcal{E}(\cdot,\cdot)$ is δ-nearstandardly concentrated. Let X be the related hyperfinite Markov chain. Then for each $F \in H$ with ${}^\circ\mathcal{E}_1^{(\delta)}(F,F) < \infty$, there exists a δ-exceptional set*

$A_1(F, \delta)$ *such that for* $s_i \in S_0 - A_1(F, \delta)$

$$L(P_i)\left\{\omega \middle| \exists t \in T_\delta^{\text{fin}} \left({}^\circ t \geq \zeta_\delta(\omega) \wedge \left(F(X^{(\delta)}(t)) \not\approx 0\right)\right)\right\} = 0.$$

Proof. Let $\{B_n \mid n \in {}^*\mathbb{N}\}$ be an increasing extension of $\{B_n \mid n \in \mathbb{N}\}$, which is the increasing sequence in Definition 3.2.2 (1). Set $B = \cup_{n \in \mathbb{N}} B_n$. Define

$$\tau_n^{(\delta)}(\omega) = \min\left\{t \in T_\delta \middle| X^{(\delta)}(\omega, t) \notin B_n\right\}, n \in {}^*\mathbb{N},$$

$$\tau^{(\delta)}(\omega) = \inf\left\{{}^\circ t \middle| X^{(\delta)}(\omega, t) \notin B\right\}.$$

It is not hard to check that

$$\tau^{(\delta)}(\omega) = \sup\left\{{}^\circ\tau_n(\omega) \middle| n \in \mathbb{N}\right\}.$$

Let $A_1(F, \delta)$ be a properly δ-exceptional set containing $B - \overline{S}_0$ and the set $A(F)$, where $A(F)$ is defined in Lemma 3.2.3 by the definition (3.2.7). Given $s_i \in S_0 - A_1(F, \delta)$, we can find an $\eta \in {}^*\mathbb{N} - \mathbb{N}$ such that

$$\tau^{(\delta)}(\omega) = {}^\circ\tau_\eta^{(\delta)}(\omega), L(P_i)\text{-a.e.} \tag{3.2.10}$$

By the definition of $A_1(F, \delta)$, we have $B - \overline{S}_0 \subset A_1(F, \delta)$. Since $A_1(F, \delta)$ is properly δ-exceptional and $s_i \notin A_1(F, \delta)$, this implies that

$$\tau(\omega) = \zeta_\delta(\omega), L(P_i)\text{-a.e.} \tag{3.2.11}$$

Combining (3.2.10) and (3.2.11), we have

$${}^\circ\tau_\eta^{(\delta)}(\omega) = \zeta_\delta(\omega), L(P_i)\text{-a.e.} \tag{3.2.12}$$

Notice that $s_i \notin A_1(F, \delta)$, $A_1(F, \delta)$ contains $A(F)$, and it is properly δ-exceptional. We have

$$L(P_i)(X(\omega, \tau_\eta^{(\delta)}(\omega)) \in A(F)) = 0.$$

Hence, we have by the definition of $A(F)$

$$L(P_i)\left\{\omega \middle| \exists t \in T_\delta^{\text{fin}} \left(t \geq \tau_\eta^{(\delta)}(\omega) \wedge \left(F(X^{(\delta)}(t)) \not\approx 0\right)\right)\right\} = 0.$$

The proposition follows from the relation (3.2.12). □

3.2.3 Quasi-Continuity

Definition 3.2.3. (1) An internal function $F : S_0 \longrightarrow {}^*\mathbb{R}$ is said to be *hyperfinite δ-quasi-continuous* if there is a hyperfinite δ-exceptional set $A \subset S_0$ such that for all infinitely close $s_i, s_j \in \overline{S}_0 - A$, we have $F(s_i) \approx F(s_j)$.

(2) An internal function $F : S_0 \longrightarrow {}^*\mathbb{R}$ is called *hyperfinite quasi-continuous* if there is an infinitesimal δ such that F is hyperfinite δ-quasi-continuous.

Remark 3.2.1. We remark that usually one uses zero capacity sets to define the quasi-continuity in regular Dirichlet spaces (refer to Sect. 4.1 of Chap. 4, or refer to [175]). Nevertheless, we utilize the exceptional sets in the definition of the hyperfinite quasi-continuity. The reason is that we have the equivalence between the concept of exceptional sets and that of zero capacity sets in regular Dirichlet space theory, e.g., Theorem 4.3.1 in [175]). However, we only have Theorem 2.4.1, Theorem 2.4.2, and Theorem 2.4.3 in Sect. 2.4, from which we understand that the conception of exceptional sets is not equivalent to that of zero capacity sets in hyperfinite Dirichlet space theory.

Given a hyperfinite δ-quasi-continuous and S-bounded function $F \in H$, let A be a δ-exceptional set satisfying $F(s_i) \approx F(s_j)$, $s_i \approx s_j$, $s_i, s_j \in \overline{S}_0 - A$. Define ${}^\circ F = f : Y \longrightarrow \mathbb{R}$ by:

(i) $f(y) = {}^\circ F(s_i)$ whenever $y \notin A^\circ$ and $s_i \in \mathrm{st}^{-1}(y) - A$;
(ii) $f(y) = 0$ whenever $y \in A^\circ$.

In order to get the condition that makes the process X to satisfy Definition 3.1.3 (iii), we impose the following condition on $\mathcal{E}(\cdot,\cdot)$.

Definition 3.2.4. (1) A hyperfinite quadratic form $\mathcal{E}(\cdot,\cdot)$ is said to have a *hyperfinite δ-quasi-continuous core* if there are a family of countable functions $\pi = \{F_n \mid n \in \mathbb{N}\} \subset H$ and a δ-exceptional set A such that $F_n(s_i) \approx F_n(s_j)$, $s_i \approx s_j$, $s_i, s_j \in \overline{S}_0 - A, \forall n \in \mathbb{N}$, and

$$(Y - A^\circ) \cap \mathcal{B}(Y) \subset \sigma\{f_n = {}^\circ F_n \mid n \in \mathbb{N}\}. \tag{3.2.13}$$

Moreover, $\mathrm{st}\{\max_{s_i \in S_0} |F_n(s_i)|\} < \infty, {}^\circ\mathcal{E}_1^{(\delta)}(F_n, F_n) < \infty, \forall n \in \mathbb{N}$. π is called a hyperfinite δ-quasi-continuous core of $\mathcal{E}(\cdot,\cdot)$.

(2) A hyperfinite quadratic form $\mathcal{E}(\cdot,\cdot)$ is said to have a *hyperfinite quasi-continuous core* if it has a hyperfinite δ-quasi-continuous core for some infinitesimal $\delta \in T$.

Lemma 3.2.4. *Let M be a set. Consider a countable subset G of M and a countable collection Λ of maps from $M \times M$ into M. Then, there exists a countable set L such that*

$$(a) \quad G \subset L \subset M. \quad (b) \quad \lambda(L \times L) \subset L, \forall \lambda \in \Lambda.$$

Proof. This is proven in Fukushima [175] Lemma 6.1.1. □

Lemma 3.2.5. *Suppose that $\mathcal{E}(\cdot,\cdot)$ has a hyperfinite δ-quasi-continuous core $\pi = \{F_n \mid n \in \mathbb{N}\}$. Then, there exists a countable set $\hat{\pi} = \{\hat{F}_n \mid n \in \mathbb{N}\}$ such that*

(1) $\hat{F}_n, \hat{F}_m \in \hat{\pi}, a \in Q \Longrightarrow |\hat{F}_n| \in \hat{\pi}, \hat{F}_n + \hat{F}_m \in \hat{\pi}, \hat{F}_n\hat{F}_m \in \hat{\pi}$, $a\hat{F}_n \in \hat{\pi}$, and $\hat{F}_n \wedge a \in \hat{\pi}$,
(2) $\pi = \{F_n \mid n \in \mathbb{N}\} \subset \hat{\pi} = \{\hat{F}_n \mid n \in \mathbb{N}\}$,
(3) every element in $\hat{\pi}$ is hyperfinite δ-quasi-continuous,

where $Q = \{a_1, a_2, \cdots, a_n, \cdots\}$ is the set of all rational numbers.

Proof. Let us denote by $Q_\delta(S_0)$ the set of hyperfinite δ-quasi-continuous functions on S_0. We define the maps

$$\Lambda = \{\lambda_{-2}, \lambda_{-1}, \lambda_0, \lambda_1, \lambda_2, \cdots\}$$

from $[H \cap Q_\delta(S_0)] \times [H \cap Q_\delta(S_0)]$ into $H \cap Q_\delta(S_0)$ by

$$\begin{aligned} \lambda_{-2}(u,v) &= |u|, \\ \lambda_{-1}(u,v) &= u+v, \\ \lambda_0(u,v) &= uv, \\ \lambda_{2i-1}(u,v) &= a_i u, \\ \lambda_{2i}(u,v) &= u \wedge a_i, i = 1, 2, \cdots \end{aligned}$$

Applying Lemma 3.2.4 to π and Λ, we get a countable set $\hat{\pi}$ satisfying (1), (2), and (3). □

Lemma 3.2.6. *Assume that the hyperfinite quadratic form $\mathcal{E}(\cdot,\cdot)$ has a δ-quasi-continuous core $\pi = \{F_n \mid n \in \mathbb{N}\}$. For any two finite measures ν_1 and ν_2 on $(Y, \mathcal{B}(Y))$, if ν_1 and ν_2 satisfy the following condition:*

$$\int_Y {}^\circ F(y)\, \nu_1(dy) = \int_Y {}^\circ F(y)\, \nu_2(dy) \quad \textit{for all} \quad F \in \hat{\pi}, \tag{3.2.14}$$

where $\hat{\pi}$ is the countable set obtained in the Lemma 3.2.5, then ν_1 and ν_2 coincide on $(Y, \sigma({}^\circ F \mid {}^\circ F \in \hat{\pi}))$.

Proof. For any $F \in \hat{\pi}, a \in Q$, $F_n = [n(F - F \wedge a)] \wedge 1 \in \hat{\pi}$ and ${}^\circ F_n \uparrow 1_{({}^\circ F > a)}$. This implies that

$$\int_Y 1_{({}^\circ F(y) > a)}\, \nu_1(dy) = \int_Y 1_{({}^\circ F(y) > a)}\, \nu_2(dy)$$

from the conditions (3.2.14). Now we can prove our Lemma by using the monotone class theorem. □

Lemma 3.2.7. *If the hyperfinite quadratic form $\mathcal{E}(\cdot,\cdot)$ has a hyperfinite δ-quasi-continuous core $\pi = \{F_n \mid n \in \mathbb{N}\}$, then π separates the points of $Y - A^\circ$ by $\mathcal{E}^{(\delta)}(\cdot,\cdot)$, where A is the δ-exceptional set in the Definition 3.2.4 (1).*

Proof. Since Y is a Hausdorff space, the singleton $\{x\}$ is a closed subset of Y. This implies that $\{x\} \in \mathcal{B}(Y)$ for any $x \in Y$. For $x, y \in Y - A^\circ$, let us assume that $F_n(s_1) \approx F_n(s_2)$ for every $F_n \in \pi, s_1 \in \mu(x), s_2 \in \mu(y)$. Then we obtain

$$\{x,y\} \subset \cap \left\{ ({}^\circ F_n)^{-1}(({}^\circ F_n)(x)) \mid F_n \in \pi \right\}.$$

It follows from the condition (3.2.13) and above observation that

$$\cap \left\{ ({}^\circ F_n)^{-1}(({}^\circ F_n)(x)) \mid F_n \in \pi \right\} = \{x\}.$$

Therefore, we get $x = y$. This shows that π separates the points of $Y - A^\circ$ by $\mathcal{E}^{(\delta)}(\cdot,\cdot)$. □

Lemma 3.2.8. *For $\delta \in T$, let $\mathcal{E}^{(\delta)}(\cdot,\cdot)$ be a hyperfinite weak coercive quadratic form with continuity constant C. Suppose that $\mathcal{E}(\cdot,\cdot)$ has a hyperfinite δ-quasi-continuous core $\pi = \{F_n \mid n \in \mathbb{N}\}$ and is δ-nearstandardly concentrated. For all $\eta \in {}^*\mathbb{N} - \mathbb{N}$, $s_i \notin A_2(\delta)$, $\hat{F}_n \in \hat{\pi}$, $t \in T_\delta^{\mathrm{fin}}$, we have*

$$\int_\Omega \hat{F}_n(X^{(\delta)}(\omega,t))\ P_i(d\omega) \approx \int_\Omega \hat{F}_n(X^{(\delta)}(\omega,t)) 1_{(t<\tau_\eta^{(\delta)}(\omega))}\ P_i(d\omega), \tag{3.2.15}$$

where $A_2(\delta)$ is any properly δ-exceptional subset of S_0 containing $B - \overline{S}_0$ and the set $\cup_{n\in\mathbb{N}} A(\hat{F}_n)$. Here, $A(\hat{F}_n)$ is the set defined in Lemma 3.2.3 by (3.2.7).

Proof. Notice that $s_i \notin A_2(\delta)$, and $A_2(\delta)$ contains $\cup_{n\in\mathbb{N}} A(\hat{F}_n)$ and is properly δ-exceptional. We have

$$L(P_i)(X(\omega, \tau_\eta^{(\delta)}(\omega)) \in A(\hat{F}_n)) = 0.$$

Hence, we have by the definition of $A(\hat{F}_n)$ that

$$L(P_i)\left\{ \omega \middle| \exists t \in T_\delta^{\mathrm{fin}} \left(t \geq \tau_\eta^{(\delta)}(\omega) \wedge \left(\hat{F}_n(X^{(\delta)}(t)) \not\approx 0 \right) \right) \right\} = 0.$$

Therefore, we get

$$\int_{\Omega} \hat{F}_n(X^{(\delta)}(\omega,t))1_{(t\geq\tau_\eta^{(\delta)}(\omega))}\ P_i(d\omega) \approx 0, s_i \notin A_2(\delta).$$

Hence, the approximation (3.2.15) holds. □

Definition 3.2.5. (1) We say that a hyperfinite quadratic form $\mathcal{E}(\cdot,\cdot)$ generates a *δ-quasi-continuous semigroup* if for all δ-quasi-continuous internal functions $f : S_0 \longrightarrow {}^*\mathbb{R}$, $T_t f$ is δ-quasi-continuous for all $t \in T_\delta^{\text{fin}}$.

(2) A hyperfinite quadratic form $\mathcal{E}(\cdot,\cdot)$ generates a *quasi-continuous semigroup* if it generates a δ-quasi-continuous semigroup for some infinitesimal $\delta \in T$.

Proposition 3.2.3. *Let Y be a regular Hausdorff space, and let $\mathcal{E}^{(\delta)}(\cdot,\cdot)$ be a hyperfinite weak coercive quadratic form for every infinitesimal $\delta\in T$. Assume that $\mathcal{E}(\cdot,\cdot)$ has a hyperfinite $\hat{\delta}$-quasi-continuous core and is $\hat{\delta}$-nearstandardly concentrated for some $\hat{\delta} \in T, \hat{\delta} \approx 0$. Moreover, suppose that there is an infinitesimal $\delta' \in T$ such that $\mathcal{E}(\cdot,\cdot)$ generates a δ-quasi-continuous semigroup for every infinitesimal $\delta \in T, \delta \geq \delta'$. Then, there is an infinitesimal $\delta_1 \in T$ and a δ_1-exceptional set $A_1(\delta_1)$ such that*

(i) For all $s_i \in S_0 - A_1(\delta_1)$ and $L(P_i)$-a.e.ω, the path $X^{(\delta_1)}(\omega,\cdot)$ has S-right and S-left limits at all $t < \zeta_{\delta_1}(\omega)$.
(ii) For all infinitely close $s_i, s_j \in \overline{S}_0 - A_1(\delta_1)$, we have

$$L(P_i)\{{}^\circ X^{(\delta_1)+}(\omega,t) \in B\} = L(P_j)\{{}^\circ X^{(\delta_1)+}(\omega,t) \in B\} \qquad (3.2.16)$$

for all finite $t \in [0,\infty)$ and all Borel sets B.
(iii) For all $s_i \in S_0 - A_1(\delta_1)$, we have

$$L(P_i)\left\{\omega \middle| \exists t \in T_{\delta_1}^{\text{fin}} \left({}^\circ t \geq \zeta_{\delta_1}(\omega) \wedge \left(\hat{F}_n(X^{(\delta_1)}(t)) \not\approx 0\right)\right)\right\} = 0,$$

for all $\hat{F}_n \in \hat{\pi}$, where $\hat{\pi} = \{\hat{F}_n \mid n \in \mathbb{N}\} \supset \pi = \{F_n \mid n \in \mathbb{N}\}$ is the countable set obtained in Lemma 3.2.5.

Proof. Step 1. Since $\mathcal{E}(\cdot,\cdot)$ has a hyperfinite $\hat{\delta}$-quasi-continuous core, $\hat{\delta}\in T, \hat{\delta} \approx 0$, there are a family of countable functions $\pi = \{F_n \mid n \in \mathbb{N}\} \subset H$ and a $\hat{\delta}$-exceptional set A such that $F_n(s_i) \approx F_n(s_j)$, $s_i \approx s_j$, $s_i, s_j \in \overline{S}_0 - A, \forall n \in \mathbb{N}$, and

$$(Y - A^\circ) \cap \mathcal{B}(Y) \subset \sigma\{f_n = {}^\circ F_n \mid n \in \mathbb{N}\}.$$

Moreover, we have

$$\mathrm{st}\{\max_{s_i \in S_0} |F_n(s_i)|\} < \infty,$$
$${}^\circ\mathcal{E}_1^{(\hat{\delta})}(F_n, F_n) < \infty, \forall n \in \mathbb{N}.$$

By Lemma 3.2.7, π separates the points of $Y - A^\circ$ by $\mathcal{E}_1^{(\hat{\delta})}(\cdot,\cdot)$. From Proposition 3.2.1, there exists an infinitesimal $\delta_0 \in T$ such that for all infinitesimal $\delta \in T_{\delta_0}$, there is a δ-exceptional set $A_0(\delta)$ such that for all $s_i \in S_0 - A_0(\delta)$, the hyperfinite Markov chain $X^{(\delta)}(t)$ has S-left and S-right limits at all $t < \zeta_\delta, L(P_i)$-a.e.

Step 2. For simplicity, we may suppose $\delta_0 \geq \delta'$ and $\delta_0 \in T_{\hat{\delta}}$. Now let us take $\delta_1 = \delta_0$. Let $\hat{\pi} = \{\hat{F}_n \mid n \in \mathbb{N}\} \supset \pi = \{F_n \mid n \in \mathbb{N}\}$ be the countable set obtained in Lemma 3.2.5. Pick up $m_0 \in {}^*\mathbb{N} - \mathbb{N}$ such that $\frac{1}{2} \leq 2^{m_0}\delta_1 \leq 1$. Set

$$\delta(1) = 2^{m_0}\delta_1, \delta(N+1)$$
$$= \frac{1}{2}\delta(N), N \geq 2, N \in \mathbb{N}.$$

Define for $N \in \mathbb{N}$

$$\Lambda(N) = \bigcup_{n=1}^{N} \{T_t\hat{F}_n \mid t \in T_{\delta(N)}^N\}.$$

Then, $\{\Lambda(N) \mid N \in \mathbb{N}\}$ is an increasing sequence.

For each $N \in \mathbb{N}, t \in T_{\delta(N)}^N, T_t\hat{F}_n \in \Lambda(N)$ is hyperfinite δ_1-quasi-continuous. Let us fix $N \in \mathbb{N}$. Then, there exists a sequence of internal sets $\{B_{m,N} \mid m \in \mathbb{N}\}$ decreasing as a function of m such that

$$\int_\Omega \hat{F}_n(X_t(\omega))\, dP_i(\omega) \approx \int_\Omega \hat{F}_n(X_t(\omega))\, dP_j(\omega),$$

for all $s_i, s_j \in \overline{S}_0 - \bigcap_{m=1}^\infty B_{m,N}, s_i \approx s_j, t \in T_{\delta(N)}^N, n = 1, 2, \cdots, N$. Moreover, $\bigcap_{m=1}^\infty B_{m,N}$ is δ_1-exceptional.

Step 3. Set

$$A_1(\delta_1) = A_1(\delta_0)$$
$$= A_0(\delta_0) \bigcup A_2(\hat{\delta}) \bigcup \left(\bigcup_{N=1}^\infty \bigcap_{m=1}^\infty B_{m,N} \right),$$

where $A_2(\hat{\delta})$ is obtained in Lemma 3.2.8 for $\delta = \hat{\delta}$. Then, $A_1(\delta_1)$ is δ_1-exceptional. Moreover, we have

$$\int_\Omega \hat{F}_n(X_t(\omega))\, dP_i(\omega) \approx \int_\Omega \hat{F}_n(X_t(\omega))\, dP_j(\omega), \qquad (3.2.17)$$

for $s_i, s_j \in \overline{S}_0 - A_1(\delta_1), s_i \approx s_j, t \in \bigcup_{N=1}^{\infty} T^N_{\delta(N)}, n \in \mathbb{N}$. Recalling the result obtained in step 1 and Lemma 3.2.8, we have from (3.2.17) that for all $\hat{F}_n \in \hat{\pi}, t \in [0, \infty), s_i, s_j \in \overline{S}_0 - A_1(\delta_1)$, $s_i \approx s_j$,

$$\int_\Omega {}^\circ\hat{F}_n({}^\circ X(t)^{(\delta_1)+})\, dL(P_i) = \int_\Omega {}^\circ\hat{F}_n({}^\circ X^{(\delta_1)+}(t))\, dL(P_j). \quad (3.2.18)$$

We deduce from the relation (3.2.18) and Lemma 3.2.6 that for all $s_i, s_j \in \overline{S}_0 - A_1(\delta_1)$, $s_i \approx s_j$ and $t \in [0, \infty)$, the result (3.2.16) holds. Hence, we have completed the proof of Proposition 3.2.3. □

3.2.4 Construction of Strong Markov Processes

Let Y be a regular Hausdorff space, and let $\mathcal{E}^{(\delta)}(\cdot,\cdot)$ be a hyperfinite weak coercive quadratic form for every infinitesimal $\delta \in T$. Assume that $\mathcal{E}(\cdot,\cdot)$ has a hyperfinite $\hat{\delta}$-quasi-continuous core and is $\hat{\delta}$-nearstandardly concentrated for some $\hat{\delta} \in T, \hat{\delta} \approx 0$. Moreover, suppose that there is an infinitesimal $\delta' \in T$ such that $\mathcal{E}(\cdot,\cdot)$ generates a δ-quasi-continuous semigroup for every infinitesimal $\delta \in T, \delta \geq \delta'$. Let $\delta_1 \in T$ be the infinitesimal and let $A_1(\delta_1)$ be the δ_1-exceptional set satisfying the results (i), (ii) and (iii) of Proposition 3.2.3.

Let $x : \Omega \times \mathbb{R}_+ \longrightarrow Y_\Delta$ be defined by

(i) if $X^{(\delta)}(\omega, 0) \notin A_1(\delta_1)$, then we have

$$x(\omega, t) = {}^\circ X^{(\delta_1)+}(\omega, t).$$

(ii) if $X^{(\delta)}(\omega, 0) \in A_1(\delta_1)$, then we have for all $t \in \mathbb{R}_+$

$$x(\omega, t) = \mathrm{st}(X^{(\delta)}(\omega, 0)).$$

In the same way as in Sect. 3.1, we denote by $A_1(\delta_1)^\circ$ the *inner standard part* of $A_1(\delta_1)$, i.e.,

$$A_1(\delta_1)^\circ = \left\{y \in Y \mid \mathrm{st}^{-1}(y) \cap S_0 \subset A_1(\delta_1)\right\}.$$

Lemma 3.2.9. *For all infinitely close $s_i, s_j \in \overline{S}_0 - A_1(\delta_1)$, all finite sequences $t_1 < t_2 < \cdots < t_n$ from $\mathbb{R}_+$, all elements $u_1, u_2, \cdots, u_n \in$*

$\hat{\pi} = \{\hat{F}_n \mid n \in \mathbb{N}\}$ *and all Borel sets* $B_1, B_2, \cdots, B_n$, *we have*

$$\int_\Omega \prod_{l=1}^{n} {}^\circ u_l({}^\circ X^{(\delta_1)+}(t_l))\ L(P_i)(d\omega) = \int_\Omega \prod_{l=1}^{n} {}^\circ u_l({}^\circ X^{(\delta_1)+}(t_l))\ L(P_j)(d\omega) \tag{3.2.19}$$

and

$$L(P_i)\left(\bigcap_{l=1}^{n} \{x(t_l) \in B_l\}\right) = L(P_j)\left(\bigcap_{l=1}^{n} \{x(t_l) \in B_l\}\right). \tag{3.2.20}$$

Proof. The case of $n = 1$ is Proposition 3.2.3 (ii) and the relation (3.2.18). If $n = 2$, let us pick $\hat{t}_1, \hat{t}_2 \in T_{\delta_1}^{\text{fin}}$ such that $\hat{t}_1 \approx t_1, \hat{t}_2 \approx t_2, \hat{t}_1 < \hat{t}_2$ and

$$\begin{aligned} &{}^\circ\!\!\int_\Omega u_1(X^{(\delta_1)}(\hat{t}_1))u_2(X^{(\delta_0)}(\hat{t}_2))\ L(P_k)(d\omega) \\ &\quad = \int_\Omega {}^\circ u_1({}^\circ X^{(\delta_1)+}(t_1)){}^\circ u_2({}^\circ X^{(\delta_1)+}(t_2))\ L(P_k)(d\omega), k = i, j. \end{aligned}$$

We have from the Markov property

$$\begin{aligned} &\int_\Omega u_1(X^{(\delta_1)}(\hat{t}_1))u_2(X^{(\delta_1)}(\hat{t}_2))\ P_k(d\omega) \\ &\quad = \int_\Omega u_1(X^{(\delta_1)}(\hat{t}_1))E_{X^{(\delta_1)}(\hat{t}_1)}u_2(X^{(\delta_1)}(\hat{t}_2 - \hat{t}_1))\ P_k(d\omega), k = i, j. \end{aligned}$$

Define a function $f : \Omega \longrightarrow \mathbb{R}$ as follows:

$$f(\omega) = {}^\circ u_1({}^\circ X^{(\delta_1)}(\omega, t_1)){}^\circ u_2({}^\circ X^{(\delta_1)}(\omega, t_2))1_{(t_2 < \zeta_{\delta_1}(\omega))}.$$

Then, the function

$$\omega \longrightarrow u_1(X^{(\delta_1)}(\hat{t}_1))u_2(X^{(\delta_1)}(\hat{t}_2))$$

is a lifting of f with respect to both P_i and P_j. Therefore, we have

$$\begin{aligned} &\int_\Omega {}^\circ u_1({}^\circ X^{(\delta_1)+}(t_1)){}^\circ u_2({}^\circ X^{(\delta_1)+}(t_2))\ L(P_i)(d\omega) \\ &\quad \approx \int_\Omega u_1(X^{(\delta_1)}(\hat{t}_1))u_2(X^{(\delta_1)}(\hat{t}_2))\ P_i(d\omega) \\ &\quad = \int_\Omega u_1(X^{(\delta_1)}(\hat{t}_1))E_{X^{(\delta_1)}(\hat{t}_1)}u_2(X^{(\delta_1)}(\hat{t}_2 - \hat{t}_1))\ P_i(d\omega) \end{aligned}$$

$$\approx \int_\Omega {}^\circ u_1({}^\circ X^{(\delta_1)}(\hat{t}_1)) \left[{}^\circ\!\left(Q^{\hat{t}_1-\hat{t}_2} u_2(X^{(\delta_1)}(\hat{t}_1))\right)\right] L(P_i)(d\omega)$$
$$\approx \int_\Omega {}^\circ u_1({}^\circ X^{(\delta_1)}(\hat{t}_1)) \left[{}^\circ\!\left(Q^{\hat{t}_1-\hat{t}_2} u_2(X^{(\delta_1)}(\hat{t}_1))\right)\right] L(P_j)(d\omega)$$
$$\approx \int_\Omega u_1(X^{(\delta_1)}(\hat{t}_1)) E_{X^{(\delta_1)}(\hat{t}_1)} u_2(X^{(\delta_1)}(\hat{t}_2 - \hat{t}_1))\ P_j(d\omega)$$
$$= \int_\Omega u_1(X^{(\delta_1)}(\hat{t}_1)) u_2(X^{(\delta_1)}(\hat{t}_2))\ P_j(d\omega)$$
$$\approx \int_\Omega {}^\circ u_1({}^\circ X^{(\delta_1)+}(t_1) {}^\circ u_2({}^\circ X^{(\delta_1)+}(t_2))\ L(P_j)(d\omega).$$

Now it is easy to show that (3.2.19) and (3.2.20) hold if $n = 2$. We can then complete the proof of Lemma 3.2.9 by mathematical induction. □

In the following, we will define quantities $\{\Pi_t\}$ and $\{\Theta_y\}$ associated with our standard Markov process. For $t \in \mathbb{R}_+$, let Π_t° be the σ-algebra generated by the sets

$$\left\{\omega \middle| x(\omega, t_1) \in B_1 \wedge \cdots \wedge x(\omega, t_n) \in B_n\right\},$$

where $0 \le t_1 < t_2 < \cdots < t_n \le t$ and $B_1, \cdots, B_n$ are Borel sets in Y. Define

$$\Pi_t = \bigcap_{s>t} \Pi_s^\circ.$$

The filtration $\{\Pi_t \mid t \in [0, \infty)\}$ is right continuous. Let us set

$$\Pi_\infty = \vee \{\Pi_t \mid t \in [0, \infty)\}.$$

If $y \in A_1(\delta_1)^\circ \cup \{\Delta\}, C \in \Pi_\infty$, let

$$\Theta_y(C) = \begin{cases} 1, & \text{if } C \text{ contains all constant paths } x(t) = y, \\ 0, & \text{else.} \end{cases}$$

If $y \notin A_1(\delta_1)^\circ$, let for all $C \in \Pi_\infty$

$$\Theta_y(C) = L(P_i)(C) \text{ for all } s_i \in \mathrm{st}^{-1}(y) - A_1(\delta_1).$$

Theorem 3.2.1. *Let Y be a regular Hausdorff space, and let $\mathcal{E}^{(\delta)}(\cdot,\cdot)$ be a hyperfinite weak coercive quadratic form for every infinitesimal $\delta \in T$. Assume that $\mathcal{E}(\cdot,\cdot)$ has a hyperfinite $\hat{\delta}$-quasi-continuous core and is $\hat{\delta}$-nearstandardly concentrated for some $\hat{\delta} \in T, \hat{\delta} \approx 0$. Moreover, suppose that there is an infinitesimal $\delta' \in T$ such that $\mathcal{E}(\cdot,\cdot)$ generates a δ-quasi-continuous semigroup for every infinitesimal $\delta \in T, \delta \ge \delta'$. Then, the associated* *hyperfinite Markov chain has a modified standard part which is a strong Markov process.*

Proof. The proof of this theorem follows exactly the same line as the proof of Theorem 3.1.1. □

In Theorem 3.2.1, we assume that there is an infinitesimal $\delta' \in T$ such that $\mathcal{E}(\cdot,\cdot)$ generates δ-quasi-continuous semigroup for every infinitesimal $\delta \in T, \delta \geq \delta'$. In practice, we have the following result:

Proposition 3.2.4. *Assume that for each S-bounded $u \in \mathcal{D}(\mathcal{E})$ there is a sequence $\{u_n \mid n \in \mathbb{N}\}$ of hyperfinite quasi-continuous functions such that ${}^\circ|u_n - u|_1 \longrightarrow 0$ as $n \longrightarrow \infty$. Then, all functions with ${}^\circ\mathcal{E}_1(v,v) < \infty$ are quasi-continuous, and the hyperfinite quadratic form $\mathcal{E}(\cdot,\cdot)$ generates a quasi-continuous semigroup.*

Proof. In a similar way as that of Albeverio et al. [25], 5.5.16 Lemma, we may prove Proposition 3.2.4 by using Proposition 1.7.1. □

Remark 3.2.2. In Theorem 3.2.1, we obtain conditions on a hyperfinite weak coercive quadratic form to guarantee that the associated hyperfinite Markov chain has a modified standard part which is a strong Markov process, using the concepts of exceptional sets and semigroups. Correspondingly, one may obtain conditions on a hyperfinite weak coercive quadratic co-form to guarantee that the associated hyperfinite Markov chain has a modified standard part which is a strong Markov process, using the concepts of co-exceptional sets and co-semigroup. The details will not be presented here, since the construction is entirely similar to the one which leads to Theorem 3.2.1.

Chapter 4
Construction of Markov Processes Associated With Quasi-Regular Dirichlet Forms

In this chapter, we consider the construction of tight strong Markov processes associated with standard quasi-regular Dirichlet forms by using the language of nonstandard analysis. In Fan [165], the construction of strong Markov processes associated with standard symmetric Dirichlet forms was considered. In this chapter, we consider nonsymmetric Dirichlet forms. We consider a standard Dirichlet form $(F(\cdot,\cdot), D(F))$ whose state space is a regular Hausdorff space Y. We impose conditions directly on the Dirichlet form $(F(\cdot,\cdot), D(F))$ to obtain associated tight strong Markov processes. The motivation comes from the desire to find the relationship between the family of Dirichlet forms and the family of strong Markov processes.

Before describing our main result of Theorem 4.1.1, let us briefly summarize the development and applications of standard Dirichlet space theory. In the symmetric case, the regular Dirichlet space theory is well developed [175, 183, 333, 334]. For the nonsymmetric case, a similar theory is also developed [44, 105, 257]. The references mentioned above consider the case where the state space is a locally compact separable space (thus essentially a finite dimensional space, like $\mathbb{R}^d$ or a d-dimensional Riemannian manifold). This theory gives us, in particular, a very good understanding about the property of diffusion processes with non-smooth diffusion coefficients [45, 112, 161, 164, 240, 331, 346, 347].

In the last decades, an extensive development of infinite dimensional stochastic analysis theory has taken place. In this case, the state space of the corresponding Dirichlet space theory is a general topological space, in particular not a locally compact separable space [2, 43, 50, 159, 160, 162, 163, 249, 295, 314–316, 326–328]. Applications in mathematical physics have been given [2, 48–50, 314].

In Albeverio et al. [25], a nonstandard symmetric Dirichlet space is proposed. In this book, we have generalized the corresponding results to nonsymmetric case in the Chaps. 1, 2, and 3. Let us recall the main outline we have achieved in this book, which will be the basis for our probabilistic construction. In Sect. 1.5, we have discussed hyperfinite Dirichlet forms and

S. Albeverio et al., *Hyperfinite Dirichlet Forms and Stochastic Processes*, Lecture Notes of the Unione Matematica Italiana 10,
DOI 10.1007/978-3-642-19659-1_4, © Springer-Verlag Berlin Heidelberg 2011

hyperfinite Markov chains. In the Chap. 2, we have developed the potential theory of hyperfinite Dirichlet forms. In Sect. 3.1 of Chap. 3, we have studied the standard parts of hyperfinite Markov chains. Suppose that S_0 is a hyperfinite subset of $^*Y$ for some topological space Y. If $X(t)$ is a hyperfinite Markov chain taking values in $S = S_0 \cup \{s_0\}$, we have got conditions that guarantee the existence of the standard part of $X(t)$. In addition, we have showed that it is a Y-valued strong Markov process. After that, we have translated the conditions into the language of hyperfinite Dirichlet forms in Sect. 3.2.

The arrangement of this chapter is as follows. We will first present our main result, and related notations and definitions in Sect. 4.1. Then, we will construct hyperfinite quadratic forms of the standard Dirichlet forms in Sect. 4.2. In the meantime, we will construct hyperfinite Markov chains as well as dual hyperfinite Markov chains of the hyperfinite quadratic forms. The main aim of rest of the chapter is then devoted to show that the hyperfinite Markov chains have the desired projection of dual strong Markov processes from Sect. 4.3 to Sect. 4.6. In the last Sect. 4.7, we will prove the necessity part of our main result (Theorem 4.1.1).

4.1 Main Result

In this chapter, we assume that the state space Y is a regular Hausdorff space or a T_3 space, which is introduced in Sect. 3.2. Let $\mathcal{B}(Y)$ be the topological Borel σ-field of Y. Let ν be a positive Radon measure on $(Y, \mathcal{B}(Y))$ (see Sect. 4.2 below for details). Let $\mathrm{supp}(\nu)$ be the smallest closed subset of Y such that $\nu(Y - \mathrm{supp}(\nu)) = 0$. We suppose that $Y = \mathrm{supp}(\nu)$ for convenience.

Definition 4.1.1. (1) A strong Markov process $(\Omega, \Pi, \Pi_t, x(t), \Theta_y)$ is called *ν-tight* if there exists an increasing sequence of compact sets $\{Y_n \mid n \in \mathbb{N}\}$ of Y such that

$$\Theta_y \left\{ \lim_{n\to\infty} \sigma_{Y-Y_n} = \zeta \right\} = 1, \nu\text{-a.e.} y \in Y, \tag{4.1.1}$$

where σ_{Y-Y_n} is the hitting time of $Y - Y_n$ and ζ is the lifetime of $x(t)$. That is,

$$\sigma_{Y-Y_n} = \inf \{t > 0 \mid x(t) \in Y - Y_n\} \tag{4.1.2}$$

and

$$\zeta(\omega) = \inf \{t \geq 0 \mid x(\omega, t) = \Delta\}.$$

(2) Two strong Markov processes $(\Omega, \Pi, \Pi_t, x(t), \Theta_y)$ and $(\Omega, \Pi, \Pi_t, x(t), \hat{\Theta}_y)$ are said to be *ν-dual* if for all $t \in \mathbb{R}_+, u, v \in L^2(Y, \nu)$, we have

$$\int_Y u(y)\Theta_t v(y)\ \nu(dy) = \int_Y v(y)\hat{\Theta}_t u(y)\ \nu(dy).$$

Let $(F(\cdot,\cdot), D(F))$ be a coercive closed form on $L^2(Y,\nu)$ as introduced in Sect. 1.6 by replacing K with $L^2(Y,\nu)$. The inner product $(\cdot,\cdot)_\nu$ of $K = L^2(Y,\nu)$ is defined by

$$(f,f)_\nu = \int_Y f^2(y)\nu(dy), \alpha \in [0,\infty).$$

Let $\{R_\alpha \mid \alpha \in (-\infty, 0)\}$ and $\{\hat{R}_\alpha \mid \alpha \in (-\infty, 0)\}$ be the resolvent and co-resolvent of $(F(\cdot,\cdot), D(F))$, respectively. Let us denote by $\{T_t \mid t \in (0,\infty)\}$ and $\{\hat{T}_t \mid t \in (0,\infty)\}$ the strong continuous contraction semigroups corresponding to $\{R_\alpha \mid \alpha \in (-\infty, 0)\}$ and $\{\hat{R}_\alpha \mid \alpha \in (-\infty, 0)\}$, respectively. Moreover, set

$$F_\alpha(f,f) = F(f,f) + \alpha(f,f)_\nu.$$

A bounded operator R on $L^2(Y,\nu)$ is called a *sub-Markovian operator* if for every $u \in L^2(Y,\nu), 0 \le u(x) \le 1, \nu\text{-a.e.} x \in Y$, we have $0 \le Ru(x) \le 1$, ν-a.e.$x \in Y$. A coercive closed form $F(\cdot,\cdot)$ on $L^2(Y,\nu)$ is called a *Dirichlet form* if it generates a semigroup $\{T_t \mid t \ge 0\}$ of sub-Markovian operators and a co-semigroup $\{\hat{T}_t \mid t \ge 0\}$ of sub-Markovian operators.

Let $(F(\cdot,\cdot), D(F))$ be a Dirichlet form on $L^2(Y,\nu)$. For an open subset B of Y, let us define

$$L(B) = \{g \in D(F) \mid g \ge 1, \nu\text{-a.e. on } B\}. \tag{4.1.3}$$

Suppose $L(B) \ne \emptyset$. From Ma and Röckner [270], Proposition 1.5 of Chap. III, there exists a unique $\gamma_1(B), \hat{\gamma}_1(B) \in L(B)$ such that for all $w \in L(B)$

$$\begin{aligned} F_1(\gamma_1(B), w) &\ge F_1(\gamma_1(B), \gamma_1(B)), \\ F_1(w, \hat{\gamma}_1(B)) &\ge F_1(\hat{\gamma}_1(B), \hat{\gamma}_1(B)). \end{aligned} \tag{4.1.4}$$

From Ma and Röckner [270], Remark 1.6 and Exercise 1.7 of Chap. III, we know

$$\begin{aligned} F_1(\gamma_1(B), \gamma_1(B)) &= F_1(\gamma_1(B), \hat{\gamma}_1(B)) \\ &= F_1(\hat{\gamma}_1(B), \hat{\gamma}_1(B)). \end{aligned} \tag{4.1.5}$$

Set

$$\Gamma_1(B) = F_1(\gamma_1(B), \gamma_1(B)) \text{ if } L(B) \neq \emptyset \text{ and } \Gamma_1(B) = \infty \text{ if } L(B) = \emptyset. \tag{4.1.6}$$

For a general subset A of Y, let us define

$$\Gamma_1(A) = \inf \{\Gamma_1(B) \mid A \subset B, B \text{ is open}\}. \tag{4.1.7}$$

We call $\Gamma_1(\cdot)$ the *1-capacity* of $F(\cdot, \cdot)$.

For a closed subset $B \subset Y$, we set

$$D(F)_B = \{f \in D(F) \mid f = 0, \nu\text{-a.e. on } Y - B\}.$$

It is easy to see that $D(F)_B$ is a closed set of $D(F)$ with the inner product $F_1(\cdot, \cdot)$.

Definition 4.1.2. An increasing sequence of closed subsets $\{G_n \mid n \in \mathbb{N}\}$ of Y is called a *nest* if $\Gamma_1(Y - G_n) \downarrow 0$.

A subset $B \subset Y$ is said to be of *zero capacity* if there exists a nest $\{G_n \mid n \in \mathbb{N}\}$ such that $B \subset \bigcap_{n=1}^{\infty}(Y - G_n)$. A function f on Y is said to be *quasi-continuous* if there exists a nest $\{G_n \mid n \in \mathbb{N}\}$ such that the restriction $f|_{G_n}$ is continuous on G_n for each $n \geq 1$.

We shall say that two ν-dual strong Markov processes $(\Omega, \Pi, \Pi_t, x(t), \Theta_y)$ and $(\Omega, \Pi, \Pi_t, x(t), \hat{\Theta}_y)$ are *properly associated* with the Dirichlet form $(F(\cdot, \cdot), D(F))$ if

$$T_t f = \Theta_t f \text{ and } \hat{T}_t f = \hat{\Theta}_t f, \nu\text{-a.e.}, \forall f \in L^2(Y, \nu), t > 0.$$

The main result of this chapter is the following:

Theorem 4.1.1. *Assume that Y is a regular Hausdorff topological space and ν is a Radon measure on Y. Let $(F(\cdot, \cdot), D(F))$ be a Dirichlet form on $L^2(Y, \nu)$. Then there are two ν-dual tight strong Markov processes $(\Omega, \Pi, \Pi_t, x(t), \Theta_y)$ and $(\Omega, \Pi, \Pi_t, x(t), \hat{\Theta}_y)$ properly associated with the Dirichlet form $(F(\cdot, \cdot), D(F))$ if and only if the following three conditions are satisfied:*

(I) There is an increasing sequence of compact subsets $\{Y_n \mid n \in \mathbb{N}\}$ of Y such that $\bigcup_{n \in \mathbb{N}} D(F)_{Y_n}$ is F_1-dense in $D(F)$.

(II) There exists a subset π_0 of $D(F)$ such that π_0 is dense in $D(F)$ with the inner product $F_1(\cdot, \cdot)$. In addition, u is quasi-continuous for each $u \in \pi_0$.

(III) *There is a countable subset π of $D(F)$ such that every element $u \in \pi$ is quasi-continuous and bounded. Moreover, we have*

$$\mathcal{B}(Y)\bigcap\left(\bigcup_{n\in\mathbb{N}} Y_n\right) \subset \sigma(u|u\in\pi),$$

where $\{Y_n \mid n \in \mathbb{N}\}$ is a sequence of compact sets in (I).

Remark 4.1.1. If $(F(\cdot,\cdot), D(F))$ satisfies conditions (I), (II) and (III) in Theorem 4.1.1, we call it *quasi-regular.*

Remark 4.1.2. Previous work of Albeverio and Ma [41] and Ma and Röckner [270] has established necessary and sufficient conditions for the existence of ν-perfect processes associated with Dirichlet forms. In these references, the state space Y is a metrizable space and ν is a σ-finite measure on Y. Thus, there is a strictly positive function $\psi \in L^2(Y,\nu)$, $0 < \psi \le 1$. Set $h = R_1\psi$, where $\{R_\alpha \mid \alpha > 0\}$ is the resolvent of $F(\cdot,\cdot)$. The authors defined the h-weighted capacity as in Röckner [313] and used it as a useful tool. In our case, Y is a regular Hausdorff space and m is a Radon measure. We do not have an h-weighted capacity, since this concept has not been developed in our context. Instead, we utilize the 1-capacity theory developed in Chap. 2, Fan [166], and Fukushima [175].

In order to prove Theorem 4.1.1, we will construct in Sect. 4.2 a hyperfinite quadratic form $(\mathcal{E}(\cdot,\cdot), \mathcal{D}(\mathcal{E}))$ from $(F(\cdot,\cdot), D(F))$. We obtain a hyperfinite Markov chain $X(t)$ associated with $\mathcal{E}(\cdot,\cdot)$, as well as the dual $\hat{X}(t)$ of $X(t)$. In Sect. 4.3, we will discuss the relation between the potential theory of $(F(\cdot,\cdot), D(F))$ and the counterpart of its hyperfinite lifting $(\mathcal{E}(\cdot,\cdot), \mathcal{D}(\mathcal{E}))$. On this basis, we will consider the path regularity of $X(t)$ and get its modified standard part $x(t)$. In fact, $x(t)$ is the wanted process. We shall show that $x(t)$ satisfies all the requirements of Theorem 4.1.1 in Sect. 4.6. At last, we prove in Sect. 4.7 that if there are two ν-dual tight Markov processes $(\Omega, \Pi, \Pi_t, x(t), \Theta_y)$ and $(\Omega, \Pi, \Pi_t, x(t), \hat{\Theta}_y)$ properly associated with $F(\cdot,\cdot)$, then $x(t)$ satisfies the conditions (I), (II) and (III) of Theorem 4.1.1. Combining Propositions 4.6.1 and 4.7.1, we shall then prove the Theorem 4.1.1.

Remark 4.1.3. From Albeverio and Ma [41], we know that even if ν is a nowhere Radon measure on a metric space, it is still possible to get the Markov process $x(t)$. In order to include this case by methods of nonstandard analysis, we would have to develop first a nonstandard measure theory associated with nowhere Radon measures on general topological spaces. After that, we could study the related stochastic analysis problems associated with a nonstandard version of nowhere Radon measure spaces. This is, however, a program which remains to be fulfilled.

4.2 Hyperfinite Lifts of Quasi-Regular Dirichlet Forms

Let Y be a Hausdorff topological space. Consider the topological Borel σ-field $\mathcal{B}(Y)$ on Y. Assume that ν is a positive Borel measure on $(Y, \mathcal{B}(Y))$ such that $\nu(K) < \infty$ for every compact set K. It is a *Radon measure* in the sense that for all $B \in \mathcal{B}(Y)$,

$$\nu(B) = \sup\{\nu(K) \mid K \subset B, K \text{ compact}\} \tag{4.2.1}$$

and for all $B \in \mathcal{B}(Y)$ with $\nu(B) < \infty$,

$$\nu(B) = \inf\{\nu(O) \mid B \subset O, O \text{ open}\}.$$

We denote by $(Y, \overline{\mathcal{B}(Y)}, \nu)$ the completion of $(Y, \mathcal{B}(Y), \nu)$. For simplicity, we suppose that $\text{supp}(\nu) = Y$.

Let ${}^*Y$ be the nonstandard extension of Y. Given an element $a \in Y$, we know that the monad of a is the subset of ${}^*Y$ defined by

$$\mu(a) = \bigcap\{{}^*O \mid a \in O \quad \text{and} \quad O \subset Y \quad \text{is open}\}.$$

A point $y \in {}^*Y$ is nearstandard if it belongs to $\mu(a)$ for some $a \in Y$. We shall say that it is nearstandard to a. The set of all nearstandard points is denoted by $Ns({}^*Y)$. Since Y is Hausdorff, each element $y \in Ns({}^*Y)$ is nearstandard to exactly one element $a \in Y$ (refer to page 48, 2.1.6. Proposition, [25]). We call a the standard part of y and denote it by $\text{st}(y)$ or ${}^\circ y$.

Let Δ be an element outside Y, $Y_\Delta = Y \cup \{\Delta\}$. For all $y \in {}^*Y - Ns({}^*Y)$, we define $\text{st}(y) = \Delta$. Now we have got the *standard part operation* as a map

$$\text{st} : {}^*Y \longrightarrow Y_\Delta.$$

A subset S_0 of ${}^*Y$ is called *rich* if it is hyperfinite and $\text{st}(S_0 \cap Ns({}^*Y)) = Y$. In the following, we will construct a rich subset S_0 in ${}^*Y$ and an internal measure m on S_0 such that $\nu = L(m) \circ \text{st}^{-1}$. By a *finite* $\overline{\mathcal{B}(Y)}$ *partition of* Y, we mean a finite collection

$$\left\{B_i \in \overline{\mathcal{B}(Y)} \middle| 1 \le i \le n\right\}$$

of non-empty sets with $Y = \bigcup_{i=1}^n B_i$ and $B_i \cap B_j = \emptyset$ if $i \neq j$. Let us denote by $\mathcal{P}$ the collection of all finite $\overline{\mathcal{B}(Y)}$ partitions of Y.

If P_1 and P_2 are elements of $\mathcal{P}$, we say that P_2 is *finer* than P_1 and we write $P_1 \le P_2$ if for each $C \in P_1, C = \bigcup\{B \in P_2 \mid B \subset C\}$.

We have the following result from 1.1 Theorem, Loeb [266].

Proposition 4.2.1. *There is a partition $P \in {}^*\mathcal{P}$ such that ${}^*P_0 \le P$ for any $P_0 \in \mathcal{P}$. That is, $P \in {}^*\mathcal{P}$ has the following properties:*

(i) There are an infinite integer $N \in {}^*\mathbb{N}$ and an internal bijection from $I = \{i \in {}^*\mathbb{N} \mid 1 \le i \le N\}$ onto P. Thus we may write $P = \{A_i \mid i \in I\}$.
(ii) If i and j are in I and $i \neq j$, then $A_i \neq \emptyset, A_j \neq \emptyset$ and $A_i \cap A_j = \emptyset$.
(iii) ${}^*Y = \bigcup\{A_i \mid i \in I\}$.
(iv) For each $B \in \mathcal{B}(Y)$, let $I_B = \{i \in I \mid A_i \subset {}^*B\}$. Then I_B is hyperfinite, and ${}^*B = \bigcup_{i \in I_B} A_i$.

Hereafter, let us fix a partition $P = \{A_i \mid i \in I\}$ of ${}^*Y$ with above properties (i) through (iv). For every $i \in I$, we pick up an element $s_i \in A_i$ and define $S_0 = \{s_i \mid i \in I\}$. Let m be the internal measure on S_0 defined by

$$m(\{s_i\}) = {}^*\nu(A_i).$$

Proposition 4.2.2. *S_0 is rich in ${}^*Y$ and $\nu = L(m) \circ \mathrm{st}^{-1}$.*

Proof. We simply refer to Albeverio et al. [25], 3.4.10 Corollary. □

Let H be the hyperfinite dimensional space of all the internal functions $f : S_0 \longrightarrow {}^*\mathbb{R}$ with the inner product

$$\langle f, g \rangle = \int_{S_0} fg \, dm.$$

Let us set $K = L^2(Y, \nu)$. Let $\tilde{H}$ be the subspace of ${}^*K$ consisting of all functions which are constant on each class $A_j, j \in I$. Obviously, H and $\tilde{H}$ are isomorphic. From now on, we will identify H with $\tilde{H}$. By using Theorem 1.6.1, we can prove the following result.

Proposition 4.2.3. *Let Y be a Hausdorff space, ν be a Radon measure on Y, and $F(\cdot,\cdot)$ be a coercive closed form on $L^2(Y, \nu)$. Then, we have*

(i) There exist a hyperfinite, rich subset S_0 of ${}^*Y$, and an internal measure m on S_0 such that H is S-dense in ${}^*K$.
(ii) There are a nonnegative quadratic form $(\mathcal{E}(\cdot,\cdot), \mathcal{D}(\mathcal{E}))$ on $L^2(S_0, m)$ and an internal time-line $T = \{k\Delta t \mid k \in {}^*\mathbb{N}\}$ representing $F(\cdot,\cdot)$ in the following sense: $\nu = L(m) \circ \mathrm{st}^{-1}$ and for all $u \in L^2(Y, \nu)$,

$$F(u, u) = \inf\{{}^\circ\mathcal{E}(v, v) \mid v \text{ is a 2-lifting of } u\}.$$

Moreover, let $\{Q^s \mid s \in T\}$ and $\{\hat{Q}^s \mid s \in T\}$ be the semigroup and co-semigroup generated by $\mathcal{E}(\cdot,\cdot)$, respectively. Then for all $t \in [0, \infty)$, $s \in T, u \in K, v \in H$ such that $t = {}^\circ s, u = st_K(v)$, we have

$$st_K Q^s v = T_t u \text{ and } st_K \hat{Q}^s v = \hat{T}_t u.$$

On the other hand, let $\{G_\alpha \mid \alpha \in {}^*(-\infty, 0)\}$ *and* $\{\hat{G}_\alpha \mid \alpha \in {}^*(-\infty, 0)\}$ *be the resolvent and co-resolvent generated by* $\mathcal{E}(\cdot,\cdot)$*, respectively. Then for all* $\beta \in (-\infty, 0), \alpha \in {}^*(-\infty, 0), u \in K, v \in H$ *such that* $\beta = {}^\circ\alpha, u = st_K(v)$*, we have*

$$st_K G_\alpha v = R_\beta u \text{ and } st_K \hat{G}_\alpha v = \hat{R}_\beta u,$$

where $\{R_\alpha \mid \alpha \in (-\infty, 0)\}$ *is the resolvent, and* $\{\hat{R}_\alpha \mid \alpha \in (-\infty, 0)\}$ *is the co-resolvent of* $(F(\cdot,\cdot), D(F))$.

Proof. From Sect. 3.2 of Albeverio et al. [25], we know that each function f in $L^2(Y, \nu)$ has an S-square integrable lifting $\tilde{f}$ in $\tilde{H}$ such that ${}^\circ\tilde{f}(x) = f({}^\circ x)$ for almost all nearstandard x. If we can show that $||{}^*f - \tilde{f}|| \approx 0$, then $\tilde{H}$ is dense in ${}^*(L^2(Y, \nu))$. By (4.2.1), it suffices to show this when f is bounded and of compact support. For such functions, the statement is an immediate consequence of Anderson's nonstandard version of Lusin's theorem. The proposition follows by Theorem 1.6.1. □

Corollary 4.2.1. *If the form* $F(\cdot,\cdot)$ *in Proposition 4.2.3 is a Dirichlet form, we can also take* $\mathcal{E}(\cdot,\cdot)$ *to be a hyperfinite quadratic form.*

Proof. We observe that from Proposition 1.5.1 it suffices to prove that $Q^{\Delta t}$ and $\hat{Q}^{\Delta t}$ are Markov operators, where

$$\langle Q^{\Delta t} u, v\rangle = \langle u, \hat{Q}^{\Delta t} v\rangle$$

and for all $u, v \in H$

$$\begin{aligned} \mathcal{E}(u,v) &= \frac{1}{\Delta t}\langle (I - Q^{\Delta t})u, v\rangle \\ &= \frac{1}{\Delta t}\langle u, (I - \hat{Q}^{\Delta t})v\rangle. \end{aligned}$$

Let P be the projection of ${}^*K$ onto H. We shall write $\{{}^*R_\alpha \mid \alpha \in {}^*(-\infty, 0)\}$ for ${}^*(\{R_\alpha \mid \alpha \in (-\infty, 0)\})$ and $\{{}^*\hat{R}_\alpha \mid \alpha \in {}^*(-\infty, 0)\}$ for ${}^*(\{\hat{R}_\alpha \mid \alpha \in (-\infty, 0)\})$, respectively. We see from Sect. 1.6 that $Q^{\Delta t}$ and $\hat{Q}^{\Delta t}$ are just $-P^*(\gamma R_\gamma)$ and $-P^*(\gamma \hat{R}_\gamma)$ for a suitable infinitesimal $\Delta t = -\frac{1}{\gamma}$, respectively. With the choice of H made in the proof of Proposition 4.2.3, this projection is just the conditional expectation with respect to the algebra generated by the partition P. Since conditional expectations preserve nonnegativity and decrease the supremum norm, the corollary follows. □

In the rest of this chapter, we shall be interested in quadratic forms generating Markov processes. That is, we shall assume that $F(\cdot,\cdot)$ is a Dirichlet form on $L^2(Y,\nu)$. We shall use the results obtained above. First, let us introduce some notations.

The infinitesimal generator A and co-generator $\hat{A}$ of $\mathcal{E}(\cdot,\cdot)$ are given by

$$A = \frac{1}{\Delta t}(I - Q^{\Delta t}) \text{ and } \hat{A} = \frac{1}{\Delta t}(I - \hat{Q}^{\Delta t}), 0 < \Delta t \leq \frac{1}{||A||} = \frac{1}{||\hat{A}||}.$$

The semigroup Q^t and co-semigroup $\hat{Q}^t, t \in T$, are defined by

$$Q^t = (Q^{\Delta t})^k \text{ and } \hat{Q}^t = (\hat{Q}^{\Delta t})^k \quad \text{for each} \quad t = k\Delta t.$$

Moreover, we have the following relation from Proposition 1.6.3,

$$\begin{aligned} F(u,u) &= \lim_{\alpha \longrightarrow -\infty} \left[-\alpha \int_Y u(y)\Big(u(y) + \alpha R_\alpha u(y)\Big)\ \nu(dy) \right] \\ &= \lim_{\substack{\alpha \to -\infty \\ \alpha \not\approx -\infty \\ \alpha \in {}^*(-\infty,0)}} {}^\circ[-\alpha\langle (I + \alpha R_\alpha)P^*u, P^*u\rangle] \\ &= \lim_{\substack{\alpha \to -\infty \\ \alpha \not\approx -\infty \\ \alpha \in {}^*(-\infty,0)}} {}^\circ[-\alpha\langle (I + \alpha G_\alpha)P^*u, P^*u\rangle] \\ &= \lim_{\substack{\alpha \to -\infty \\ \alpha \not\approx -\infty \\ \alpha \in {}^*(-\infty,0)}} {}^\circ[{}^{(\alpha)}\mathcal{E}(u,u)] \end{aligned} \tag{4.2.2}$$

for every $u \in D(F)$, where P is the projection of ${}^*K$ onto H. Therefore, we see from the proof of Theorem 1.6.1 that

$$\frac{1}{\Delta t}\langle (I - Q^{\Delta t})P^*u, P^*u\rangle \approx F(u,u),$$

i.e.,

$${}^\circ\mathcal{E}(P^*u, P^*u) = F(u,u) < \infty \quad \text{for each} \quad u \in D(F). \tag{4.2.3}$$

We recall from Definition 1.4.2 that the domain $\mathcal{D}(\mathcal{E})$ of $\mathcal{E}(\cdot,\cdot)$ is the set of all $u \in H$ such that

(i) ${}^\circ\mathcal{E}_1(u,u) = {}^\circ[\mathcal{E}(u,u) + \langle u,u\rangle] < \infty$.
(ii) For all infinite $\alpha \in {}^*(-\infty,0), \mathcal{E}(u + \alpha G_\alpha u, u + \alpha G_\alpha u) \approx 0$ and $\mathcal{E}(u + \alpha\hat{G}_\alpha u, u + \alpha\hat{G}_\alpha u)$.

We remark that $\mathcal{E}(\cdot,\cdot)$ does not necessarily satisfy the hyperfinite weak sector condition. Hence, the results of Proposition 1.4.2 need not hold. However, it is easy to see from the proof of Proposition 1.4.2 the following: if ${}^\circ\mathcal{E}(u,u) < \infty$ and for all infinite $\alpha < 0$, ${}^{(\alpha)}\mathcal{E}(u,u) \approx \mathcal{E}(u,u)$ and ${}^{(\alpha)}\hat{\mathcal{E}}(u,u) \approx \mathcal{E}(u,u)$, then we have $u \in \mathcal{D}(\mathcal{E})$.

Therefore, we have proved the following using (4.2.2) and (4.2.3):

Lemma 4.2.1. *For every $u \in D(F)$, we have $P^*u \in \mathcal{D}(\mathcal{E})$.*

Now, we are in the position to construct the dual processes $X(t)$ and $\hat{X}(t)$ and their transition matrices associated with $\mathcal{E}(\cdot,\cdot)$. Remember that m is an internal measure on the hyperfinite set $S_0 = \{s_1, s_2, \cdots, s_N\}$. We shall write m_i for $m(s_i)$ if it is convenient.

Fix $j \in I$. Let $u_j : S_0 \longrightarrow {}^*\mathbb{R}$ be the function satisfying $u_j(s_l) = \delta_{lj}, l \in I$. We define

$$b_{ij} = Q^{\Delta t} u_j(s_i) \text{ and } \hat{b}_{ij} = \hat{Q}^{\Delta t} u_j(s_i), j \in I.$$

It is easy to see that for any internal function $u : S_0 \longrightarrow {}^*\mathbb{R}$, we have

$$u(\cdot) = \sum_{j=1}^{N} u(s_j) u_j(\cdot).$$

Hence, we obtain

$$\begin{aligned} Q^{\Delta t} u(s_i) &= \sum_{j=1}^{N} u(s_j) Q^{\Delta t} u_j(s_i) \\ &= \sum_{j=1}^{N} b_{ij} u(s_j), \\ Au(s_i) &= \frac{1}{\Delta t} \left(u(s_i) - \sum_{j=1}^{N} u(s_j) b_{ij} \right). \end{aligned}$$

Similarly, we have

$$\begin{aligned} \hat{Q}^{\Delta t} u(s_i) &= \sum_{j=1}^{N} u(s_j) \hat{Q}^{\Delta t} u_j(s_i) \\ &= \sum_{j=1}^{N} \hat{b}_{ij} u(s_j), \\ \hat{A}u(s_i) &= \frac{1}{\Delta t} \left(u(s_i) - \sum_{j=1}^{N} u(s_j) \hat{b}_{ij} \right). \end{aligned}$$

Let us denote $u(s_i)$ by $u(i)$. Then, we have

$$\begin{aligned}\mathcal{E}(u,v) &= \frac{1}{\Delta t}\sum_{i=1}^{N} Au(i)v(i)m(i)\\ &= \frac{1}{\Delta t}\sum_{i=1}^{N}\left[u(i)-\sum_{j=1}^{N} b_{ij}u(j)\right]v(i)m(i)\\ &= \frac{1}{\Delta t}\sum_{i=1}^{N}\sum_{j=1}^{N}(\delta_{ij}-b_{ij})u(j)v(i)m(i)\\ &= \frac{1}{\Delta t}\sum_{i=1}^{N}\sum_{j=1}^{N}(\delta_{ij}-\hat{b}_{ij})v(j)u(i)m(i).\end{aligned}$$

Since the duality of $Q^{\Delta t}$ and $\hat{Q}^{\Delta t}$, we have for all $i,j \in I$,

$$m(i)b_{ij} = m(j)\hat{b}_{ji}.$$

Recall that s_0 is a point outside S_0, and we put $S = S_0 \cup \{s_0\}$. We assume $m(\{s_0\}) = 0$ and $\mathrm{st}(s_0) = \Delta$. Let $Q = \{q_{ij} \mid 0 \le i,j \le N\}$ be an $(N+1)\times(N+1)$ matrix satisfying the following conditions:

$$\begin{aligned} q_{00} &= 1,\\ q_{0i} &= 0 \quad \text{for} \quad i \in I,\\ q_{ij} &= b_{ij}, i,j \in I,\\ q_{i0} &= 1-\sum_{j=1}^{N} b_{ij}, i \in I.\end{aligned}$$

It is easy to see that

$$\sum_{j=0}^{N} q_{ij} = 1 \quad \text{for all} \quad i \in I \cup \{0\}.$$

We construct a hyperfinite Markov chain $\{X(t) \mid t \in T\}$ associated with Q and m in the following manner. Let Ω be the set of all internal functions $\omega : T \longrightarrow S$. Let X be the coordinate function $X(\omega,t) = \omega(t)$. We take P to be the measure generated by

$$P([\omega]_0) = m(X(\omega,0))$$

and

$$P([\omega]_{k\Delta t}) = m\{X(\omega,0)\}\prod_{n=0}^{k-1} q(\omega(n\Delta t),\omega((n+1)\Delta t)),$$

where $q(s_i, s_j)$ is just q_{ij} and $[\omega]_t$ is defined by

$$[\omega]_t = \{\omega' \mid X(\omega', s) = X(\omega, s) \quad \text{for all} \quad s \leq t\}.$$

Let $\mathcal{F}_t$ be the internal algebra on Ω generated by the sets $[\omega]_t, \omega \in \Omega$. Let us define the probability measure P_i as follows:

$$P_i([\omega]_{k\Delta t}) = \delta_{i\omega(0)} \prod_{n=0}^{k-1} q(\omega(n\Delta t), \omega((n+1)\Delta t)).$$

It is easy to verify that the following Markov property of $X(t)$ holds. If $X(\omega, t) = s_i$, then we have

$$P\left(\omega' \in [\omega]_t \middle| X(\omega', t + \Delta t) = s_j\right) = q_{ij} P([\omega]_t).$$

Similarly, let $\hat{Q} = \{\hat{q}_{ij} \mid 0 \leq i, j \leq N\}$ be an $(N+1) \times (N+1)$ matrix satisfying the following conditions:

$$\begin{aligned}
\hat{q}_{00} &= 1, \\
\hat{q}_{0i} &= 0 \quad \text{for} \quad i \in I, \\
\hat{q}_{ij} &= \hat{b}_{ij}, i, j \in I, \\
\hat{q}_{i0} &= 1 - \sum_{j=1}^{N} \hat{b}_{ij}, i \in I.
\end{aligned}$$

It is easy to see that

$$\sum_{j=0}^{N} \hat{q}_{ij} = 1 \quad \text{for all} \quad i \in I \cup \{0\}.$$

We may construct the dual hyperfinite Markov chain $\{\hat{X}(t) = X(t) \mid t \in T\}$ and the corresponding internal measure $\hat{P}, \hat{P}_i, i \in I$.

If $\delta \in T$, let T_δ be the sub-line

$$T_\delta = \{k\delta \mid k \in {}^*\mathbb{N}_0\}.$$

We write $X^{(\delta)}$ for the restriction $X|T_\delta$. For each $t \in T_\delta$, let $\mathcal{F}_t^{(\delta)}$ be the internal algebra on Ω generated by the sets

$$[\omega]_t^{(\delta)} = \left\{\omega' \in \Omega \middle| X^{(\delta)}(\omega', s) = X^{(\delta)}(\omega, s) \text{ for all } s \in T_\delta, s \leq t\right\}.$$

In the following of this chapter, we shall use the notations $A^{(\delta)}$, $q_{ij}^{(\delta)}$, $\mathcal{E}^{(\delta)}(\cdot,\cdot)$, and $\mathcal{D}(\mathcal{E}^{(\delta)})$ introduced in Sect. 1.5.

For $\delta \in T, \alpha \in {}^*\mathbb{R}, \alpha \geq 0$, let us introduce the notation

$$\mathcal{E}_\alpha^{(\delta)}(u,u) = \mathcal{E}^{(\delta)}(u,u) + \alpha\langle u,u\rangle.$$

4.3 Relation with Capacities

Let $(F(\cdot,\cdot), D(F))$ be a Dirichlet form on $L^2(Y,\nu)$. We consider the 1-capacity $\Gamma_1(\cdot)$ defined in the relations (4.1.3), (4.1.4), (4.1.5), (4.1.6), and (4.1.7) in Sect. 4.1. We get the following results from Ma and Röckner [270], Chap. III.

Lemma 4.3.1. *Let B be an open subset of Y such that $L(B) \neq \emptyset$, where $L(B)$ is defined by the definition (4.1.3) in Sect. 4.1. Then we have the following results:*

(1) There exists a unique element $\gamma_1(B) \in L(B)$ such that

$$\Gamma_1(B) = F_1(\gamma_1(B), \gamma_1(B)). \tag{4.3.1}$$

(2) $0 \leq \gamma_1(B) \leq 1$, ν-a.e. and $\gamma_1(B) = 1$, ν-a.e. on B.

Proposition 4.3.1. *$\Gamma_1(\cdot)$ has the properties:*

(1) $A \subset B \Longrightarrow \Gamma_1(A) \leq \Gamma_1(B)$.
(2) $A_n \uparrow \Longrightarrow \Gamma_1(\cup_{n\in\mathbb{N}} A_n) = \sup_{n\in\mathbb{N}} \Gamma_1(A_n)$.
(3) A, B open $\Longrightarrow \Gamma_1(A \cup B) + \Gamma_1(A \cap B) \leq \Gamma_1(A) + \Gamma_1(B)$.
(4) $\Gamma_1(\cup_{n\in\mathbb{N}} A_n) \leq \sum_{n\in\mathbb{N}} \Gamma_1(A_n)$.

In the same way as in Sects. 2.2 and 2.3, we introduce the capacity theory associated with $(\mathcal{E}^{(\delta)}(\cdot,\cdot), \mathcal{D}(\mathcal{E}^{(\delta)}))$ and state some properties without proof.

For $\delta \in T, \alpha \in {}^*\mathbb{R}_+$, and an internal subset A of S_0, we define

$$\begin{aligned} \sigma_A^{(\delta)}(\omega) &= \min\{t \in T_\delta \mid X(\omega,t) \in A\}, \\ e_\alpha^{(\delta)}(A)(i) &= E_i\left[(1+\alpha\delta)^{-\sigma_A^{(\delta)}/\delta}\right], \\ \mathcal{L}(A) &= \{G \mid S_0 \longrightarrow {}^*\mathbb{R} \text{ is internal and } G|_A = 1_A\}. \end{aligned}$$

Using the terminology of Sects. 2.2 and 2.3, we call $e_\alpha^{(\delta)}(A)$ the equilibrium α-potential of A associated with $\mathcal{E}^{(\delta)}(\cdot,\cdot)$. Let us set

$$\mathrm{Cap}_\alpha^{(\delta)}(A) = \mathcal{E}_\alpha^{(\delta)}\left(e_\alpha^{(\delta)}(A), e_\alpha^{(\delta)}(A)\right).$$

For an arbitrary subset B of S_0, we define

$$\mathrm{Cap}_\alpha^{(\delta)}(B) = \inf\left\{\mathrm{Cap}_\alpha^{(\delta)}(A) \middle| B \subset A, A \text{ is internal}\right\}.$$

We call $\mathrm{Cap}_\alpha^{(\delta)}(\cdot)$ the *α-capacity* associated with $\mathcal{E}^{(\delta)}(\cdot,\cdot)$. We have studied the potential theory of hyperfinite quadratic forms in Chap. 2. We shall use the same notations of Chap. 2 in the following.

We first remark that the weak sector condition

$$|F_1(x,y)| \leq C\sqrt{F_1(x,x)}\sqrt{F_1(y,y)} \text{ for all } x,y \in D(F)$$

is equivalent to the following:

$$|F(x,y)| \leq C'\sqrt{F_1(x,x)}\sqrt{F_1(y,y)} \text{ for all } x,y \in D(F),$$

for some nonnegative constant $C' \in [0,\infty)$. Obviously, we can always suppose that $C = C'$. Hence, we have

$$|F(x,y)| \leq C\sqrt{F_1(x,x)}\sqrt{F_1(y,y)} \text{ for all } x,y \in D(F). \tag{4.3.2}$$

Proposition 4.3.2. *Let C satisfy the inequality (4.3.2). For all open subset $B \subset Y$, we have for all $\delta \in T, \delta \approx 0$ the following*

$$\mathrm{Cap}_1^{(\delta)}({}^*B \cap S_0) \leq (C+1)^2 \Gamma_1(B).$$

Proof. For simplicity, we may suppose that $L(B) \neq \emptyset$. From Lemma 4.3.1, there is an element $\gamma_1(B) \in D(F)$ satisfying (4.3.1). For convenience, let us assume that $\gamma_1(B)(x) = 1$ for all $x \in B$. For all $\delta \in T, \delta \approx 0$, we have from Proposition 2.2.1

$$\begin{aligned}\mathrm{Cap}_1^{(\delta)}({}^*B \cap S_0) &= \mathcal{E}_1^{(\delta)}(e_1^{(\delta)}(1_{{}^*B}), e_1^{(\delta)}(1_{{}^*B})) \\ &= \mathcal{E}_1^{(\delta)}(e_1^{(\delta)}({}^*B), \gamma_1(B)).\end{aligned} \tag{4.3.3}$$

On the other hand, let γ be given by (1.6.5) in the proof of Theorem 1.6.1, Chap. 1. Then, we have

$$\begin{aligned}\mathcal{E}^{(\delta)}(e_1^{(\delta)}({}^*B), \gamma_1(B)) &= -\gamma\langle (I + {}^*(\gamma R_\gamma))e_1^{(\delta)}({}^*B), \gamma_1(B)\rangle \\ &= {}^{(\gamma)}F(e_1^{(\delta)}({}^*B), \gamma_1(B)).\end{aligned} \tag{4.3.4}$$

From the inequality (4.3.2) and Corollary 1.6.1 (i), we get

$$
\begin{aligned}
&{}^{(\gamma)}F(e_1^{(\delta)}({}^*B),\gamma_1(B)) = F((-\gamma R_\gamma)e_1^{(\delta)}({}^*B),\gamma_1(B)) \\
&\quad \le C\sqrt{F_1((-\gamma R_\gamma)e_1^{(\delta)}({}^*B),(-\gamma R_\gamma)e_1^{(\delta)}({}^*B))}\sqrt{F_1(\gamma_1(B),\gamma_1(B))}. \qquad (4.3.5)
\end{aligned}
$$

From Corollary 1.6.1 (ii), we have

$$
\begin{aligned}
&F_1((-\gamma R_\gamma)e_1^{(\delta)}({}^*B),(-\gamma R_\gamma)e_1^{(\delta)}({}^*B)) \\
&= F((-\gamma R_\gamma)e_1^{(\delta)}({}^*B),(-\gamma R_\gamma)e_1^{(\delta)}({}^*B)) + \langle(-\gamma R_\gamma)e_1^{(\delta)}({}^*B),(-\gamma R_\gamma)e_1^{(\delta)}({}^*B)\rangle \\
&\le {}^{(\gamma)}F(e_1^{(\delta)}({}^*B),e_1^{(\delta)}({}^*B)) + \langle e_1^{(\delta)}({}^*B),e_1^{(\delta)}({}^*B)\rangle \\
&= -\gamma\langle(I+{}^*(\gamma R_\gamma))e_1^{(\delta)}({}^*B),e_1^{(\delta)}({}^*B)\rangle + \langle e_1^{(\delta)}({}^*B),e_1^{(\delta)}({}^*B)\rangle \\
&= \mathcal{E}_1^{(\delta)}(e_1^{(\delta)}({}^*B),e_1^{(\delta)}({}^*B)). \qquad (4.3.6)
\end{aligned}
$$

Therefore, we have from the relations (4.3.3), (4.3.4), (4.3.5), and (4.3.6) that

$$
\begin{aligned}
{}^\circ\left(\mathrm{Cap}_1^{(\delta)}({}^*B\cap S_0)\right) &= {}^\circ\left(\mathcal{E}_1^{(\delta)}(e_1^{(\delta)}(1_{{}^*B}),e_1^{(\delta)}(1_{{}^*B}))\right) \\
&\le (C+1)^2F_1(\gamma_1(B),\gamma_1(B)).
\end{aligned}
$$

This finishes the proof of Proposition 4.3.2. □

Corollary 4.3.1. *Let $A\subset Y$ satisfy the condition $\Gamma_1(A)=0$. Then for all $\delta\in T,\delta\approx 0$, we have $\mathrm{Cap}_1^{(\delta)}({}^*A\cap S_0)\approx 0$.*

Proof. The proof follows easily from Proposition 4.3.2. □

Corollary 4.3.2. *Let $\{G_n\mid n\in\mathbb{N}\}$ be a sequence of increasing closed subsets of Y satisfying $\Gamma_1(Y-G_n)\downarrow 0$. Then for all $\delta\in T,\delta\approx 0$, we have*

$$
{}^\circ\left(\mathrm{Cap}_1^{(\delta)}\left(S_0-\bigcup_{n=1}^{\infty}({}^*G_n\cap S_0)\right)\right)=0. \qquad (4.3.7)
$$

Proof. From Proposition 4.3.2, we have

$$
\begin{aligned}
{}^\circ\left(\mathrm{Cap}_1^{(\delta)}(S_0-{}^*G_n\cap S_0)\right) &= {}^\circ\left(\mathrm{Cap}_1^{(\delta)}({}^*(Y-G_n)\cap S_0)\right) \\
&\le (C+1)^2\Gamma_1(Y-G_n) \\
&\longrightarrow 0, n\to\infty.
\end{aligned}
$$

Therefore, (4.3.7) holds. □

Proposition 4.3.3. *Let $\{G_n\mid n\in\mathbb{N}\}$ be a sequence of increasing closed subsets of Y satisfying $\Gamma_1(Y-G_n)\downarrow 0$. Then $S_0-\bigcup_{n=1}^{\infty}{}^*G_n\cap S_0$ is δ-exceptional for all $\delta\in T,\delta\approx 0$.*

Proof. The proof follows easily from Corollary 4.3.2 and Theorem 2.4.1. □

4.4 Path Regularity of Hyperfinite Markov Chains

Similarly as in Definition 3.2.1, we introduce the following

Definition 4.4.1. Let Z be a subset of Y. We say that a subset π of $L^2(Y,\nu)$ *separates points* of Z, if for any two different points x and y in Z, there exists $u \in \pi$ such that $u(x) \neq u(y)$.

Lemma 4.4.1. *If the hypothesis (III) in Theorem 4.1.1 holds, then π separates the points of $\bigcup_{n\in\mathbb{N}} Y_n$.*

Proof. The proof can be carried through in the same way as the proof of Lemma 3.2.7. □

Lemma 4.4.2. *Assume that Y is a regular Hausdorff space and the hypothesis (III) of Theorem 4.1.1 holds. Let $\delta \in T, \delta \approx 0$. For every $\omega \in \Omega$, if the path $X^{(\delta)}(\omega,t)$ fails to have an S-left or S-right limit at some $t < \tau^{(\delta)}(\omega)$, then so does $(P^*u)(X^{(\delta)}(\omega,t))$ for some $u \in \pi$, where the stopping time $\tau^{(\delta)}(\omega)$ is defined by*

$$\tau^{(\delta)}(\omega) = \inf\left\{ {}^\circ t \middle| X^{(\delta)}(\omega,t) \notin \bigcup_{n\in\mathbb{N}} ({}^*Y_n \cap S_0) \right\}.$$

Proof. Let us set $B_n = {}^*Y_n \cap S_0, n \in \mathbb{N}$ and $B = \cup_{n\in\mathbb{N}} B_n$. For each $n \in \mathbb{N}$, we define

$$\tau_n^{(\delta)}(\omega) = \min\left\{ t \in T \middle| X^{(\delta)}(\omega,t) \notin B_n \right\},$$

$$\zeta_\delta(\omega) = \inf\left\{ {}^\circ t \middle| X^{(\delta)}(\omega,t) \notin \overline{S}_0 \right\},$$

i.e., ζ_δ is the lifetime of $X^{(\delta)}(\omega,t)$. It is easy to show that

$$\tau^{(\delta)}(\omega) \leq \zeta_\delta(\omega),$$

$$\tau^{(\delta)}(\omega) = \sup\left\{ {}^\circ\tau_n^{(\delta)}(\omega) \middle| n \in \mathbb{N} \right\}.$$

Let ω be a fixed point in Ω. Fix $t < \tau^{(\delta)}(\omega), t \in [0,\infty)$. Then, there is an element $N \in \mathbb{N}$ such that $t < \tau_N^{(\delta)}(\omega)$. Given a sequence $\{t_n \mid n \in \mathbb{N}\}$ from T such that the standard part ${}^\circ t_n$ increases strictly to t, we can show that the sequence

$$\left\{ {}^\circ X^{(\delta)}(\omega,t_n) \middle| n \in \mathbb{N} \right\}$$

has a cluster point in Y_N in same way as the proof of Lemma 3.2.1. The rest follows in a similar way. □

Proposition 4.4.1. *Assume that Y is a regular Hausdorff space and the hypothesis (III) of Theorem 4.1.1 holds. Let $\delta \in T, \delta \approx 0$. Then there exists a δ-exceptional set $A_0(\delta)$ such that for all $s_i \in S_0 - A_0(\delta)$, the hyperfinite Markov chain $X^{(\delta)}(t)$ has S-left and S-right limits at all $t < \zeta_\delta, L(P_i)$-a.e.*

Proof. The result can be proved by using the method of Proposition 3.2.1, and the results of Lemmas 4.2.1 and 4.4.2. □

4.5 Quasi-Continuity and Nearstandard Concentration

Let $(F(\cdot,\cdot), D(F))$ be a Dirichlet form on $L^2(Y,\nu)$. First of all, we present some properties of $E(\cdot,\cdot)$.

Lemma 4.5.1. *Let $\{f_k \mid k \in \mathbb{N}\}$ be a sequence of quasi-continuous functions on Y. Then there exists a nest $\{G_n \mid n \in \mathbb{N}\}$ such that for all $k \in \mathbb{N}$, $f_k \in C(\{G_n\})$, where*

$$C(\{G_n\}) = \{f \mid f|_{G_n} \quad \textit{is continuous for each} \quad n \in \mathbb{N}\}.$$

Proof. The proof is similar to the one in Fukushima [175] Theorem 3.1.2. □

Lemma 4.5.2. *(1) If the condition (I) in Theorem 4.1.1 is satisfied, we have*

$$F_1(u-u_n, u-u_n) \longrightarrow 0 \quad as \quad n \longrightarrow \infty, \forall u \in D(F),$$

where $u_n \in D(F)_{Y_n}$ is the projection of u.

(2) Suppose that the condition (II) in Theorem 4.1.1 holds. Then, every element $f \in D(F)$ admits a quasi-continuous version.

Proof. (1) Let $\overline{F}(\cdot,\cdot)$ be the symmetric part of $F(\cdot,\cdot)$, i.e.,

$$\overline{F}(u,v) = \frac{1}{2}\Big(F(u,v) + F(v,u)\Big), u, v \in D(F).$$

Set

$$\overline{F}_1(u,v) = \overline{F}(u,v) + (u,v), u, v \in D(F).$$

For every $u \in D(F)$, we have the following decomposition with respect to the inner product $\overline{F}_1(\cdot,\cdot)$

$$u = (u-u_n) + u_n, u_n \in D(F)_{Y_n}, u_n - u \bot D(F)_{Y_n}, \forall n \in \mathbb{N}.$$

Let $v_n = u - u_n$. Then, we have

$$\begin{aligned}\overline{F}_1(v_n, v_{n+m}) &= \overline{F}_1(v_{n+m}, v_{n+m}) + \overline{F}_1(v_n - v_{n+m}, v_{n+m}) \\ &= \overline{F}_1(v_{n+m}, v_{n+m}) + \overline{F}_1(u_{n+m} - u_n, u - u_{n+m}) \\ &= \overline{F}_1(v_{n+m}, v_{n+m}) + \overline{F}_1(-u_n, u - u_{n+m}).\end{aligned}$$

Since $u_n \in D(F)_{Y_n} \subset D(F)_{Y_{n+m}}, \overline{F}_1(-u_n, u - u_{n+m}) = 0$. Hence, we have

$$\overline{F}_1(v_n, v_{n+m}) = \overline{F}_1(v_{n+m}, v_{n+m}).$$

Therefore, we have

$$\begin{aligned}F_1(v_n - v_{n+m}, v_n - v_{n+m}) &= F_1(v_n, v_n) - F_1(v_{n+m}, v_{n+m}) \\ &\longrightarrow 0, n \to \infty.\end{aligned}$$

This means that $\{v_n \mid n \in \mathbb{N}\}$ is a $F_1(\cdot, \cdot)$ Cauchy sequence. Hence, there is an element $v_\infty \in D(F)$ such that

$$F_1(v_n - v_\infty, v_n - v_\infty) \longrightarrow 0, n \longrightarrow \infty.$$

It is easy to see that

$$F_1(v_\infty, v) = 0, \forall v \in \bigcup_{n \in \mathbb{N}} D(F)_{Y_n}.$$

Since $\bigcup_{n\in\mathbb{N}} D(F)_{Y_n}$ is F_1-dense in $D(F)$, we know that $v_\infty = 0$. This proves (1).

(2) For this proof, we refer to Albeverio and Ma [41], 4.1 Lemma and Fukushima [175], Theorem 3.1.3. □

Lemma 4.5.3. *Suppose that the condition (I) in Theorem 4.1.1 is satisfied. Then* $(F_1(\cdot, \cdot), D(F))$ *is a separable Hilbert space.*

Proof. Since Y_n is compact, $(F_1(\cdot, \cdot), D(F)_{Y_n})$ is separable. Hence $(F(\cdot, \cdot), D(F))$ is separable. □

Lemma 4.5.4. *Suppose that the hypotheses (I), (II) and (III) in Theorem 4.1.1 hold. Then, there exists a countable set* $\hat{\pi}$ *such that*

$$\begin{aligned}&(1) \quad \pi \subset \hat{\pi} \subset D(F) \cap Q_b(Y), &(4.5.1)\\ &(2) \quad u, v \in \hat{\pi}, a \in Q \Longrightarrow |u| \in \hat{\pi}, u + v \in \hat{\pi}, uv \in \hat{\pi}, \\ &\qquad av \in \hat{\pi} \text{ and } u \wedge a \in \hat{\pi}, &(4.5.2)\\ &(3) \quad \hat{\pi} \text{ is } F_1\text{-dense in } D(F),\end{aligned}$$

where $Q = \{a_1, a_2, \cdots, a_n, \cdots\}$ *is the set of all rational numbers and* $Q_b(Y)$ *is the space of all bounded quasi-continuous functions defined on* Y. *Moreover, there is a nest* $\{Z_n \mid n \in \mathbb{N}\}$ *such that for every* $u \in \hat{\pi}$, $u \in C(\{Z_n\})$.

Proof. From Lemmas 4.5.2 and 4.5.3, we can assume that π is F_1-dense in $D(F)$. Define the maps

$$\Lambda = \{\lambda_{-2}, \lambda_{-1}, \lambda_0, \lambda_1, \lambda_2, \cdots\}$$

from $[D(F) \cap Q_b(Y)] \times [D(F) \times Q_b(Y)]$ into $D(F) \cap (Q_b(Y)$ by

$$\begin{aligned}
\lambda_{-2}(u,v) &= |u|,\\
\lambda_{-1}(u,v) &= u+v,\\
\lambda_0(u,v) &= uv,\\
\lambda_{2i-1}(u,v) &= a_i u,\\
\lambda_{2i}(u,v) &= u \wedge a_i, i=1,2,\cdots
\end{aligned}$$

Applying Lemma 3.2.4 to π and Λ, we get a countable set $\hat{\pi}$ satisfying the relations (4.5.1) and (4.5.2). Obviously, $\hat{\pi}$ is F_1-dense in $D(F)$. The existence of $\{Z_n \mid n \in \mathbb{N}\}$ follows from Lemma 4.5.1. □

Lemma 4.5.5. *Assume that the hypotheses (I), (II) and (III) in Theorem 4.1.1 hold. For any two finite measures ν_1 and ν_2 on $(Y, \mathcal{B}(Y))$ satisfying the following conditions:*

$$\int_Y u(y)\ \nu_1(dy) = \int_Y u(y)\ \nu_2(dy) \quad \textit{for all} \quad u \in \hat{\pi},$$

where $\hat{\pi}$ is the countable set obtained in the Lemma 4.5.4, we have that ν_1 and ν_2 coincide on $(Y, \sigma(u|u \in \hat{\pi}))$.

Proof. The proof is similar to the one of Lemma 3.2.6. □

We recall that we have introduced the hyperfinite quasi-continuity and related concepts of $(\mathcal{E}(\cdot,\cdot), \mathcal{D}(\mathcal{E}))$ in Sect. 3.2. We then have

Proposition 4.5.1. *Let us assume that $f : Y \longrightarrow \mathbb{R}$ is a quasi-continuous function. Then for all $\delta \in T, \delta \approx 0, P^*f$ is hyperfinite δ-quasi-continuous.*

Proof. The proof follows easily from Corollary 4.3.2 and Theorem 2.4.1. □

Corollary 4.5.1. *Let us assume that the condition (II) in Theorem 4.1.1 holds. Then for every element $f \in D(F)$ and all infinitesimal $\delta \in T$, there is a δ-exceptional set $\underline{A}(\delta) \subset S_0$ such that $(P^*f)(s_i) \approx (P^*f)(s_j)$ for all infinitely close $s_i, s_j \in \overline{S}_0 - A(\delta)$.*

Proof. From Lemma 4.5.2 (2), there is a quasi-continuous version $\tilde{f}$ of f. Hence, $P^*\tilde{f}$ is hyperfinite δ-quasi-continuous, $\forall \delta \in T, \delta \approx 0$. Since $P^*f \approx P^*\tilde{f}$, we prove Corollary 4.5.1 by using Lemma 2.1.1 (i). □

Proposition 4.5.2. *Let Y be a regular Hausdorff space. If the conditions (I), (II), and (III) in Theorem 4.1.1 are satisfied, then there exist a δ_0-exceptional set $A_1(\delta_0)$ of $X(t)$ for some infinitesimal δ_0 and some $\beta \in T_{\delta_0}, \frac{1}{2} < \beta \leq 1$ such that*

(i) For all $s_i \in S_0 - A_1(\delta_0)$, the path $X^{(\delta_0)}(\omega, t)$ has S-right and S-left limits at all $t < \zeta_{\delta_0}(\omega), L(P_i)$-a.e.$\omega$.
(ii) Let us set $\hat{Q}(\delta, \beta) = \bigcup_{n=1}^{\infty} T_{2^{-n}\beta}$. For all infinitely close $s_i, s_j \in \overline{S}_0 - A_1(\delta_0)$, $s \in \hat{Q}(\delta_0, \beta)$, $u \in \hat{\pi}$, we have

$$
{}^{\circ}((P^* T_{\mathrm{st}(s)} u)(s_i)) = {}^{\circ}\!\!\int_{\Omega} (P^* u)(X^{(\delta_0)}(s))\ P_i(d\omega)
$$
$$
= {}^{\circ}\!\!\int_{\Omega} (P^* u)(X^{(\delta_0)}(s))\ P_j(d\omega). \tag{4.5.3}
$$

(iii) For all $s_i \in S_0 - A_1(\delta_0), u \in \hat{\pi}$, ${}^{\circ}[P^*(T_t u)(s_i)] : [0, \infty) \longrightarrow \mathbb{R}$ is continuous with respect to t. Moreover, for all $s_i \in S_0 - A_1(\delta_0), u \in \hat{\pi}$, $s \in T_{\delta_0}^{\mathrm{fin}}$, we have

$$
P^*(T_{\mathrm{st}(s)} u)(s_i) \approx Q^s(P^* u)(s_i).
$$

Proof. Step 1. For $s \in T^{\mathrm{fin}}$, let $t = \mathrm{st}(s)$. If $u \in \hat{\pi}$, $P^* u$ is a 2-lifting of u. Therefore, $Q^s(P^* u)$ is a 2-lifting of $T_t u$ from Proposition 4.2.3. For all $k \in \mathbb{N}$, we have

$$
\mu\left\{ s_i \in S_0 \middle| |Q^s(P^* u)(s_i) - P^*(T_t u)(s_i)| > \frac{1}{k} \right\} < \frac{1}{k}. \tag{4.5.4}
$$

Therefore, (4.5.4) holds for some infinite $k \in {}^*\mathbb{N} - \mathbb{N}$ also. This means that

$$
\left\{ s_i \middle| Q^s(P^* u)(s_i) \not\approx P^*(T_t u)(s_i) \right\}
$$

is contained in an internal set of infinitesimal measure. Having this in mind, we shall show our proposition step by step.

Step 2. Suppose that $\hat{\pi} = \{u_n \mid n \in \mathbb{N}\}$. Let $\{P^* u_n \mid n \in {}^*\mathbb{N}\}$ be an internal extension of $\{P^* u_n \mid n \in \mathbb{N}\}$. For $\varepsilon \in T^1, N \in {}^*\mathbb{N}$, define

$$
B_1(\varepsilon) = \bigcup_{n \in \mathbb{N}} \left\{ s_i \middle| \exists s \in T_{\varepsilon}^{\mathrm{fin}} \left(Q^s(P^* u_n)(s_i) \not\approx P^*(T_{\mathrm{st}(s)} u_n)(s_i) \right) \right\},
$$
$$
B_1(\varepsilon, N) = \bigcup_{n=1}^{N} \left\{ s_i \middle| \exists s \in T_{\varepsilon}^{N} \left(Q^s(P^* u_n)(s_i) \not\approx P^*(T_{\mathrm{st}(s)} u_n)(s_i) \right) \right\}, \tag{4.5.5}
$$
$$
\hat{B}_1(\varepsilon, N) = \bigcup_{n=1}^{N} \left\{ s_i \middle| \exists s \in T_{\varepsilon}^{N} \left(|Q^s(P^* u_n)(s_i) - P^*(T_{\mathrm{st}(s)} u_n)(s_i)| > \frac{1}{N} \right) \right\}.
$$

Fix $\varepsilon \not\approx 0, \varepsilon \in T^1$. For each $N \in \mathbb{N}$, we have

$$\mu(\hat{B}_1(\varepsilon, N)) < \frac{1}{N}.$$

Consider the following internal set

$$\left\{\varepsilon \in T^1 \middle| \mu(\hat{B}_1(\varepsilon, N(\varepsilon)) < \frac{1}{N(\varepsilon)}\right\}, N(\varepsilon) = \left[\frac{1}{\varepsilon}\right] + 1.$$

By saturation, there is an infinitesimal $\varepsilon_1 \in T^1$ such that

$$\mu\left(\hat{B}_1(\varepsilon_1, N(\varepsilon_1))\right) < \frac{1}{N(\varepsilon_1)}.$$

In view of our Lemma 2.1.1, there is an infinitesimal $\delta_1 \in T_{\varepsilon_1}$ such that $\hat{B}_1(\varepsilon_1, N(\varepsilon_1))$ is δ-exceptional for all $\delta \geq \delta_1, \delta \in T_{\varepsilon_1}$. Therefore, $\hat{B}_1(\delta, N(\delta))$ is δ-exceptional for all $\delta \in T_{\varepsilon_1}, \delta \geq \delta_1$. Noticing that

$$B_1(\delta) \subset B_1(\delta, N(\delta)) \subset \hat{B}_1(\delta, N(\delta)),$$

we know that $B_1(\delta)$ is δ-exceptional for all $\delta \in T_{\varepsilon_1}, \delta \geq \delta_1$.

Step 3. We take $n_0 \in {}^*\mathbb{N} - \mathbb{N}$ satisfying $\frac{1}{2} < 2^{n_0}\delta_1 \leq 1$. Let $\beta = 2^{n_0}\delta_1$ and $\hat{Q}(\beta) = \bigcup_{n=1}^{\infty} T_{2^{-n}\beta}$. Define

$$B_2 = \bigcup_{u \in \hat{\pi}} \left\{ s_i \in \overline{S}_0 \middle| \exists s \in \hat{Q}(\beta) \left(Q^s(P^*u)(s_i) \not\approx Q^s(P^*u)(s_j) \text{ for some } s_j \approx s_i\right)\right\}.$$

Since $P^*(T_{\mathrm{st}(s)}u)$ is hyperfinite δ-quasi-continuous for all $\delta \in T_1, \delta \approx 0$, and $Q^s(P^*u) \approx P^*(T_{\mathrm{st}(s)}u)$, it follows from Lemma 2.1.1 that there is an infinitesimal $\delta_2 \in T$ such that B_2 is δ-exceptional for all $\delta \geq \delta_2, \delta \in T$.

Step 4. We pick an infinite number $m_0 \leq n_0$ such that $\delta_0 \hat{=} 2^{n_0 - m_0}\delta_1 \approx 0$ and $\delta_0 \geq \delta_2$. Let $A_0(\delta_0)$ be the δ_0-exceptional set obtained in Proposition 4.4.1. Then, there exists a δ_0-exceptional set $A_1(\delta_0)$ which contains $A_0(\delta_0), B_1(\delta_0)$ and B_2. Obviously, $A_1(\delta_0)$ satisfies (i). For $s_i \in S_0 - A_1(\delta_0), u \in \hat{\pi}, s \in T_{\delta_0}^{\mathrm{fin}}$, we have from the relations (4.5.5) that

$$Q^s(P^*u)(s_i) \approx P^*(T_{\mathrm{st}(s)}u)(s_i),$$

since $T_{\delta_0} \subset T_{\delta_1} \subset T_{\varepsilon_1}$. Thus, $\overset{\circ}{(}P^*(T_t u)(s_i)) : [0, \infty) \longrightarrow \mathbb{R}$ is continuous. This completes the proof of Proposition 4.5.2. □

Lemma 4.5.6. *Suppose that the condition (I) in Theorem 4.1.1 holds. For every $u \in D(F), s \in [0,\infty)$, the set*

$$B(u,s) = \left\{ s_i \in S_0 - \bigcup_{n\in\mathbb{N}} ({}^*Y_n \cap S_0) \middle| P^*(T_s u)(s_i) \not\approx 0 \right\}$$

is Δt-exceptional.

Proof. From Lemma 4.5.2, we have

$$F_1(u-u_n, u-u_n) \longrightarrow 0 \quad \text{as} \quad n \longrightarrow \infty, \forall u \in D(F),$$

where $u_n \in D(F)_{Y_n}$ is the projection of u. For each $k \in \mathbb{N}$, let us take $n_k \in \mathbb{N}$ such that

$$\delta_k = \sqrt{F_1(T_s u - (T_s u)_{n_k}, T_s u - (T_s u)_{n_k})} < 2^{-k},$$

where $(T_s u)_n \in D(F)_{Y_n}$ is the projection of $T_s u$. For simplicity, we may suppose that $(T_s u)_{n_k}|_{Y-Y_{n_k}} = 0$. Define

$$B_k(u,s) = \left\{ s_i \in S_0 - {}^*Y_{n_k} \cap S_0 \middle| |P^*(T_s u)(s_i)| \geq \sqrt{\delta_k} \right\}, k \in \mathbb{N}.$$

Noticing that

$$|P^*(T_s u)(s_i) - P^*(T_s u)_{n_k}(s_i)| \geq \sqrt{\delta_k}, s_i \in B_k(u,s),$$

we have

$$\begin{aligned}
&P\left(\omega \middle| \exists t \leq 1\,(X(\omega,t) \in B_k(u,s))\right) \\
&\leq P\left[\omega \middle| \exists t \leq 1\left(|P^*(T_s u)(X(\omega,t)) - P^*(T_s u)_{n_k}(X(\omega,t))| \geq \sqrt{\delta_k}\right)\right] \\
&\leq P\left[\omega \middle| \exists t \leq 1\left(P^*(T_s u)(X(\omega,t)) - P^*(T_s u)_{n_k}(X(\omega,t)) \geq \sqrt{\delta_k}\right)\right] \\
&\quad + P\left[\omega \middle| \exists t \leq 1\left(P^*(T_s u)(X(\omega,t)) - P^*(T_s u)_{n_k}(X(\omega,t)) \leq -\sqrt{\delta_k}\right)\right].
\end{aligned} \tag{4.5.6}$$

Let us set

$$\begin{aligned}
A &= \{i \in S_0 \mid P^*(T_s u)(s_i) - P^*(T_s u)_{n_k}(s_i) \geq \sqrt{\delta_k}\}, \\
\sigma_A(\omega) &= \min\{t \in T \mid X(t) \in A\}, \\
e_1(A)(i) &= E_i\left[(1+\Delta)^{-\sigma_A/\Delta t}\right].
\end{aligned}$$

Then, we have

$$\begin{aligned}
&P\left[\omega \middle| \exists t \leq 1 \left(P^*(T_s u)(X(\omega,t)) - P^*(T_s u)_{n_k}(X(\omega,t)) \geq \sqrt{\delta_k}\right)\right] \\
&\quad = P(\omega | \sigma_A(\omega) \leq 1) \\
&\quad = \int_{S_0} P_i \left\{\omega \middle| (1+\Delta t)^{-\sigma_A/\Delta t} \geq (1+\Delta t)^{-1/\Delta t}\right\} dm(i) \\
&\quad \leq \int_{S_0} E_i \left[(1+\Delta t)^{-\sigma_A/\Delta t} (1+\Delta t)^{1/\Delta t}\right] dm(i) \\
&\quad = (1+\Delta t)^{1/\Delta t} \int_{S_0} e_1(A)(i)\, dm(i) \\
&\quad \leq (1+\Delta t)^{1/\Delta t}\, \mathcal{E}_1(e_1(A), e_1(A)),
\end{aligned} \tag{4.5.7}$$

where the reason for the last step holding is (2.3.1) in Sect. 2.3.

From Theorem 2.2.1, we have

$$\begin{aligned}
\mathcal{E}_1(e_1(A), e_1(A)) &\leq \mathcal{E}_1\left(e_1(A), \frac{1}{\sqrt{\delta_k}}\left(P^*(T_s u) - P^*(T_s u)_{n_k}\right)\right) \\
&= F^{(\gamma)}\left(e_1(A), \frac{1}{\sqrt{\delta_k}}\left(P^*(T_s u) - P^*(T_s u)_{n_k}\right)\right) \\
&\quad + \frac{1}{\sqrt{\delta_k}} \int_{S_0} e_1(A)(i)\left(P^*(T_s u) - P^*(T_s u)_{n_k}\right)(i) dm(i) \\
&= F\left((-\gamma R_\gamma) e_1(A), \frac{1}{\sqrt{\delta_k}}\left(P^*(T_s u) - P^*(T_s u)_{n_k}\right)\right) \\
&\quad + \frac{1}{\sqrt{\delta_k}} \int_{S_0} e_1(A)(i)\left(P^*(T_s u) - P^*(T_s u)_{n_k}\right)(i) dm(i),
\end{aligned} \tag{4.5.8}$$

where γ is given by (1.6.5) in the proof of Theorem 1.6.1. From the inequality (4.3.2) in Sect. 4.3, we have

$$\begin{aligned}
&F\left((-\gamma R_\gamma) e_1(A), \frac{1}{\sqrt{\delta_k}}\left(P^*(T_s u) - P^*(T_s u)_{n_k}\right)\right) \\
&\leq C \frac{1}{\sqrt{\delta_k}} \sqrt{F_1\left((-\gamma R_\gamma) e_1(A), (-\gamma R_\gamma) e_1(A)\right)} \\
&\times \sqrt{F_1\left(P^*(T_s u) - P^*(T_s u)_{n_k}, P^*(T_s u) - P^*(T_s u)_{n_k}\right)}.
\end{aligned} \tag{4.5.9}$$

In the same way as for the relation (4.3.6) in Sect. 4.3, we can show that

$$F_1\left((-\gamma R_\gamma) e_1(A), (-\gamma R_\gamma) e_1(A)\right) \leq \mathcal{E}_1(e_1(A), e_1(A)).$$

Hence, we have from the inequality (4.5.9) that

$$F\left((-\gamma R_\gamma)e_1(A), \frac{1}{\sqrt{\delta_k}}\left(P^*(T_s u) - P^*(T_s u)_{n_k}\right)\right)$$
$$\leq C\frac{1}{\sqrt{\delta_k}}\sqrt{\mathcal{E}_1(e_1(A), e_1(A))}$$
$$\times\sqrt{F_1\left(P^*(T_s u) - P^*(T_s u)_{n_k}, P^*(T_s u) - P^*(T_s u)_{n_k}\right)}. \tag{4.5.10}$$

From the relations (4.5.8) and (4.5.10), we have

$$\mathcal{E}_1(e_1(A), e_1(A)) \leq (C+1)\frac{1}{\sqrt{\delta_k}}\sqrt{\mathcal{E}_1(e_1(A), e_1(A))}$$
$$\times\sqrt{F_1\left(P^*(T_s u) - P^*(T_s u)_{n_k}, P^*(T_s u) - P^*(T_s u)_{n_k}\right)}.$$

Therefore, we have

$$\mathcal{E}_1(e_1(A), e_1(A)) \leq \frac{(C+1)^2}{\delta_k}F_1\left(P^*(T_s u) - P^*(T_s u)_{n_k}, P^*(T_s u) - P^*(T_s u)_{n_k}\right)$$
$$\leq (C+1)^2\delta_k. \tag{4.5.11}$$

From the relations (4.5.7) and (4.5.11), we have

$$P\left[\omega\middle|\exists t \leq 1\left(P^*(T_s u)(X(t)) - P^*(T_s u)_{n_k}(X(t)) \geq \sqrt{\delta_k}\right)\right]$$
$$\leq (1+\Delta t)^{1/\Delta t}(C+1)^2\delta_k. \tag{4.5.12}$$

Similarly, we have

$$P\left[\omega\middle|\exists t \leq 1\left(P^*(T_s u)(X(\omega,t)) - P^*(T_s u)_{n_k}(X(\omega,t)) \leq -\sqrt{\delta_k}\right)\right]$$
$$\leq (1+\Delta t)^{1/\Delta t}(C+1)^2\delta_k. \tag{4.5.13}$$

From the relations (4.5.6), (4.5.12), and (4.5.13), we have

$$P\left(\omega\middle|\exists t \leq 1\Big(X(\omega,t) \in B_k(u,s)\Big)\right) \leq 2\Big(1+\Delta t\Big)^{1/\Delta t}(C+1)^2\delta_k$$
$$\leq 2\Big(1+\Delta t\Big)^{1/\Delta t}(C+1)^2 2^{-k}. \tag{4.5.14}$$

For each $K \in \mathbb{N}$, we have $B(u,s) \subset \bigcup_{k=K}^{\infty} B_k(u,s)$. It follows from the inequality (4.5.14) that $B(u,s)$ is Δt-exceptional. □

Lemma 4.5.7. *Suppose that the condition (I) in Theorem 4.1.1 is satisfied. Then, there exists a Δt-exceptional set A_2 containing all $B(u,s), u \in \hat{\pi}$, $s \in Q_+$, where Q_+ is the family of all positive rational numbers.*

Proof. This follows easily from Lemma 2.1.1 and Lemma 4.5.6. □

In Sect. 4.4, we have defined the lifetime ζ_δ of $X^{(\delta)}(\omega,t)$. That is, for $\delta \in T$, we have

$$\zeta_\delta(\omega) = \inf\left\{{}^\circ t \mid X^{(\delta)}(\omega,t) \notin \overline{S}_0\right\}.$$

Now, we define the right standard part ${}^\circ X^{(\delta)+}$ of $X^{(\delta)}$ as follows. If $t < \zeta_\delta(\omega)$, let

$${}^\circ X^{(\delta)+}(\omega,t) = S\text{-}\lim_{s\downarrow t} X^{(\delta)}(\omega,s)$$

whenever this limit exists and for all $t_1 < t$, the $S\text{-}\lim_{s\downarrow t_1} X^{(\delta)}(\omega,s)$ exist also, and ${}^\circ X^{(\delta)+}(\omega,t) = \Delta$ else. If $t \geq \zeta_\delta(\omega)$, we define ${}^\circ X^{(\delta)+}(\omega,t) = \Delta$.

Proposition 4.5.3. *Assume that Y is a regular Hausdorff space. If the conditions (I), (II), and (III) in Theorem 4.1.1 are satisfied, then there exists a properly δ_0-exceptional set $A(\delta_0)$ of $X(t)$ for some infinitesimal δ_0 and some $\beta \in T_{\delta_0}, \frac{1}{2} < \beta \leq 1$ such that*

(i) For all $s_i \in S_0 - A(\delta_0)$, the path $X^{(\delta_0)}(\omega,t)$ has S-right and S-left limits at all $t < \zeta_{\delta_0}(\omega), L(P_i)$-a.e.$\omega$.
(ii) For all infinitely close $s_i, s_j \in \overline{S}_0 - A(\delta_0), u \in \hat{\pi}, t \in T_{\delta_0}^{\text{fin}}$, we have

$$\begin{aligned}\int_\Omega u({}^\circ X^{(\delta_0)}(\omega,t))\ L(P_i)(d\omega) &= \int_\Omega u({}^\circ X^{(\delta_0)}(\omega,t))1_{(t<\zeta_{\delta_0})}\ L(P_i)(d\omega)\\ &= \int_\Omega u({}^\circ X^{(\delta_0)}(\omega,t))1_{(t<\zeta_{\delta_0})}\ L(P_j)(d\omega)\\ &= \int_\Omega u({}^\circ X^{(\delta_0)}(\omega,t))\ L(P_j)(d\omega). \quad (4.5.15)\end{aligned}$$

(iii) For all infinitely close $s_i, s_j \in \overline{S}_0 - A(\delta_0)$, all finite $t \in [0,\infty)$, and every Borel set $B \in \mathcal{B}(Y)$, we have

$$L(P_i)\left({}^\circ X^{(\delta_0)+}(\omega,t) \in B\right) = L(P_j)\left({}^\circ X^{(\delta_0)+}(\omega,t) \in B\right). \quad (4.5.16)$$

(iv) *For all* $s_i \in S_0 - A(\delta_0), u \in \hat{\pi}, {}^\circ[P^*(T_t u)(s_i)] : [0,\infty) \longrightarrow \mathbb{R}$ *is continuous with respect to* t. *Moreover, for all* $s_i \in S_0 - A(\delta_0), u \in \hat{\pi}, s \in T_{\delta_0}^{\text{fin}}$, *we have*

$$P^*(T_{\text{st}(s)}u)(s_i) \approx Q^s(P^*u)(s_i).$$

Proof. Let A_2 be the Δt-exceptional set obtained in Lemma 4.5.7 and let $A_1(\delta_0)$ be the δ_0-exceptional set obtained in Proposition 4.5.2, $\delta_0 \in T, \delta_0 \approx 0$. Let $A(\delta_0)$ be a properly δ_0-exceptional set containing $A_1(\delta_0)$, A_2, and $S_0 - \bigcup_{n\in\mathbb{N}}({}^*Z_n \cap S_0)$, where $\{Z_n \mid n \in \mathbb{N}\}$ is the nest obtained in Lemma 4.5.4.

Let $\{Y_n \mid n \in \mathbb{N}\}$ be the increasing sequence of compact sets in (I). Let us set $B_n = {}^*Y_n \cap S_0, n \in \mathbb{N}$, and $B = \cup_{n\in\mathbb{N}} B_n$. Let $\{B_n \mid n \in {}^*\mathbb{N}\}$ be an increasing extension of $\{B_n \mid n \in \mathbb{N}\}$. Define

$$\tau_n^{(\delta_0)}(\omega) = \min\left\{t \in T_{\delta_0} \middle| X^{(\delta_0)}(\omega,t) \notin B_n\right\}, n \in {}^*\mathbb{N},$$

$$\tau^{(\delta_0)}(\omega) = \inf\left\{{}^\circ t \middle| X^{(\delta_0)}(\omega,t) \notin B\right\}.$$

We have

$$\tau^{(\delta_0)}(\omega) = \sup\left\{{}^\circ\tau_n^{(\delta_0)}(\omega) \middle| n \in \mathbb{N}\right\}. \tag{4.5.17}$$

Given $s_i, s_j \in \overline{S}_0 - A(\delta_0), s_i \approx s_j$, we can find an $\eta \in {}^*\mathbb{N} - \mathbb{N}$ such that

$$\tau^{(\delta_0)}(\omega) = {}^\circ\tau_\eta^{(\delta_0)}(\omega), L(P_i) \text{ and } L(P_j)\text{-a.e.}$$

Since Y_n is compact, $n \in \mathbb{N}$, we have from (4.5.17) that $\tau^{(\delta_0)}(\omega) \leq \zeta_{\delta_0}(\omega)$, $\forall \omega \in \Omega$. Let us pick an internal stopping time $\sigma(\omega)$ such that

$$L(P_k)({}^\circ\sigma = \zeta_{\delta_0}) = 1, k = i,j, \quad \text{and} \quad \tau_\eta^{(\delta_0)}(\omega) \leq \sigma(\omega), \forall \omega \in \Omega.$$

If $t \in T_{\delta_0}^{\text{fin}}, u \in \hat{\pi}$, we have

$$\begin{aligned}\int_\Omega (P^*u)(X^{(\delta_0)}(t))\ P_i(d\omega) &= \int_\Omega (P^*u)(X^{(\delta_0)}(t))1_{(t<\tau_\eta^{(\delta_0)})}\ P_i(d\omega) \\ &\quad + \int_\Omega (P^*u)(X^{(\delta_0)}(t))1_{(t\geq\tau_\eta^{(\delta_0)})}\ P_i(d\omega).\end{aligned} \tag{4.5.18}$$

From Proposition 4.5.2 (iii) and Lemma 4.5.7, we have

$$\int_\Omega (P^*u)(X^{(\delta_0)}(t))1_{(t\geq\tau_\eta^{(\delta_0)})}\ P_k(d\omega) \approx 0, k = i,j. \tag{4.5.19}$$

It follows from the relations (4.5.3), (4.5.18), and (4.5.19) that for all $t \in T_{\delta_0}^{\text{fin}}$

$$\begin{aligned}
{}^{\circ}\!\!\int_{\Omega} (P^* u)(X^{(\delta_0)}(t)) 1_{(t<\tau_\eta^{(\delta_0)})} \ P_i(d\omega) &= {}^{\circ}\!\!\int_{\Omega} (P^* u)(X^{(\delta_0)}(t)) \ P_i(d\omega) \\
&= {}^{\circ}\!\!\int_{\Omega} (P^* u)(X^{(\delta_0)}(t)) \ P_j(d\omega) \\
&= {}^{\circ}\!\!\int_{\Omega} (P^* u)(X^{(\delta_0)}(t)) 1_{(t<\tau_\eta^{(\delta_0)})} \ P_j(d\omega).
\end{aligned} \tag{4.5.20}$$

Therefore, we have from the relations (4.5.19) and (4.5.20) that for all $t \in \hat{Q}(\delta_0, \beta)$

$$\begin{aligned}
\int_{\Omega} u({}^{\circ}X^{(\delta_0)}(t)) 1_{(t<\zeta_{\delta_0})} \ L(P_i)(d\omega) &= {}^{\circ}\!\!\int_{\Omega} (P^* u)(X^{(\delta_0)}(t)) 1_{(t<\sigma)} \ P_i(d\omega) \\
&= {}^{\circ}\!\!\int_{\Omega} (P^* u)(X^{(\delta_0)}(t)) 1_{(t<\tau_\eta^{(\delta_0)})} \ P_i(d\omega) \\
&= {}^{\circ}\!\!\int_{\Omega} (P^* u)(X^{(\delta_0)}(t)) \ P_i(d\omega) \\
&= {}^{\circ}\!\!\int_{\Omega} (P^* u)(X^{(\delta_0)}(t)) \ P_j(d\omega) \\
&= {}^{\circ}\!\!\int_{\Omega} (P^* u)(X^{(\delta_0)}(t)) 1_{(t<\tau_\eta^{(\delta_0)})} \ P_j(d\omega) \\
&= {}^{\circ}\!\!\int_{\Omega} (P^* u)({}^{\circ}X^{(\delta_0)}(t)) 1_{(t<\sigma)} \ P_j(d\omega) \\
&= \int_{\Omega} u({}^{\circ}X^{(\delta_0)}(t)) 1_{(t<\zeta_{\delta_0})} \ L(P_j)(d\omega).
\end{aligned} \tag{4.5.21}$$

From Proposition 4.5.2 (iii), we see that (4.5.15) holds for all $t \in T_{\delta_0}^{\text{fin}}$. For all infinitely close $s_i, s_j \in \overline{S}_0 - A(\delta_0), t \in [0, \infty), u \in \hat{\pi}$, we have from (4.5.21) that

$$\int_{\Omega} u({}^{\circ}X^{(\delta_0)+}(t)) \ L(P_i)(d\omega) = \int_{\Omega} u({}^{\circ}X^{(\delta_0)+}(t)) \ L(P_j)(d\omega). \tag{4.5.22}$$

Because ${}^{\circ}X^{(\delta_0)+}(\omega, t) \in \bigcup_{n=1}^{\infty} Y_n \bigcup \{\Delta\}, L(P_k)$-a.e.$\omega, k = i, j$, we get (4.5.16) from (4.5.22) and Lemma 4.5.5. This completes the proof of Proposition 4.5.3.

□

4.6 Construction of Strong Markov Processes

In this section, we suppose that the conditions (I), (II) and, (III) in Theorem 4.1.1 are satisfied. Let $A(\delta_0)$ be the properly δ_0-exceptional set of $X(t)$ in Proposition 4.5.3, $\delta_0 \approx 0$. Let us set

$$\tau^{(\delta_0)}(\omega) = \inf \left\{ {}^{\circ}t \middle| X^{(\delta_0)}(\omega, t) \notin \bigcup_{n \in \mathbb{N}} ({}^{*}Y_n \cap S_0) \right\},$$

where $\{Y_n \mid n \in \mathbb{N}\}$ is the sequence of compact sets in (I) of Theorem 4.1.1. Let $x : \Omega \times \mathbb{R}_+ \longrightarrow Y_\Delta$ be defined by

(i) if $X(\omega, 0) \notin A(\delta_0)$, then $x(\omega, t) = {}^{\circ}X^{(\delta_0)+}(\omega, t)$ if $t < \tau^{(\delta_0)}(\omega)$ and $x(\omega, t) = \Delta$ if $t \geq \tau^{(\delta_0)}(\omega)$.
(ii) if $X(\omega, 0) \in A(\delta_0)$, then $x(\omega, t) = \mathrm{st}(X(\omega, 0))$ for all $t \in \mathbb{R}_+$.

We call x the *modified standard part* of $X(\omega, t)$.

In the same way as in Sect. 3.1, we denote by $A(\delta_0)^{\circ}$ the *inner standard part* of $A(\delta_0)$, i.e.,

$$A(\delta_0)^{\circ} = \left\{ y \in Y \mid \mathrm{st}^{-1}(y) \cap S_0 \subset A(\delta_0) \right\}.$$

Lemma 4.6.1. *For all infinitely close $s_i, s_j \in \overline{S}_0 - A(\delta_0)$, all finite sequences $t_1 < t_2 < \cdots < t_n$ from $\mathbb{R}_+$, all elements $u_1, u_2, \cdots, u_n \in \hat{\pi}$, and all Borel sets $B_1, B_2, \cdots, B_n$, we have*

$$\int_{\Omega} \prod_{l=1}^{n} u_l({}^{\circ}X^{(\delta_0)+}(t_l)) \; L(P_i)(d\omega) = \int_{\Omega} \prod_{l=1}^{n} u_l({}^{\circ}X^{(\delta_0)+}(t_l)) \; L(P_j)(d\omega) \tag{4.6.1}$$

and

$$L(P_i) \left(\bigcap_{l=1}^{n} \{x(t_l) \in B_l\} \right) = L(P_j) \left(\bigcap_{l=1}^{n} \{x(t_l) \in B_l\} \right). \tag{4.6.2}$$

Proof. The case of $n = 1$ is part of Proposition 4.5.3. If $n = 2$, let us pick $\hat{t}_1, \hat{t}_2 \in T_{\delta_0}^{\mathrm{fin}}$ such that $\hat{t}_1 \approx t_1, \hat{t}_2 \approx t_2, \hat{t}_1 < \hat{t}_2$, and

$$\begin{aligned} &{}^{\circ}\!\!\int_{\Omega} (P^{*}u_1)(X^{(\delta_0)}(\hat{t}_1))(P^{*}u_2)(X^{(\delta_0)}(\hat{t}_2)) \; L(P_k)(d\omega) \\ &= \int_{\Omega} u_1({}^{\circ}X^{(\delta_0)+}(t_1))u_2({}^{\circ}X^{(\delta_0)+}(t_2)) \; L(P_k)(d\omega), k = i, j. \end{aligned}$$

We have from the Markov property

$$\int_\Omega (P^*u_1)(X^{(\delta_0)}(\hat{t}_1))(P^*u_2)(X^{(\delta_0)}(\hat{t}_2))\ P_k(d\omega)$$
$$= \int_\Omega (P^*u_1)(X^{(\delta_0)}(\hat{t}_1))E_{X^{(\delta_0)}(\hat{t}_1)}(P^*u_2)(X^{(\delta_0)}(\hat{t}_2 - \hat{t}_1))\ P_k(d\omega), k = i, j.$$

Define a function $f : \Omega \longrightarrow \mathbb{R}$ as follows:

$$f(\omega) = u_1({}^\circ X^{(\delta_0)}(\omega, t_1))u_2({}^\circ X^{(\delta_0)}(\omega, t_2))1_{(t_2 < \tau^{\delta_0}(\omega))}.$$

Then, the function

$$\omega \mapsto (P^*u_1)(X^{(\delta_0)}(\hat{t}_1))P^*u_2(X^{(\delta_0)}(\hat{t}_2))$$

is a lifting of f with respect to both P_i and P_j. Therefore, we have

$$\int_\Omega u_1({}^\circ X^{(\delta_0)+}(t_1))u_2({}^\circ X^{(\delta_0)+}(t_2))\ L(P_i)(d\omega)$$
$$\approx \int_\Omega (P^*u_1)(X^{(\delta_0)}(\hat{t}_1))(P^*u_2)(X^{(\delta_0)}(\hat{t}_2))\ P_i(d\omega)$$
$$= \int_\Omega (P^*u_1)(X^{(\delta_0)}(\hat{t}_1))E_{X^{(\delta_0)}(\hat{t}_1)}(P^*u_2)(X^{(\delta_0)}(\hat{t}_2 - \hat{t}_1))\ P_i(d\omega)$$
$$\approx \int_\Omega u_1({}^\circ X^{(\delta_0)}(\hat{t}_1)) \left[{}^\circ\!\left(Q^{\hat{t}_1 - \hat{t}_2}(P^*u_2)(X^{(\delta_0)}(\hat{t}_1))\right)\right]\ L(P_i)(d\omega)$$
$$= \int_\Omega u_1({}^\circ X^{(\delta_0)}(t_1))(T_{t_2 - t_1}u_2)({}^\circ X^{(\delta_0)}(t_1))\ L(P_i)(d\omega)$$
$$= \int_\Omega u_1({}^\circ X^{(\delta_0)}(t_1))(T_{t_2 - t_1}u_2)({}^\circ X^{(\delta_0)}(t_1))\ L(P_j)(d\omega)$$
$$\approx \int_\Omega u_1({}^\circ X^{(\delta_0)}(\hat{t}_1)) \left[{}^\circ\!\left(Q^{\hat{t}_2 - \hat{t}_1}(P^*u_2)(X^{(\delta_0)}(\hat{t}_1))\right)\right]\ L(P_j)(d\omega)$$
$$\approx \int_\Omega (P^*u_1)(X^{(\delta_0)}(\hat{t}_1))E_{X^{(\delta_0)}(\hat{t}_1)}(P^*u_2)(X^{(\delta_0)}(\hat{t}_2 - \hat{t}_1))\ P_j(d\omega)$$
$$= \int_\Omega (P^*u_1)(X^{(\delta_0)}(\hat{t}_1))(P^*u_2)(X^{(\delta_0)}(\hat{t}_2))\ P_j(d\omega)$$
$$\approx \int_\Omega u_1({}^\circ X^{(\delta_0)+}(t_1)u_2({}^\circ X^{(\delta_0)+}(t_2))\ L(P_j)(d\omega).$$

Now, it is easy to show that (4.6.1) and (4.6.2) hold if $n = 2$. We can complete the proof of Lemma 4.6.1 by mathematical induction. □

In the following, we will define quantities $\{\Pi_t\}$, $\{\Theta_y\}$, and $\{\hat{\Theta}_y\}$ associated with our standard Markov processes. For $t \in \mathbb{R}_+$, let Π_t° be the σ-algebra generated by the sets

$$\left\{\omega \middle| x(\omega, t_1) \in B_1 \wedge \cdots \wedge x(\omega, t_n) \in B_n\right\},$$

where $0 \leq t_1 < t_2 < \cdots < t_n \leq t$ and $B_1, \cdots, B_n$ are Borel sets in Y. Define

$$\Pi_t = \bigcap_{s>t} \Pi_s^\circ.$$

The filtration $\{\Pi_t \mid t \in [0, \infty)\}$ is right continuous. Let us set

$$\Pi_\infty = \vee \{\Pi_t \mid t \in [0, \infty)\}.$$

If $y \in A(\delta_0)^\circ \cup \{\Delta\}, C \in \Pi_\infty$, let

$$\Theta_y(C) = \begin{cases} 1, & \text{if } C \text{ contains all constant paths } x(t) = y, \\ 0, & \text{else.} \end{cases}$$

If $y \notin A(\delta_0)^\circ$, let for all $C \in \Pi_\infty$

$$\Theta_y(C) = L(P_i)(C) \text{ for all } s_i \in \mathrm{st}^{-1}(y) - A(\delta_0).$$

Similarly, we can define $\{\hat{\Theta}_y\}$.

Proposition 4.6.1. *$(\Omega, \Pi, \Pi_t, x(t), \Theta_y)$ and $(\Omega, \Pi, \Pi_t, x(t), \hat{\Theta}_y)$ are ν-tight dual strong Markov processes which are properly associated with the Dirichlet form $(F(\cdot,\cdot), D(F))$.*

Proof. Obviously, $(\Omega, \Pi, \Pi_t, x(t), \Theta_y)$ and $(\Omega, \Pi, \Pi_t, x(t), \hat{\Theta}_y)$ satisfy the conditions (ii), (iii) and (iv) of Definition 3.1.1 by our construction. Moreover, they are ν-tight and dual with respect to ν. We shall prove that they satisfy (i) and (v) of Definition 3.1.1.

Given a Radon probability measure μ on Y, we define two new measures μ_0 and μ_1 on Y by

$$\mu_0(B) = \mu(B - A(\delta_0)^\circ),$$
$$\mu_1(B) = \mu(B \cap A(\delta_0)^\circ).$$

By Corollary 3.1.1, we can find an internal measure λ on S_0 satisfying

$$\mu_0 = L(\lambda) \circ \mathrm{st}^{-1},$$
$$L(\lambda)(A(\delta_0)) = 0. \tag{4.6.3}$$

We first prove that $(\Omega, \Pi, \Pi_t, x(t), \Theta_y)$ satisfies Definition 3.1.1 (i):

(i) Given a Borel set $E \subset Y$, and $t \in \mathbb{R}_+$, we must show that

$$y \mapsto \Theta_y(x(t) \in E)$$

is μ-measurable for all finite Radon measures μ. Since for $\alpha \in [0,1]$

$$\{y \in A(\delta_0)^\circ \mid \Theta_y\{x(t) \in E\} > \alpha\} = \begin{cases} A(\delta_0)^\circ \cap E, & \alpha \in [0,1) \\ \emptyset, & \alpha = 1 \end{cases}$$

is universally measurable by Lemma 3.1.2, it suffices to show that

$$Y - A(\delta_0)^\circ \ni y \mapsto \Theta_y(x(t) \in E)$$

is μ_0-measurable.

Fix $u \in \hat{\pi}$. From (4.5.20), (4.5.21), and (4.5.22) in Sect. 4.5, we can choose $\hat{t} \approx t$ so large that

$$^\circ\!\!\int_\Omega (P^* u)(X^{(\delta_0)}(\hat{t}))\ P_\lambda(d\omega) = \int_\Omega u({}^\circ X^{(\delta_0)+}(t))\ L(P_\lambda)(d\omega).$$

Define a function $f : Y \longrightarrow \mathbb{R}$ as follows:

(1) If $y \notin A(\delta_0)^\circ$, let us define

$$\begin{aligned} f(y) &= \int_\Omega u({}^\circ X^{(\delta_0)+}(t))\ L(P_i)(d\omega) \\ &= \Theta_t u(y) \end{aligned}$$

for some $s_i \in \mathrm{st}^{-1}(y) \cap S_0 - A(\delta_0)$;

(2) If $y \in A(\delta_0)^\circ$, define $f(y)$ arbitrarily.

The function

$$s_i \mapsto \int_\Omega u(X^{(\delta_0)}(\hat{t})) P_i(d\omega)$$

is a lifting of f with respect to λ. Therefore, $y \mapsto \Theta_t u(y)$ is μ_0-measurable. For $a \in Q$, we can show that the map

$$y \mapsto \Theta_y\{u(x(t)) > a\}$$

is μ_0-measurable, since $u_n = [n(u - u \wedge a)] \wedge 1 \in \hat{\pi}$ and $u_n \uparrow 1_{(u>a)}$. Hence,

$$y \mapsto \Theta_y\{x(t) \in E\}$$

is μ_0-measurable for all $E \in \mathcal{B}(Y)$ by a monotone class Theorem.

Similarly, we may prove that $(\Omega, \Pi, \Pi_t, x(t), \hat{\Theta}_y)$ also satisfies Definition 3.1.1 (i).

(v) In order to show that $(\Omega, \Pi, \Pi_t, x(t), \Theta_y)$ satisfies (3.1.9) in Sect. 3.1, it suffices to prove that for all $\{\Pi_t\}$ stopping times σ, $u_1, u_2 \in \hat{\pi}$ and $s \in [0, \infty)$, we have

$$\int_Y E_{\Theta_y}\left(u_1(x(\sigma))u_2(x(\sigma+s))\right)\ \mu(dy) \\ = \int_Y E_{\Theta_y}\left(u_1(x(\sigma))E_{\Theta_{x(\sigma)}}u_2(x(s))\right)\ \mu(dy). \tag{4.6.4}$$

First we notice that since the paths of x are constant Θ_{μ_1}-a.e., we have

$$\int_Y E_{\Theta_y}\left(u_1(x(\sigma))u_2(x(\sigma+s))\right)\ \mu_1(dy) \\ = \int_Y E_{\Theta_y}\left(u_1(x(\sigma))E_{\Theta_{x(\sigma)}}u_2(x(s))\right)\ \mu_1(dy).$$

Equation (4.6.4) will hold if we can prove

$$\int_Y E_{\Theta_y}\left(u_1(x(\sigma))u_2(x(\sigma+s))\right)\ \mu_0(dy) \\ = \int_Y E_{\Theta_y}\left(u_1(x(\sigma))E_{\Theta_{x(\sigma)}}u_2(x(s))\right)\ \mu_0(dy). \tag{4.6.5}$$

In the sequel, we will show that (4.6.5) holds. Let ν be the nonstandard representation of μ_0 given in the construction (4.6.3). We pick an internal stopping time τ such that ${}^\circ\tau = \sigma$, $L(P_\nu)$-a.e and $\hat{s} \in T^{\text{fin}}_{\delta_0}$, $\hat{s} \geq s$, $\hat{s} \approx s$ such that

$$\int_\Omega u_1(x(\sigma))u_2(x(\sigma+s))\ L(P_\nu)(d\omega) \\ \approx \int_\Omega (P^* u_1)(X^{(\delta_0)}(\tau))(P^* u_2)(X^{(\delta_0)}(\tau+\hat{s}))\ P_\nu(d\omega).$$

Then, we have

$$\int_Y E_{\Theta_y}\left(u_1(x(\sigma))u_2(x(\sigma+s))\right)\ \mu_0(dy) \\ = \int_\Omega u_1(x(\sigma))u_2(x(\sigma+s))\ \Theta_{\mu_0}(d\omega) \\ = \int_{S_0}\int_\Omega u_1(x(\sigma))u_2(x(\sigma+s))\ L(P_i)(d\omega)\ L(\lambda)(ds_i)$$

$$= \int_{\Omega} u_1(x(\sigma))u_2(x(\sigma+s))\ L(P_\lambda)(d\omega)$$

$$= \int_{\overset{\circ}{\Omega}} (P^* u_1)(X^{(\delta)}(\tau))(P^* u_2)(X^{(\delta)}(\tau+\hat{s}))\ P_\lambda(d\omega)$$

$$= \int_{\overset{\circ}{\Omega}} (P^* u_1)(X^{(\delta)}(\tau))E_{X^{(\delta)}(\tau)}(P^* u_2)(X^{(\delta)}(\hat{s}))\ P_\lambda(d\omega)$$

$$= \int_{\Omega} u_1(x(\sigma))E_{\Theta_{x(\sigma)}}(u_2(s))\ L(P_\lambda)(d\omega)$$

$$= \int_{\Omega} u_1(x(\sigma))E_{\Theta_{x(\sigma)}}(u_2(s))\ \Theta_{\mu_0}(d\omega)$$

$$= \int_{Y} E_{\Theta_y}\left[u_1(x(\sigma))E_{\Theta_{x(\sigma)}}(u_2(s))\right]\ \mu_0(dy)$$

This is (4.6.5). Similarly, we can also prove that $(\Omega, \Pi, \Pi_t, x(t), \hat{\Theta}_y)$ satisfies (3.1.9) in Sect. 3.1. □

4.7 Necessity for Existence of Dual Tight Markov Processes

Lemma 4.7.1. *Let $(F(\cdot,\cdot), D(F))$ be a Dirichlet form on $L^2(Y,\nu)$. If the condition (I) holds, then $\nu(Y - \bigcup_{n\in\mathbb{N}} Y_n) = 0$.*

Proof. Since $\bigcup_{n\in\mathbb{N}} D(F)_{Y_n}$ is F_1-dense in $D(F)$, it is a dense subset of $L^2(Y,\nu)$. This implies that $f = 0, \nu$-a.e. on $Y - \bigcup_{n\in\mathbb{N}} Y_n$, $\forall f \in L^2(Y,\nu)$. Since ν is a Radon measure, we deduce that $\nu(Y - \bigcup_{n\in\mathbb{N}} Y_n) = 0$. □

Proposition 4.7.1. *Let $(F(\cdot,\cdot), D(F))$ be a Dirichlet form on $L^2(Y,\nu)$. If there exist ν-tight dual strong Markov processes $(\Omega, \Pi, \Pi_t, x(t), \Theta_y)$ and $(\Omega, \Pi, \Pi_t, x(t), \hat{\Theta}_y)$ which are properly associated with $F(\cdot,\cdot)$, then conditions (I), (II) and (III) of Theorem 4.1.1 hold.*

Proof. (I) Let $\{R_\alpha \mid \alpha \in (-\infty, 0)\}$ be the resolvent of $x(t)$. Then, the family $\{R_1 f \mid f \in L^2(Y,\nu)\}$ is a F_1-dense subset of $D(F)$. Let $\{Y_n \mid n \in \mathbb{N}\}$ be an increasing sequence of compact sets satisfying the condition (4.1.1) in Sect. 4.1. Denote by σ_{Y-Y_n} the hitting time of $Y - Y_n$ defined by the definition (4.1.2) in Sect. 4.1. For $f \in L^2(Y,\nu)$, set

$$R_1^n f(y) = E_{\Theta_y} \int_0^{\sigma_{Y-Y_n}} e^{-t} f(x_t)\ dt.$$

Noticing that $R_1^n f \in D(F)_{Y_n}$ is the projection of $R_1 f$, we have

$$\begin{aligned} &F_1(R_1^n f - R_1^{n+m} f, R_1^n f - R_1^{n+m} f) \\ &\quad = F_1(R_1^n f, R_1^n f) - F_1(R_1^{n+m} f, R_1^{n+m} f) \longrightarrow 0, n \longrightarrow \infty. \end{aligned}$$

Moreover, it follows from the condition (4.1.1) in Sect. 4.1 that

$$R_1^n f \longrightarrow R_1 f, \nu\text{-a.e.}$$

Hence, we have

$$F_1(R_1^n f - R_1 f, R_1^n f - R_1 f) \longrightarrow 0, n \to \infty.$$

Therefore, $\bigcup_{n\in\mathbb{N}} D(F)_{Y_n}$ is F_1-dense in $D(F)$.

(II) For all $f \in L^2(Y,\nu), R_1 f$ is quasi-continuous by Theorem 4.3.3, Fukushima [175]. Take $\pi_0 = \{R_1 f \mid f \in L^2(Y,\nu)\}$.

(III) Let us set $Z = \bigcup_{n\in\mathbb{N}} Y_n$. Then, we know that the relative topology of Z is second countable. We pick a countable family of open sets such that $\{A_n \cap Z \mid n \in \mathbb{N}\}$ forms a basis for the relative topology of Z and $A \cap Z$ is contained in a compact subset of Y. For each $n \in \mathbb{N}$, we define

$$u_n(y) = E_{\Theta_y} \int_0^{\sigma_{Y-A_n}} e^{-t} 1_{A_n}(x_t)\, dt,$$

where σ_{Y-A_n} is the hitting time of $Y - A_n$. From Lemma 4.4.2 of Fukushima [175], we have that u_n is quasi-continuous. Moreover, u_n satisfies

$$\begin{aligned} u_n(y) > 0 \quad &\text{for} \quad y \in A_n, \\ u_n(y) = 0 \quad &\text{for} \quad y \in Y - A_n. \end{aligned}$$

This means that $\pi \hat{=} \{u_n \mid n \in \mathbb{N}\}$ satisfies (III) in Theorem 4.1.1. □

Combining Propositions 4.6.1 and 4.7.1, we complete the proof of Theorem 4.1.1.

Chapter 5
Hyperfinite Lévy Processes

So far, we have concentrated in this book on studying arbitrary Markov processes which admit a modification whose paths are right-continuous with left limits (i.e. *càdlàg* Markov or *Feller processes*, cf. e.g. [307, Chap. 3, Theorem 2.7]), and their relations to energy forms and Hamiltonians – from the perspective of nonstandard analysis.

Independently from this theory, Tom Lindstrøm [263] developed a different approach of finding nonstandard representations for a special class of Feller processes and their infinitesimal generators, viz. stochastically continuous processes with stationary and independent increments, for short: *Lévy processes*.

In terms of the state space, we will have to narrow the scope of our investigation: The state space should now be at least a separable Hilbert space, and in this chapter, we will only treat Lévy processes on $\mathbb{R}^d$.

Nevertheless, it is rather surprising to see how confining oneself to this, still fairly large, class of Markov processes reduces the amount of technicalities involved in developing a theory of internal hyperfinite representations of these processes, their semigroups and their infinitesimal generators.

Lindstrøm's theory was carried further by Albeverio and Herzberg [14] who proved that the jump part of any Lévy process can be conceived of, in a natural and rigorous sense, as the hyperfinite sum of independent hyperfinite Poisson processes. In addition, Lindstrøm [264] introduced nonlinear stochastic integration with respect to (hyperfinite) Lévy processes, Herzberg [205] developed a theory of pathwise integration with respect to hyperfinite Lévy processes with bounded-variation jump part, and Herzberg and Lindstrøm [206] proved an internal jump-diffusion decomposition. Whilst there are also other approaches to Lévy processes from a nonstandard-analysis viewpoint, due to Albeverio and Herzberg [15] and Ng [288], Lindstrøm's theory of hyperfinite Lévy processes is the most promising for potential theory and other applications, in particular in equilibrium asset pricing (see [204]).

S. Albeverio et al., *Hyperfinite Dirichlet Forms and Stochastic Processes*,
Lecture Notes of the Unione Matematica Italiana 10,
DOI 10.1007/978-3-642-19659-1_5, © Springer-Verlag Berlin Heidelberg 2011

In this chapter, we will present the most important results in Lindstrøm's original paper [263] and their extensions by Albeverio and Herzberg [14]. The thrust of the first sections of this Chapter is Lindstrøm's work, though we shall provide much more detailed proofs and choose a slightly different approach that does not put as much emphasis on the jump-diffusion decomposition. *En passant*, we shall give detailed proofs of some of the results presented in Albeverio et al. [25], as well as outlining the link between hyperfinite Lévy processes (which are going to be defined in Sect. 5.2) and certain hyperfinite nonnegative quadratic forms.

Finally, we will briefly discuss a few applications of the theory presented in this chapter.

5.1 Standard Lévy Processes

There is a vast literature on standard Lévy processes. A selection of references can be found in the monograph by Applebaum [65] and papers by Albeverio, Mandrekar, and Rüdiger [46], Marinelli, Prévôt and Röckner [277], Albeverio and Rüdiger [52] as well as by Albeverio, Rüdiger, and Wu [55].

First, we shall give the precise definition of a Lévy process. As previously, $\mathbb{R}_+$ will be short hand for the half-open interval $[0, +\infty)$.

Definition 5.1.1. Consider a probability space $(\Omega, \mathcal{C}, P)$ and let $d \in \mathbb{N}$. A stochastic process $x : \Omega \times \mathbb{R}_+ \to \mathbb{R}^d$ is called a *Lévy process* if and only if

1. $x_0 = 0$ P-almost surely,
2. $x_t - x_s$ is independent of $\sigma(x_u \mid u \leq s)$,
3. the law of $x_t - x_s$ equals the law of x_{t-s}, and
4. P-almost all paths of $(x_t)_{t\in\mathbb{R}_+}$ are right-continuous with left limits (càdlàg).

Much of the appeal of the area of Lévy processes is due to representation results of the following kind (cf. e.g. [307, 323]):

Theorem 5.1.1 (Lévy-Khintchine formula). *Consider a probability space $(\Omega, \mathcal{C}, P)$ and let $d \in \mathbb{N}$. A stochastic process $x : \Omega \times \mathbb{R}_+ \to \mathbb{R}^d$ is a Lévy process if and only if there are a symmetric $d \times d$-matrix with non-negative entries C, a vector $\gamma \in \mathbb{R}^d$, and a Radon measure ν on $\mathbb{R}^d \setminus \{0\}$ satisfying*

1. $\int_{B_1(0)} |y|^2 \, \nu(dy) < +\infty$,
2. $\nu[\complement B_1(0)] < +\infty$,

such that for all $y \in \mathbb{R}^d$ and $t \geq 0$,

$$E\left[e^{iy\cdot x_t}\right] = \exp\left(\begin{array}{c} ity\cdot\gamma - \frac{t}{2} y\cdot Cy + \\ t\int_{\mathbb{R}^d\setminus\{0\}} \left(e^{ity\cdot z} - 1 - iy\cdot z\chi_{\overline{B_1(0)}}\right) \nu(dz) \end{array} \right). \tag{5.1.1}$$

Here, $B_\rho(x)$ – as usual – is the open ball of radius ρ centered at x, for any $x \in \mathbb{R}^d$ and $\rho \in \mathbb{R}_{>0}$.

In internal contexts, however, we will also write $B_\rho(x)=\{y \in {}^*\mathbb{R}^d \mid |y-x| < \rho\}$ for $\rho \in {}^*\mathbb{R}_{>0}$ and $x \in {}^*\mathbb{R}^d$ (wherein $|z| = \sqrt{z \cdot z}$, for arbitrary $z \in {}^*\mathbb{R}^d$, is the Euclidean norm of z, and $\cdot : (x, y) \mapsto x \cdot y$ denotes the inner product on ${}^*\mathbb{R}^d$). Speaking rather informally, we suppress the "*" when referring to the *-image of the operator $B_\cdot(\cdot) : (\rho, x) \mapsto B_\rho(x)$.

A proof of this result is beyond the scope of this book, but it can be found, in the language of Fourier transforms, e.g. in the monographs by Bertoin [86] or Sato [323]. One can also formulate an analogous result for infinitesimal generators of Lévy processes – a proof of the equivalence of the formulations by means of Fourier transforms and infinitesimal generators can be found in the volume by Revuz and Yor [307].

In order to give a more handy formulation of Theorem 5.1.1 and later results, we will introduce the following terminology.

Definition 5.1.2. If $x_\cdot = (x_t)_{t\geq 0}$ is the Lévy process associated with a continuous and translation-invariant Markovian semigroup $p_\cdot = (p_t)_{t\geq 0}$, then the measure ν, the vector γ and the matrix C of (5.1.1) will also be referred to as the *Lévy measure* of x, the drift vector of x and the covariance matrix of x, respectively.

Definition 5.1.3. A Radon measure ν on $\mathbb{R}^d \setminus \{0\}$ satisfying

1. $\int_{B_1(0)} |y|^2 \; \nu(dy) < +\infty$,
2. $\nu[\complement B_1(0)] < +\infty$

(equivalently, $\int (1 \wedge |y|^2) \; \nu(dy) < +\infty$), is called a *Lévy measure.* A triple (γ, C, ν) with $\gamma \in \mathbb{R}^d$, $C \in \mathbb{R}^{d\times d}_{\geq 0}$ symmetric, and ν a Lévy measure, will be referred to as a *generating triplet.*

There is a one-to-one correspondence between continuous translation-invariant Markovian semigroups and Lévy processes in the following sense.

Consider a probability space $(\Omega, \mathcal{C}, P)$ on which there is a Lévy process $x_\cdot$ defined. Then P_{x_t}, the distribution of the random variable x_t under P, which is a probability measure on $\mathbb{R}^d$, can be shown to give rise to a space-translation invariant Markov kernel by setting $p_t(z, B) = P_{x_t}[x_t \in B - z]$. Moreover, one can prove $(p_t)_{t\in\mathbb{R}_+}$ to be a translation-invariant Markovian semigroup which is continuous as a map $t \mapsto p_t$ (with respect to the vague, or weak* topology). The converse also holds true: For every translation-invariant continuous Markovian – i.e. translation-invariant Feller – semigroup $p_\cdot$ there is a Lévy process $x_\cdot$ on a probability space $(\Omega, \mathcal{C}, P)$ such that $p_t = P_{x_t}$ for all $t \geq 0$. For this, see e.g. Bauer [77].

As hinted at previously, one can now rephrase the Lévy-Khintchine formula:

Theorem 5.1.2 (Lévy-Khintchine formula for infinitesimal generators). *Consider a Markovian semigroup $(p_t)_{t\in\mathbb{R}_+}$ on $\mathbb{R}^d$, $d \in \mathbb{N}$. Suppose $t \mapsto p_t$ is continuous (in the vague or weak* topology). Then the semigroup $p.$ is space-translation invariant (in the sense that $p_t f(\cdot + z) = p_t f$ for all $z \in \mathbb{R}^d$ and any nonnegative measurable $f : \mathbb{R}^d \to \mathbb{R}$) if and only if the infinitesimal generator ℓ of $(p_t)_{t\in\mathbb{R}_+}$ can be written as*

$$\ell : f \mapsto \frac{1}{2}\sum_{i,j=1}^{d} C_{i,j}\partial_i\partial_j f + \sum_{i=1}^{d} \gamma_i \partial_i f + \int_{\mathbb{R}^d\setminus\{0\}} (f(\cdot + y) - f)\ \nu(dy) \quad (5.1.2)$$

where $C \in {\mathbb{R}_+}^{d\times d}$ is a symmetric $d \times d$-matrix with nonnegative entries, $\gamma \in \mathbb{R}^d$, and ν is a Radon measure on $\mathbb{R}^d \setminus \{0\}$ satisfying

1. $\int_{B_1(0)} |y|^2\ \nu(dy) < +\infty$,
2. $\nu[\complement B_1(0)] < +\infty$.

Via the Ionescu-Tulcea-Kolmogorov construction, $p.$ is a translation-invariant semigroup if and only if there is a Lévy process $x = (x_t)_{t\geq 0}$ on a probability space $(\Omega, \mathcal{C}, P)$ such that $P_{x_t} = p_t$ for all $t \geq 0$.

(For this result, see e.g. [307].)

The latter formulation of the Lévy-Khintchine formula can also be seen as the link between the theory of Lévy processes and certain Dirichlet forms:

Remark 5.1.1. To all infinitesimal generators ℓ of $\mathbb{R}^d$-valued Lévy processes and measures m on the Borel σ-algebra of $\mathbb{R}^d$, one can immediately associate a bilinear form by setting

$$E(f, g) = \int_{\mathbb{R}^d} -f(x) \cdot \ell g(x)\ m(dx)$$

for all $f \in L^2(\mathbb{R}^d, m)$ and $g \in D(\ell)$, wherein $D(\ell)$ denotes the *domain of ℓ in $L^2(\mathbb{R}^d, m)$*, i.e. the preimage of $L^2(\mathbb{R}^d, m)$ under ℓ. This form E can be completed to become a closed coercive form under certain conditions, viz. that (1) m is *$p.$-supermedian* (in the sense that $\int p_t f\ dm \leq \int f\ dm$ for all $t > 0$ and nonnegative Borel-measurable f), and (2) $p.$ is *m-symmetric* (which means $\int g \cdot p_t f\ dm = \int f \cdot p_t g\ dm$ for all m-square integrable f, g and $t > 0$) – conversely, one can uniquely recover the corresponding infinitesimal generator from the Dirichlet form. In the framework of symmetric forms, this can be found in Fukushima, Ôshima and Takeda [183, Theorem 1.3.1]. In the setting of nonsymmetric forms, this is explained, e.g., in Ma and Röckner [270, Proposition II.4.3, final remarks in Sect. II.4.a].

Analogous representation formulae can be proven for Dirichlet forms, too (cf. [271]). For closability questions of minimally defined symmetric Dirichlet forms with jump processes, we refer to Albeverio and Song [56]. See also Albeverio and Rüdiger [51] for subordination methods, in order to get symmetric Dirichlet forms that are not necessarily of diffusion type. For other work on non-diffusion forms, see, e.g., Jacob [218–220] or Jacob and Schilling [222].

5.2 Hyperfinite Lévy Processes: Definitions and Characterizations

Keeping the above results about Lévy processes in a standard setting in mind, we shall now introduce the key notions of the nonstandard theory of Lévy processes.

As before, the set $T = \{n\Delta t \mid n \in {}^*\mathbb{N}\}$, for some positive $\Delta t \approx 0$, will serve as our time-line. Again we are working on an internal probability space $(\Omega, 2^{\Omega}, \mu)$, 2^{Ω} denoting the internal power set of Ω, and μ an internal finitely-additive probability measure on Ω.

The following definition was first given by Lindstrøm [263, Definitions 1.1, 1.3].

Definition 5.2.1. [263, Definitions 1.1, 1.3] Consider an internal stochastic process $X : \Omega \times T \to {}^*\mathbb{R}^d$ and a hyperfinite set $A \subset {}^*\mathbb{R}^d$.

- X is called a *hyperfinite random walk with increments from* A *and transition probabilities* $\{p_a\}_{a\in A}$ if and only if
  1. $X_0 = 0$ on Ω,
  2. For all $t \in T$, the *increments* $\Delta X_0 = X_{\Delta t} - X_0, \cdots, \Delta X_t = X_{t+\Delta t} - X_t$ form a hyperfinite set of *-independent internal random variables, and
  3. For all $t \in T$ and for all $a \in A$,

$$\mu\{\Delta X_t = a\} = p_a.$$

- X is called a *hyperfinite Lévy process* if and only if
  1. X is a hyperfinite random walk and
  2. $L(\mu)[\bigcap_{t\in T\cap \mathrm{Fin}({}^*\mathbb{R})}\{X_t \in \mathrm{Fin}({}^*\mathbb{R}^d)\}] = 1$.

Remark 5.2.1. Note that since T is not hyperfinite, we will never have that Ω is hyperfinite unless $|A| = 1$ (in which case X is deterministic). However, when we constrain the time horizon to a finite $t \in T \cap \mathrm{Fin}({}^*\mathbb{R})$, the set of all paths up to time t will be hyperfinite.

Remark 5.2.2. To any hyperfinite Lévy process, one can associate, in analogy to Remark 5.1.1, a bilinear, not necessarily symmetric form on ${}^*\mathbb{R}^d$ by setting

$$\forall f, g \in \mathcal{D}(\mathcal{E}) \quad \mathcal{E}(f,g) = \int_{{}^*\mathbb{R}^d} f(x) \cdot \frac{1}{\Delta t} \left(g(x) - \sum_{a \in A} g(x+a)\, p_a \right) dx$$

(where $\mathcal{D}(\mathcal{E})$ depends on the expression on the right hand side via Definition 1.4.2). In order to show that this form is nonnegative, we combine the Cauchy-Bunyakovski-Schwarz inequality (transferred to the nonstandard universe) with the translation invariance of the *-Lebesgue measure and the fact that $\{a\} \mapsto p_a$ defines a probability measure on A:

$$\begin{aligned}
&\int_{{}^*\mathbb{R}^d} f(x) \cdot \frac{1}{\Delta t} \sum_{a \in A} f(x+a)\, p_a \, dx \\
&\quad = \sum_{a \in A} \int_{{}^*\mathbb{R}^d} f(x) \cdot \frac{1}{\Delta t} f(x+a)\, p_a \, dx \\
&\quad \le \sum_{a \in A} \left(\int_{{}^*\mathbb{R}^d} f(x)^2 \, dx \right)^{1/2} \left(\int_{{}^*\mathbb{R}^d} \left(\frac{1}{\Delta t} f(x+a)\, p_a \right)^2 dx \right)^{1/2} \\
&\quad \le \sum_{a \in A} \left(\int_{{}^*\mathbb{R}^d} f(x)^2 \, dx \right)^{1/2} \left(\int_{\underbrace{a + {}^*\mathbb{R}^d}_{= {}^*\mathbb{R}^d}} f(y)^2 \, dy \right)^{1/2} p_a \frac{1}{\Delta t} \\
&\quad = \frac{1}{\Delta t} \int_{{}^*\mathbb{R}^d} f(x)^2 \, dx.
\end{aligned}$$

This implies $\mathcal{E}(f,f) \ge 0$ for all f in the domain $\mathcal{D}(\mathcal{E})$ of $\mathcal{E}$. So the bilinear form $\mathcal{E}$ is indeed nonnegative.

Conversely, if we are given a form $\mathcal{E}$ which is derived from such a set of increments A and transition probabilities $(p_a)_{a \in A}$, we can reconstruct A and $(p_a)_{a \in A}$ by means of elementary linear algebra in ${}^*\mathbb{R}^d$ (the corresponding operator is given explicitly in (1.1.2) of Sect. 1.1).

Thus, by studying hyperfinite Lévy processes, one also obtains results on certain nonnegative quadratic forms. We shall see in this chapter that for hyperfinite Lévy processes, the construction of standard parts and the classification of infinitesimal generators are technically less demanding than the proofs of the analogous results for general hyperfinite Markov chains. Furthermore, hyperfinite Lévy processes and their generators admit more explicit representation results: the hyperfinite Lévy-Khintchine formula (Theorem 5.4.1) and the representation of hyperfinite Lévy processes as superpositions of hyperfinite Poisson processes (Theorem 5.6.1).

Our goal for this section is to prove a necessary and sufficient criterion for a hyperfinite random walk to be a hyperfinite Lévy process. For this sake, we shall first of all approximate hyperfinite Lévy processes by hyperfinite Lévy processes with finite increments (in the sense that $A \subset \mathrm{Fin}({}^*\mathbb{R}^d)$), and then prove integrability results for hyperfinite Lévy processes with finite increments.

Notation 5.2.1. Unless indicated otherwise, X will always be a hyperfinite random walk with increments from A and transition probabilities $\{p_a\}_{a\in A}$. We shall also use the following abbreviation:

$$\forall k \in {}^*\mathbb{R} \quad q_k = \frac{1}{\Delta t} \sum_{\substack{a\in A \\ |a|>k}} p_a,$$

whence $q_k \Delta t = \mu\{\Delta X_t \in \complement\overline{B_k(0)}\}$ for any $t \in T$.

In preceding chapters of this monograph, A denoted an internal operator. Since we will not be concerned with operators in this chapter, we are free to use A to denote the hyperfinite increment set.

Notation 5.2.2. By $\mathcal{F}_t$, for $t \in T$, we shall denote the internal algebra generated by the internal random variables $X_0, \ldots, X_t$.

Notation 5.2.3. As another notational convention, we let $E[Z] = E_\mu[Z]$, for internal random variables Z, denote the expectation with respect to the internal probability measure μ. Similarly, $E[z] = E_{L(\mu)}[z]$, for $\mathbb{R}^m$-valued Loeb-measurable random variables z, will denote the expectation with respect to $L(\mu)$. Analogously, $E[Z|\mathcal{C}]$, for an internal algebra $\mathcal{C}$ of a hyperfinite probability space Ω, denotes the conditional expectation of Z given $\mathcal{C}$, and $E[z|L(\mathcal{C})]$ will denote the (standard) conditional expectation of z given the Loeb extension of $\mathcal{C}$.

We may adopt this notation, since we will not consider any standard Dirichlet forms (also usually denoted by E) for the rest of this chapter.

Remark 5.2.3. For all $u \geq s \in T$, ΔX_u is *-independent of all ΔX_v with $v < s$ and hence *-independent of all $X_w = \sum_{v<w} \Delta X_v$ for $w < s$. Therefore the internal random variable $X_t - X_s = \sum_{s\leq u<t} \Delta X_u$, for $s < t \in T$ will always be *-independent of the internal algebra generated by the internal random variables X_w for $w < s$. This is to say that $X_t - X_s$ is independent of $\mathcal{F}_s$ whenever $s, t \in T$ with $s < t$.

Following common practice, we will often simply write "independent" when *-independent is really meant.

Remark 5.2.4. If $u, t \in T$ with $u > t$, then the internal law of $X_u - X_t = \sum_{v<u-t} \Delta X_{t+v}$ under μ is the same as the internal law of $\sum_{v<u-t} \Delta X_v = X_{u-t}$ under μ. For, all the ΔX_t, $t \in T$ are independent and each distributed according to the law $(p_a)_{a\in A}$ of ΔX_0.

Remarks 5.2.3 and 5.2.4 have prepared us for the proof of $({}^\circ q_k)_{k\in\mathbb{N}} \in (\mathbb{R}_+ \cup \{+\infty\})^{\mathbb{N}}$ being decreasing to zero whenever X is a hyperfinite Lévy process.

Lemma 5.2.1. *[263, Lemma 3.1] If X is a hyperfinite Lévy process, then ${}^\circ q_k \downarrow 0$ as $k \to \infty$.*

Proof. Assume the negation of the Lemma's assertion were true, then there is an $\varepsilon > 0$ such that $q_k > \varepsilon$ for all $k \in \mathbb{N}$. By overflow, there must be a $K \in {}^*\mathbb{N}\setminus\mathbb{N}$ such that $q_K > \varepsilon$. Using the definition of q_k and recalling the fact that the increments of X are *-independent, we can compute the probability that X will make a change that is greater than k in norm before time $t+\Delta t$ as

$$\mu\left[\bigcup_{s\leq t}\{|\Delta X_s| > k\}\right] = (q_k\Delta t)^{\frac{t}{\Delta t}}.$$

But on the other hand, one has for noninfinitesimal $t \in T$,

$${}^\circ\left(q_k\Delta t\right)^{\frac{t}{\Delta t}} \leq {}^\circ\left(1-\varepsilon\Delta t\right)^{\frac{t}{\Delta t}} = e^{-\varepsilon t} < 1,$$

since $(1-x/N)^N \longrightarrow e^{-x}$ as $N \to \infty$ holds locally uniformly for all $x \in \mathbb{R}$, which implies $(1-y/N)^N \approx e^{-y}$ for all finite $y \in {}^*\mathbb{R}$ and infinite $N \in {}^*\mathbb{R}$. Therefore the probability of X making a jump of infinite size before $t+\Delta t$ is noninfinitesimal, contradicting the almost sure finiteness condition imposed on all hyperfinite Lévy processes X. □

Definition 5.2.2. Let $k \in {}^*\mathbb{R}_{>0}$. We define the *truncated random walks* $X^{\leq k}$ and $X^{>k}$ by

$$X^{\leq k} : \Omega \times T \to {}^*\mathbb{R}^d, \quad (\omega, t) \mapsto \sum_{\substack{s<t \\ |\Delta X_s(\omega)|\leq k}} \Delta X_s(\omega),$$

$$X^{>k} : \Omega \times T \to {}^*\mathbb{R}^d, \quad (\omega, t) \mapsto \sum_{\substack{s<t \\ |\Delta X_s(\omega)|> k}} \Delta X_s(\omega).$$

Remark 5.2.5. Obviously, $X = X^{\leq k} + X^{>k}$ on all of $\Omega \times T$, and both $X^{\leq k}$ and $X^{>k}$ are hyperfinite random walks (with increments from $\{0\}\cup A\cap\overline{B_k(0)}$ and $\{0\}\cup A\setminus\overline{B_k(0)}$, respectively).

Lemma 5.2.2. *[263, Lemma 3.2] If X is a hyperfinite Lévy process, then so are $X^{\leq k}$ and $X^{>k}$ for all sufficiently large finite $k \in {}^*\mathbb{R}_+$.*

Proof. We observe that the difference of two hyperfinite Lévy processes is again a Lévy process. Hence the first part of Remark 5.2.5 implies that we only have to verify that $X^{>k}$ is a hyperfinite Lévy process for sufficiently large finite k, and $X - X^{>k} = X^{\leq k}$ will be one as well. By the second part of

Remark 5.2.5, all we have to check is whether $X^{>k}$ will be finite for all finite times on a set of probability 1 if we choose $k \in \mathrm{Fin}({}^*\mathbb{R})$ large enough.

By Lemma 5.2.1, we may choose $k \in \mathbb{N}$ such that q_k is finite. Now consider, for arbitrary $m > k$, the internal process

$$X^{(k,m]} : \Omega \times T \to {}^*\mathbb{R}^d, \quad (\omega, t) \mapsto \sum_{\substack{s<t \\ k<|\Delta X_s(\omega)|\leq m}} \Delta X_s(\omega).$$

The set of increments of $X^{(k,m]}$ is obviously contained in $\mathrm{Fin}({}^*\mathbb{R}^d)$. Hence we have

$$\bigcup_{s\leq t} \left\{ X_s^{(k,m]} \notin \mathrm{Fin}\left({}^*\mathbb{R}^d\right) \right\} \subset \left\{ \mathrm{card}\ \left\{ s<t \;\middle|\; \Delta X_s^{(k,m]} \neq 0 \right\} \in {}^*\mathbb{N}\setminus\mathbb{N} \right\}$$
$$\subset \{\mathrm{card}\ \{s<t \mid |\Delta X_s|>k\} \in {}^*\mathbb{N}\setminus\mathbb{N}\} \quad (5.2.1)$$

for all $t \in T$. But on the other hand, $\mu\{|\Delta X_s| > k\} = q_k \Delta t$ for all $s \in T$ and all increments are *-independent, therefore we may perform the following combinatorial calculation:

$$\begin{aligned} &L(\mu)\left[\{\mathrm{card}\ \{s<t \mid |\Delta X_s|>k\} \in \mathbb{N}\}\right] \\ &= \sum_{n=0}^{\infty} L(\mu)\left[\{\mathrm{card}\ \{s<t \mid |\Delta X_s|>k\} = n\}\right] \\ &= \sum_{n=0}^{\infty} {}^{\circ}\!\left(\binom{t/\Delta t}{n} (q_k\Delta t)^n (1-q_k\Delta t)^{\frac{t}{\Delta t}-n}\right). \end{aligned} \quad (5.2.2)$$

If $\frac{t}{\Delta t}$ is finite, then we immediately get

$$L(\mu)\left[\{\mathrm{card}\ \{s<t \mid |\Delta X_s|>k\} \in \mathbb{N}\}\right] = 1,$$

and if $\frac{t}{\Delta t}$ is infinite, then (5.2.2) yields

$$\begin{aligned} &L(\mu)\left[\{\mathrm{card}\ \{s<t \mid |\Delta X_s|>k\} \in \mathbb{N}\}\right] \\ &= \sum_{n=0}^{\infty} {}^{\circ}\!\left(\frac{t^n}{\Delta t^n}\frac{1}{n!}e^{-q_k t}(q_k\Delta t)^n\right) \\ &= {}^{\circ}e^{-q_k t}\sum_{n=0}^{\infty} {}^{\circ}\!\left(t^n\frac{1}{n!}{q_k}^n\right) = e^{-{}^{\circ}q_k t}\sum_{n=0}^{\infty}\left(t^n\frac{1}{n!}{{}^{\circ}q_k}^n\right) = 1 \end{aligned}$$

since $(1-q_k t/h)^{h-n} \approx e^{-q_k t}$ for all infinite h – in particular for $h = t/\Delta t$ – and finite $n \in \mathbb{N}$ (this, in turn, follows from the fact that $(1-x/N)^{N-n} \longrightarrow e^{-x}$ as $N \to \infty$ locally uniformly for all $x \in \mathbb{R}$ and for any $n \in \mathbb{N}$). Thus, in view of relation (5.2.1), we have proven that $\bigcup_{s\leq t}\{X_s^{(k,m]} \notin \mathrm{Fin}({}^*\mathbb{R}^d)\}$ has $L(\mu)$-probability zero.

Summarizing the first part of this proof, we state that on a set of probability 1, the internal random variable $X_s^{(k,m]}$ is finite for all $s \leq t$.

Now we turn to the hyperfinite random walk $X^{>k}$ again. Note that, again by a simple combinatorial argument based on the independence and stationarity of the increments of $X^{(k,m]}$ and $X^{>k}$, as well as on the definition of q_m, one has

$$L(\mu)\left[\bigcap_{s\leq t}\left\{X_s^{(k,m]} = X_s^{>k}\right\}\right] = {}^\circ\prod_{s<t}\mu\left\{|\Delta X_s| \leq m\right\}$$
$$= {}^\circ\left(1 - q_m\Delta t\right)^{t/\Delta t} = {}^\circ e^{-q_m t} = e^{-{}^\circ q_m t}$$

(where once again we have used the locally uniform convergence of $(1 - x/N)^N \longrightarrow e^{-x}$ as $N \to \infty$ for all x). But ${}^\circ q_m \longrightarrow 0$ as $m \to \infty$, hence $L(\mu)[\bigcap_{s\leq t}\{X_s^{(k,m]} = X_s^{>k}\}] \longrightarrow 1$ as $m \to \infty$. Therefore the probability for the assertion that for some $m \in \mathbb{N}$, $X_s^{(k,m]} = X_s^{>k}$ for all $s \in T$, equals one. Since with probability 1, $X_s^{(k,m]}$ is finite for all $s \leq t$ (as was shown in the first part of this proof), we obtain that with probability 1, $X_s^{>k}$ is also finite for all $s \leq t$. Thus $X^{>k}$ is not merely a hyperfinite random walk, but a hyperfinite Lévy process. This suffices to complete the proof, as was explained at the outset. □

The preceding argument also yields the following

Corollary 5.2.1. *[263, Corollary 3.3] Suppose X is a hyperfinite random walk such that $X^{\leq m}$ is a hyperfinite Lévy process for all sufficiently large finite m and assume that ${}^\circ q_m \longrightarrow 0$ as $m \to \infty$. Then X is a hyperfinite Lévy process.*

Proof. As was shown in the proof of Lemma 5.2.2, the convergence assertion ${}^\circ q_m \longrightarrow 0$ as $m \to \infty$ suffices to prove that $X^{>k}$ will be a hyperfinite Lévy process. But by assumption, so is $X^{\leq k}$. Since we have $X = X^{\leq k} + X^{>k}$ and because the sum of two hyperfinite Lévy processes also is a hyperfinite Lévy process, we conclude that X must be one as well. □

We have now almost reached the end of our study of approximations of hyperfinite Lévy processes by hyperfinite Lévy processes with finite increments.

Lemma 5.2.3. *[263, Proposition 3.4] If X is a hyperfinite Lévy process, $t \in T$ finite, and $\varepsilon \in \mathbb{R}_{>0}$, then there exists a hyperfinite Lévy process $\hat{X} = \hat{X}^{(\varepsilon,t)}$ with finite increments such that*

$$\mu\left[\bigcap_{s\leq t}\left\{X_s = \hat{X}_s\right\}\right] > 1-\varepsilon.$$

Proof. Since ${}^\circ q_k \longrightarrow 0$ as $k \to \infty$ by Lemma 5.2.1, we can choose $k \in \mathbb{R}_{>0}$ such that ${}^\circ e^{-q_k t} > 1-\varepsilon$. By Lemma 5.2.2, $X^{\leq k}$ is a hyperfinite Lévy process and obviously has finite increments. Moreover, similarly to the proof of Lemma 5.2.2, we can show that

$$\begin{aligned} L(\mu)\left[\bigcap_{s\leq t}\left\{X_s = X_s^{\leq k}\right\}\right] &= {}^\circ\prod_{s<t}\mu\left\{|\Delta X_s| \leq k\right\} = {}^\circ\left(1-q_k\Delta t\right)^{t/\Delta t} \\ &= {}^\circ e^{-q_k t} > 1-\varepsilon \end{aligned}$$

by our initial choice of k. Thus we only have to put $\hat{X} = X^{\leq k}$ to bring the proof to a close. □

As some of the readers might already have suspected, hyperfinite Lévy processes with finite increments get their importance from their S-integrability properties.

Theorem 5.2.1. *[263, Theorem 2.3] If X is a hyperfinite Lévy process with finite increments, then the random variable $|X_t|^p$ is S-integrable for all finite $p \in {}^*\mathbb{R}_{\geq 0}$ and finite $t \in T$.*

The proof will employ the following Lemma:

Lemma 5.2.4. *[263, Lemma 2.1] Consider a finite Loeb measure space $(\Omega, L(\mathcal{A}), L(\mu))$, and let $F : \Omega \to {}^*\mathbb{R}^d$ be $\mathcal{A}$-measurable and internal. If $\int_\Omega |F|^p \, d\mu$ is finite for some finite $p \in {}^*\mathbb{R}_{>0}$, then $|F|^q$ is S-integrable for all $q \in {}^*\mathbb{R}_{>0}$ with $q < p$ and $q \not\approx p$.*

Proof. First, $\int_\Omega |F|^q \, d\mu$ is finite since

$$\int_\Omega |F|^q \, d\mu \leq \mu\left\{|F| \leq 1\right\}\cdot 1 + \underbrace{\int_{\{|F|>1\}} |F|^q \, d\mu}_{\leq \int_{\{|F|>1\}} |F|^p \, d\mu}.$$

Now consider an $A \in \mathcal{A}$ such that $\mu(A) \approx 0$. It remains to prove that $\int_A |F|^q \, d\mu \approx 0$. Hölder's inequality, when transferred to the nonstandard universe, yields

$$\begin{aligned} \int_A |F|^q \cdot 1 \, d\mu &\leq \left(\int_A |F|^p \, d\mu\right)^{q/p} \cdot \mu(A)^{1-\frac{q}{p}} \\ &\leq \left(\int_\Omega |F|^p \, d\mu \vee 1\right) \cdot \mu(A)^{1-\frac{q}{p}}. \end{aligned}$$

Hence, noting that $1 - \frac{q}{p} > 0$, but $1 - \frac{q}{p} \not\approx 0$ (otherwise $q \approx p$, since p is finite), and recalling that $\int_\Omega |F|^p \, d\mu$ is finite, we conclude that $\int_A |F|^q \, d\mu$ must be infinitesimal. □

Proof. Ushering in the proof of Theorem 5.2.1, we first remark that we may assume that X_t is not almost surely zero for all $t \in T$ (otherwise there is nothing to prove). Hence (adopting the convention $\min \emptyset = {}^*\infty$) the stopping time

$$\tau_k = \min \{t \in T \mid |X_t| \geq k\},$$

is not equal to ${}^*\infty$ everywhere for any $k \in {}^*\mathbb{R}$.

For all infinite k, τ_k will be almost surely infinite, since X is a hyperfinite Lévy process. Therefore, by underspill, we must have $\mu\{\tau_k > 1\} > \frac{1}{2}$ for all sufficiently large finite k. In particular, there will be a finite noninfinitesimal k which is strictly greater than all the increments of X and satisfies $\mu\{\tau_k > 1\} > \frac{1}{2}$. Fixing such a k, we set $\alpha = E[e^{-\tau_k}]$. The choice of k ensures that $1 \not\approx \alpha < 1$.

Inductively, we shall define a sequence of new stopping times via $\sigma_0 = 0$, $\sigma_1 = \tau_k$ and

$$\forall n \in \mathbb{N} \quad \sigma_n = \min \left\{t \in T \;\middle|\; t > \sigma_{n-1}, \quad \left|X_t - X_{\sigma_{n-1}}\right| \geq k\right\}.$$

The next step is to see that the family of random variables

$$\sigma_n - \sigma_{n-1} = \min \left\{s \in T \setminus \{0\} \;\middle|\; \left|X_{s+\sigma_{n-1}} - X_{\sigma_{n-1}}\right| \geq k\right\},$$

indexed by $n \in {}^*\mathbb{N}$, is *-independent, and that each random variable $\sigma_n - \sigma_{n-1}$ will be distributed exactly like τ_k. Whilst this can be proven by applying the transfer principle to the corresponding standard result (which, in turn, usually is proven by conditioning on σ_{n-1}), we prefer to give a more self-contained argument here. Note that for all $n \in {}^*\mathbb{N}$ and $s, s_0, \ldots, s_{n-1} \in T$,

$$\begin{aligned}
&\{\sigma_n - \sigma_{n-1} = s\} \cap \bigcap_{i<n} \{\sigma_i = s_i\} \qquad (5.2.3)\\
&= \underbrace{\left\{\left|X_{s+s_{n-1}} - X_{s_{n-1}}\right| \geq k\right\} \cap \bigcap_{u<s} \left\{\left|X_{u+s_{n-1}} - X_{s_{n-1}}\right| < k\right\}}_{=:A_s} \\
&\cap \bigcap_{i<n} \{\sigma_i = s_i\}.
\end{aligned}$$

This set is the intersection of the event $\bigcap_{i<n}\{\sigma_i = s_i\}$ (an element of $\mathcal{F}_{t_{n-1}}$) with A_s, an element of the internal algebra generated by the internal random

variables $X_{v+\Delta t} - X_v$, $t_{n-1} \leq v < s$. However, each $X_{v+\Delta t} - X_v$ for $v \geq t_{n-1}$ is independent of $\mathcal{F}_v$ (cf. Remark 5.2.3) and thus also independent of $\mathcal{F}_{t_{n-1}} \subset \mathcal{F}_v$. Therefore A_s must be independent of $\mathcal{F}_{t_{n-1}}$ as well, which via (5.2.3) implies

$$\mu\left[\{\sigma_n - \sigma_{n-1} = s\} \cap \bigcap_{i<n} \{\sigma_i = s_i\}\right] = \mu\left[A_s\right] \mu\left[\bigcap_{i<n} \{\sigma_i = s_i\}\right]. \quad (5.2.4)$$

Moreover, by conditioning on the random variable $X_{s_{n-1}}$ (which can only attain a hyperfinite number of values) and taking advantage of $X_{u+s_{n-1}} - X_{s_{n-1}}$ being independent from $\mathcal{F}_{s_{n-1}}$ for any $u \in T$, we see that $\mu[A_s] = \mu\{\tau_k = s\}$. Exploiting (5.2.4), we thereby finally arrive at

$$\mu\left[\{\sigma_n - \sigma_{n-1} = s\} \cap \bigcap_{i<n} \{\sigma_i = s_i\}\right] = \mu\{\tau_k = s\} \mu\left[\bigcap_{i<n} \{\sigma_i = s_i\}\right]. \quad (5.2.5)$$

A consequence of this result is that

$$\begin{aligned}\mu\{\sigma_n - \sigma_{n-1} = s\} &= \sum_{t \in T \cup \{^*\infty\}} \mu\left[\{\sigma_n - \sigma_{n-1} = s\} \cap \{\sigma_{n-1} = t\}\right] \\ &= \sum_{t \in T \cup \{^*\infty\}} \mu\{\tau_k = s\} \mu\{\sigma_{n-1} = t\} \\ &= \mu\{\tau_k = s\}\end{aligned}$$

(note that Ω is hyperfinite, therefore only hyperfinitely many of the values $\mu\{\sigma_{n-1} = t\}$, $t \in T \cup \{^*\infty\}$, will be non-zero), hence $\sigma_n - \sigma_{n-1}$ has the same distribution as τ_k.

Reinserting this into (5.2.5) yields

$$\begin{aligned}&\mu\left[\{\sigma_n - \sigma_{n-1} = s\} \cap \bigcap_{i<n} \{\sigma_i = s_i\}\right] \\ &= \mu\{\sigma_n - \sigma_{n-1} = s\} \mu\left[\bigcap_{i<n} \{\sigma_i = s_i\}\right] \quad (5.2.6)\end{aligned}$$

and thus, after generalising over s and $s_0, \ldots, s_{n-1}$, the independence of $\sigma_n - \sigma_{n-1}$ from $\sigma_0, \ldots, \sigma_{n-1}$ – which is equivalent to the independence from $\sigma_0, \sigma_1 - \sigma_0, \ldots, \sigma_{n-1} - \sigma_{n-2}$.

Having proven that the $\sigma_n - \sigma_{n-1}$ are independent and distributed according to the law of τ_k, we may next write

$$E\left[e^{-\sigma_n}\right] = E\left[e^{-\tau_k}\right]\prod_{i=1}^{n} E\left[e^{-\sigma_n+\sigma_{n-1}}\right] = E\left[e^{-\tau_k}\right]^{n+1} = \alpha^{n+1} \qquad (5.2.7)$$

for every $n \in {}^*\mathbb{N}$. On the other hand, k is larger than all increments of X, and by definition of σ_n, one will always have $|X_{\sigma_\ell - \Delta t} - X_{\sigma_{\ell-1}}| < k$. Therefore

$$\left|X_{\sigma_\ell} - X_{\sigma_{\ell-1}}\right| \leq |X_{\sigma_\ell} - X_{\sigma_\ell - \Delta t}| + \left|X_{\sigma_\ell - \Delta t} - X_{\sigma_{\ell-1}}\right| < 2k$$

for all $\ell \in {}^*\mathbb{N}$. Using this estimate as well as the definition of σ_n, we see that if $|X(\omega, t)| \geq 2nk$ for some $t \in T$ and $\omega \in \Omega$, then we must have that $t > \sigma_n(\omega)$, for otherwise

$$|X_t| \leq |X_{\sigma_n}| \leq \underbrace{|X_{\sigma_n} - X_0|}_{=\left|\sum_{i=1}^{n}(X_{\sigma_i} - X_{\sigma_{i-1}})\right|} \leq \sum_{i=1}^{n} \left|X_{\sigma_i} - X_{\sigma_{i-1}}\right| < n \cdot 2k.$$

Thus, in light of (5.2.7), we gain

$$\mu\{|X_t| \geq 2nk\} \leq \mu\{\sigma_n < t\} \leq \frac{E\left[e^{-\sigma_n}\right]}{e^{-t}} = e^t \alpha^{n+1}$$

for all $n \in \mathbb{N}$ and $t \in T$. By means of this inequality, we derive that for arbitrary $\varepsilon \in {}^*\mathbb{R}_{>0}$ and $t \in T$,

$$\begin{aligned} E\left[e^{\varepsilon|X_t|}\right] &= \sum_{n\in{}^*\mathbb{N}} \int_{\{2(n-1)k \leq |X_t| < 2nk\}} e^{\varepsilon|X_t|}\, d\mu \\ &\leq \sum_{n\in{}^*\mathbb{N}} \mu\{2(n-1)k \leq |X_t|\}\, e^{\varepsilon\cdot 2nk} \leq \sum_{n\in{}^*\mathbb{N}} e^{2\varepsilon nk} e^t \alpha^n. \end{aligned}$$

(We point out that, thanks to the hyperfiniteness of Ω, the above sum $\sum_{n\in{}^*\mathbb{N}} \int_{\{2(n-1)k \leq |X_t| < 2nk\}} e^{\varepsilon|X_t|}\, d\mu$ will only feature hyperfinitely many non-zero addends.)

In particular, if we choose $\varepsilon \in \mathbb{R}_{>0}$ so small that ${}^\circ e^{2\varepsilon k}\alpha < 1$ (which is always possible since ${}^\circ\alpha < 1$ and k is finite), we get

$$E\left[e^{\varepsilon|X_t|}\right] \leq e^t \cdot e^{2\varepsilon k}\alpha \cdot \frac{1}{1 - e^{2\varepsilon k}\alpha} \in \mathrm{Fin}({}^*\mathbb{R})$$

for all finite $t \in T$.

Finally, if we remark that for all finite p there is always a finite $N_0 \in \mathbb{N}$ such that $e^{\varepsilon|X_t|} > |X_t|^p$ on $\{|X_t| > N_0\}$, we conclude that $E[|X_t|^p]$ is finite. Applying Lemma (5.2.4) completes the proof of the theorem. □

We introduce the following quantities, which may be called normalized increment mean and normalized increment variance, respectively.

Definition 5.2.3.

$$m_X = \frac{1}{\Delta t} E\left[\Delta X_0\right] = \frac{1}{\Delta t} \sum_{a \in A} a p_a, \quad {\sigma_X}^2 = \frac{1}{\Delta t} E\left[\left|\Delta X_0\right|^2\right] = \frac{1}{\Delta t} \sum_{a \in A} |a|^2 p_a.$$

Lemma 5.2.5. *[263, Lemma 1.2, Corollary 2.4]*

- *For all $t \in T$ and arbitrary X,*

$$E\left[\left|X_t\right|^2\right] = {\sigma_X}^2 \cdot t + \left|m_X\right|^2 \cdot t\left(t - \Delta t\right).$$

- *Moreover, if X is a hyperfinite random walk with finite increments, then X will be a hyperfinite Lévy process if and only if both m_X and ${\sigma_X}^2$ are finite.*

Proof. For the first part of the Lemma, observe that whenever $s \neq u$, ΔX_s and ΔX_u are independent and therefore

$$E\left[\Delta X_s \cdot \Delta X_u\right] = E\left[\Delta X_s\right] \cdot E\left[\Delta X_u\right] = \left|E\left[\Delta X_0\right]\right|^2 = \left|m_X\right|^2 \Delta t^2,$$

whence this quantity is independent of u, s. This yields

$$\begin{aligned} E\left[\left|X_t\right|^2\right] &= E\left[\left(\sum_{s<t} \Delta X_s\right) \cdot \left(\sum_{u<t} \Delta X_u\right)\right] \\ &= \sum_{s'<t} E\left[\left|\Delta X_{s'}\right|^2\right] + \sum_{\substack{u,s<t \\ u \neq s}} E\left[\Delta X_s \cdot \Delta X_u\right] \\ &= \frac{t}{\Delta t} E\left[\left|\Delta X_0\right|^2\right] + \left|m_X\right|^2 \Delta t^2 \cdot \left(\frac{t^2}{\Delta t^2} - \frac{t}{\Delta t}\right) \\ &= {\sigma_X}^2 \cdot t + \left|m_X\right|^2 \left(t^2 - t\Delta t\right), \end{aligned}$$

bringing the proof of the first part of the Lemma to a close.

We shall now prove the criterion of the second part of the Lemma.

Assume first that m_X and ${\sigma_X}^2$ are finite. Then, so is $E[|X_t|^2]$ for all finite $t \in T$ by the first part of the Lemma.

Next we point out that the process $(X_t - m_X t)_{t \in T}$ is not only a hyperfinite random walk, but also a martingale, since due to the independence and stationarity of the increments, $E[\Delta X_t - m_X \Delta t | \mathcal{F}_t] = E[\Delta X_t] - m_X \Delta t = E[\Delta X_0] - m_X \Delta t = 0$ for all $t \in T$. Furthermore, $E[|X_t - m_X t|^2]$ $E[|X_t - m_X t|^2]$ is finite for all finite $t \in T$, since on the one hand

$E[|X_t|^2]$ is finite for all finite $t \in T$, and on the other hand (due to the Cauchy-Bunyakovski-Schwarz inequality), one has

$$\begin{aligned} E\left[|X_t - m_X t|^2\right] &\leq E\left[|X_t|^2\right] + |m_X|^2 t^2 + 2\,|m_X \cdot E\,[X_t]|\,t \\ &\leq E\left[|X_t|^2\right] + |m_X|^2 t^2 + 2\,|m_X| \cdot \underbrace{|E\,[X_t]|}_{\leq E[|X_t|] \leq 1 + E\left[|X_t|^2\right]} \cdot t \\ &\leq E\left[|X_t|^2\right](1 + 2\,|m_X|\,t) + |m_X|^2 t^2 + 2\,|m_X|\,t, \end{aligned}$$

whilst m_X is finite by assumption.

Thus, we have shown that $(X_t - m_X t)_{t \in T}$ is an internal martingale such that $E[|X_t - m_X t|^2]$ is finite for any finite $t \in T$. Hence we may employ the subsequent Lemma 5.2.6 and conclude that with probability 1, $X_t - m_X t$ is finite for all finite $t \in T$. In light of the finiteness of m_X, this means that with probability 1, X_t is finite for all finite $t \in T$.

In order to prove that the criterion also works in the converse direction, we recall Theorem 5.2.1 to find that $E[|X_t - m_X t|^2]$ is finite for all finite $t \in T$. But, being aware of the first part of the present Lemma, this can only be true if both m_X and ${\sigma_X}^2$ are finite. □

Lemma 5.2.6. *[25, p. 119] If M is an internal martingale on an arbitrary hyperfinite adapted Loeb probability space $(\Omega, (\mathcal{F}_t)_{t\in T}, L(\mu))$, then $(|M_t|^2)_{t\in T}$ is a submartingale, and if $E[|M_t|^2]$ is finite for all finite $t \in T$ (i.e. M is a λ^2-martingale), then there is a subset $\Omega_0 \subset \Omega$ of $L(\mu)$-measure 1 such that $M_t(\omega)$ is finite for all $\omega \in \Omega_0$ and finite $t \in T$.*

Proof. Consider a finite $t \in T$, and assume for a contradiction that there was a set $\tilde{B}$ of Loeb measure $p > 0$ such that for all $\omega \in \tilde{B}$, there is an $s \leq t$ such that $M_s(\omega)$ is infinite, i.e. we assume

$$p = L(\mu)\left[\bigcup_{s \leq t}\left\{M_s \notin \mathrm{Fin}\left({}^*\mathbb{R}^d\right)\right\}\right] > 0.$$

Then, by $\aleph_1$-saturation, there will be an $N \in {}^*\mathbb{N}\backslash\mathbb{N}$ such that $\mu[\bigcup_{s\leq t}\{|M_s| \geq N\}] \geq p - \frac{1}{N}$. This can be written, when we introduce the stopping time

$$\tau = \min\{s \in T \mid |M_s| \geq N\},$$

as $\mu\{\tau \leq t\} \geq p - \frac{1}{N}$, i.e. $L(\mu)\{\tau \leq t\} > 0$. Hence the process $(|M_t - m_X t|^2)_{t\in T}$ is the square of a martingale and therefore a submartingale. The submartingale property of squares of internal martingales follows e.g. from applying the Transfer Principle to Jensen's inequality for conditional expectations [77, p. 121, Satz 15.3] or from simply observing that for all martingales M, one has

$$\forall t \in T \quad E\left[\Delta\left(|M_t|^2\right)\middle|\mathcal{F}_t\right] - E\left[|\Delta M_t|^2\middle|\mathcal{F}_t\right] = 2E\left[M_t \Delta M_t | \mathcal{F}_t\right] = 2M_t \underbrace{E\left[\Delta M_t | \mathcal{F}_t\right]}_{=0} = 0$$

and therefore $E[\Delta(|M_t|^2)|\mathcal{F}_t] = E[|\Delta M_t|^2|\mathcal{F}_t] \geq 0$ for any $t \in T$.

If we apply the transferred Optional Stopping Theorem to this submartingale and the stopping times $\tau \wedge t \leq t$, we get

$$E\left[|M_t|^2\right] \geq E\left[|M_{\tau\wedge t}|^2\right]. \tag{5.2.8}$$

On the other hand,

$$E\left[|M_{\tau\wedge t}|^2\right] \geq \underbrace{\mu\{\tau \leq t\}}_{\not\approx 0} \cdot \underbrace{\min_{\{\tau\leq t\}} |M_{\tau\wedge t}|^2}_{\in {}^*\mathbb{R}_{>0}\setminus\mathbb{R}}.$$

(The minimum $\min_{\{\tau\leq t\}} |M_{\tau\wedge t}|^2$ will be attained as the infimum of an internal function on a hyperfinite set, and it will be infinite since $|M_{\tau\wedge t}|^2 \geq N^2 \in {}^*\mathbb{N}\setminus\mathbb{N}$ on $\{\tau \leq t\}$.) So, $E[|M_{\tau\wedge t}|^2]$ and therefore $E[|M_t|^2]$ will be infinite, a contradiction to what was supposed in the Lemma's statement. Thus we have brought our initial assumption to absurdity. Hence for all finite $t \in T$, we do have the identity $L(\mu)[\bigcap_{s\leq t}\{M_s \in \mathrm{Fin}({}^*\mathbb{R}^d)\}] = 1$. After choosing $\varepsilon \in T \setminus \mathrm{st}^{-1}\{0\}$, we now conclude the proof by remarking that

$$L(\mu)\left[\bigcap_{s\in T\cap\mathrm{Fin}({}^*\mathbb{R})} \left\{M_s \in \mathrm{Fin}\left({}^*\mathbb{R}^d\right)\right\}\right] = L(\mu)\left[\bigcap_{n\in\mathbb{N}}\bigcap_{s\leq n\varepsilon} \left\{M_s \in \mathrm{Fin}\left({}^*\mathbb{R}^d\right)\right\}\right], \tag{5.2.9}$$

where the right hand side is the probability of a countable intersection of events of measure 1. □

So far, we have only truncated hyperfinite random walks X to processes such as $X^{\leq k}$ and $X^{>k}$. We would like to consider more general truncation schemes. This also prompts us to generalize the notion of q_k.

Definition 5.2.4. For all internal $B \subset {}^*\mathbb{R}^d$,

$$\hat{\nu}\,[B] = \frac{1}{\Delta t} \sum_{a \in A \cap B} p_a$$

Note that $q_k = \hat{\nu}[\complement \overline{B_k(0)}]$.

Lemma 5.2.7. *[263, Proposition 4.1] If X is a hyperfinite Lévy process and $B \subset {}^*\mathbb{R}^d$ is internal such that $B \cap \mathrm{st}^{-1}\{0\} = \emptyset$, then $\hat{\nu}[B]$ is finite.*

Proof. Since $\hat{\nu}[B \cap \complement \overline{B_k(0)}] \leq q_k \downarrow 0$ as $k \to \infty$ by virtue of Lemma 5.2.1, it suffices to prove that $\hat{\nu}[B \cap \overline{B_k(0)}]$ is finite for some $k \in \mathbb{N}$. Thus we may assume without loss of generality that $B \subset B_k(0)$ for some finite k. If this k is large enough, then Lemma 5.2.2 tells us that $X^{\leq k}$ is a hyperfinite Lévy process. On the other hand, $X^{\leq k}$ has increments that are finite (bounded by k), thus by the previous Lemma 5.2.5, $\sigma_{X^{\leq k}}{}^2$ is finite. Recalling that B is internal and does not contains any infinitesimal elements by assumption, we may apply underspill to find an $\varepsilon \in \mathbb{R}_{>0}$ such that $B \cap B_\varepsilon(0) = \emptyset$. Then

$$\varepsilon^2 \hat{\nu}[B] = \frac{1}{\Delta t} \sum_{a \in A \cap B} \varepsilon^2 p_a \leq \frac{1}{\Delta t} \sum_{a \in A \cap B} |a|^2 p_a \leq \frac{1}{\Delta t} \sum_{a \in A \cap B_k(0)} |a|^2 p_a = \sigma_{X^{\leq k}}{}^2$$

where we have exploited that $B \subset B_k(0)$. Since ε is noninfinitesimal and $\sigma_{X^{\leq k}}{}^2$ is finite as remarked earlier in this proof, we have found a finite bound $\sigma_{X^{\leq k}}{}^2/\varepsilon^2$ on $\hat{\nu}[B]$. □

As previously announced, we shall study generalizations of $X^{\leq k}$:

Definition 5.2.5. Let $\Lambda \subset {}^*\mathbb{R}^d$ be internal. We define the *Λ-truncated random walk* X^Λ by

$$X^\Lambda : \Omega \times T \to {}^*\mathbb{R}^d, \quad (\omega, t) \mapsto \sum_{\substack{s<t \\ \Delta X_s(\omega) \in \Lambda}} \Delta X_s\,(\omega)$$

We will now generalize Lemma 5.2.2:

Lemma 5.2.8. *[263, Corollary 4.2] Let X be a hyperfinite Lévy process and assume $\Lambda \subset {}^*\mathbb{R}^d$ to be internal. If either $\Lambda \cap \mathrm{st}^{-1}\{0\} = \emptyset$ or $\complement\Lambda \cap \mathrm{st}^{-1}\{0\} = \emptyset$, then both X^Λ and $X^{\complement\Lambda}$ are hyperfinite Lévy processes.*

Corollary 5.2.2. *If X is a hyperfinite Lévy process, then $X^{\leq k}$ and $X^{>k}$ are hyperfinite Lévy processes for all noninfinitesimal k.*

Proof (Proof of Lemma 5.2.8). The proof strongly resembles the one of Lemma 5.2.2. It suffices to treat the case where $\Lambda \cap \mathrm{st}^{-1}\{0\} = \emptyset$ and prove that X^Λ is a hyperfinite Lévy process. For, $X - X^\Lambda = X^{\complement\Lambda}$, and the difference of two hyperfinite Lévy processes is again a hyperfinite Lévy process. But if $\Lambda \cap \mathrm{st}^{-1}\{0\} = \emptyset$, then Lemma 5.2.7 yields that $\hat{\nu}[\Lambda]$ is finite. Hence the value $\hat{\nu}[\Lambda]$ may play exactly the rôle which was filled by q_k in the proof of Lemma 5.2.2. Next define, for finite m, $\Lambda_m = \Lambda \cap \overline{B_m(0)}$. Then X^{Λ_m} can play the rôle that was filled by $X^{(k,m]}$ in the proof of Lemma 5.2.2. With these modifications (using $\hat{\nu}[\Lambda]$ instead of q_k, and replacing X^{Λ}, $X^{\complement\Lambda}$, X^{Λ_m} by $X^{>k}$, $X^{\leq k}$, $X^{(k,m]}$, respectively), we can now copy the proof of Lemma 5.2.2 to see that there is a set of probability 1 on which $X_t^{\Lambda_m}$ is finite for all finite $t \in T$. Thus, X^{Λ_m} already is a hyperfinite Lévy process. But $X^{\Lambda_m} = (X^{\Lambda})^{\leq m}$. By means of Corollary 5.2.1, we deduce that X^{Λ} is a hyperfinite Lévy process as well. □

Now we are ready to state and prove the central result of this section.

Theorem 5.2.2. *[263, Theorem 4.3] The hyperfinite random walk X is a hyperfinite Lévy process if and only if all of the following conditions are satisfied:*

1. *For all $k \in \mathrm{Fin}({}^*\mathbb{R}) \setminus \mathrm{st}^{-1}\{0\}$, $\frac{1}{\Delta t}\sum_{|a|\leq k} a p_a \in \mathrm{Fin}({}^*\mathbb{R}^d)$.*
2. *For all $k \in \mathrm{Fin}({}^*\mathbb{R})$, $\frac{1}{\Delta t}\sum_{|a|\leq k} |a|^2 p_a \in \mathrm{Fin}({}^*\mathbb{R}^d)$.*
3. *$\lim_{k\to\infty} {}^\circ\left(\frac{1}{\Delta t}\sum_{|a|>k} p_a\right) = 0$.*

Proof. In case X is a hyperfinite Lévy process, we may apply Lemma 5.2.8 and see that $X^{\leq k}$ is a hyperfinite Lévy process – and even one with increments that are finite (bounded by k) – for all finite noninfinitesimal k. Now Lemma 5.2.5 already yields the finiteness of both $\frac{1}{\Delta t}\sum_{|a|\leq k} a p_a$ – thereby proving the first condition in the statement of the theorem – and $\frac{1}{\Delta t}\sum_{|a|\leq k} |a|^2 p_a$ for all finite noninfinitesimal k. Since $k \mapsto \frac{1}{\Delta t}\sum_{|a|\leq k} a p_a$ is increasing in k, this suffices to prove the second condition as well. The third condition, finally, is simply the conclusion of Lemma 5.2.1.

To prove that the converse implication also holds, we note that by Lemma 5.2.5, the first two conditions in the statement of the theorem imply that $X^{\leq k}$ must be a hyperfinite Lévy process for all finite noninfinitesimal k. Now the third condition entitles us to apply Corollary 5.2.1 – which readily assures us of X indeed being a hyperfinite Lévy process. □

5.3 Standard Parts of Hyperfinite Lévy Processes

Definition 5.3.1. Let $F : T \to \mathbb{R}^d$ be internal, and consider $s \in \mathbb{R}_{\geq 0}$ and $c \in {}^*\mathbb{R}^d$. Then c is called the *S-right limit* of F at s, denoted $c = S\text{-}\lim_{{}^\circ t \downarrow r} F(t)$, whenever for all $\varepsilon \in \mathbb{R}_{>0}$ there exists a real number $\delta \in \mathbb{R}_{>}0$ such that $|F(t) - c| < \varepsilon$ for all $t \in T \cap {}^*[s, s+\delta)$.

Analogously, c is called the *S-left limit* of F at s, denoted $c = S\text{-}\lim_{{}^\circ t\uparrow r} F(t)$, whenever for all $\varepsilon \in \mathbb{R}_{>0}$ there is some $\delta \in \mathbb{R}_{>}0$ such that $|F(t) - c| < \varepsilon$ for all $t \in T \cap {}^*(s-\delta, s]$.

The value c is called an *S-one-sided limit* of F at s if it is either the S-left limit or the S-right limit of F at s.

Remark 5.3.1. All S-one-sided limits are obviously unique – this would even hold if F was taking values in ${}^*Y$, Y being an arbitrary Hausdorff space, rather than ${}^*\mathbb{R}^d$.

In order to establish the existence of one-sided limits for almost all paths of a hyperfinite Lévy process, we shall use the following well-known result from the theory of λ^2-martingales:

Theorem 5.3.1. *[25, Proposition 4.2.10] If M is an internal ${}^*\mathbb{R}$-valued $(\mathcal{F}_t)_{t\in T}$-martingale such that $E[{M_t}^2]$ is finite for all finite $t \in T$, then almost all of its paths $M.(\omega)$ have one-sided limits for all $s \in \mathbb{R}_{\geq 0}$.*

Proof. Since countable intersections of events of probability 1 have again probability 1, it suffices to show that for all finite $R \in T$, almost all of the paths $M.(\omega)$, $\omega \in \Omega$, have S-one-sided limits for all $s \in \mathbb{R} \cap [0, R]$.

The next part of the proof resembles, to some extent, the proof of Doob's martingale convergence theorems via the "upcrossing inequalities". For, by Lemma 5.2.6 we know that almost all paths of M remain finite in finite time, hence the conclusion of the Lemma can only fail if with positive probability, the paths of M are "infinitely oscillating", i.e. if the event $E = \bigcup_{a,b\in\mathbb{Q}} E_{a,b}$, wherein

$$E_{a,b} = \left\{\omega \in \Omega \;\middle|\; \exists (t_n)_{n\in\mathbb{N}} \in T^{\mathbb{N}} \forall n \in \mathbb{N} \left(M_{t_{2n-1}}(\omega) \leq a < b \leq M_{t_{2n}}(\omega)\right)\right\},$$

has strictly positive (Loeb) probability. By σ-additivity, this can only be true if $L(\mu)[E_{a,b}]$ for some $a < b \in \mathbb{Q}$.

Define inductively the following ${}^*\mathbb{N}_0$-sequence of stopping times: $\tau_0 = 0$ and

$$\begin{aligned} \forall n \in {}^*\mathbb{N} \;\; \tau_{2n-1} &= \min\{t \in T \mid t > \tau_{2n-2}, \quad M_t \leq a\} \wedge R, \\ \tau_{2n} &= \min\{t \in T \mid t > \tau_{2n-1}, \quad M_t \geq b\} \wedge R. \end{aligned}$$

Obviously, $(\tau_n(\omega))_{n\in{}^*\mathbb{N}}$ is strictly increasing and takes values in $T \cap [0, R)$, for all $\omega \in \Omega$, until it hits R (where it will remain henceforth). Therefore, if we let $\gamma = R/\Delta t$, then we must have $\tau_\gamma(\omega) = R$ for all $\omega \in \Omega$. On the other hand, the transferred Optional Stopping Theorem yields that $(M_{\tau_n})_{n\in{}^*\mathbb{N}}$ is a martingale and hence the identity

$$E\left[M_{\tau_n}\left(M_{\tau_{n+1}}-M_{\tau_n}\right)\middle|\mathcal{F}_{\tau_n}\right]=M_{\tau_n}\underbrace{E\left[\left(M_{\tau_{n+1}}-M_{\tau_n}\right)\middle|\mathcal{F}_{\tau_n}\right]}_{=0}=0$$

for all $n \in {}^*\mathbb{N}$ follows. This in turn implies, via $(M_{\tau_{n+1}}-M_{\tau_n})^2 = {M_{\tau_{n+1}}}^2 - {M_{\tau_n}}^2 - 2M_{\tau_n}(M_{\tau_{n+1}}-M_{\tau_n})$, the relation

$$E\left[\left(M_{\tau_{n+1}}-M_{\tau_n}\right)^2\middle|\mathcal{F}_{\tau_n}\right]=E\left[{M_{\tau_{n+1}}}^2-{M_{\tau_n}}^2\middle|\mathcal{F}_{\tau_n}\right]$$

for arbitrary $n \in {}^*\mathbb{N}$, from which we obtain

$$\begin{aligned}E\left[{M_R}^2\right]=E\left[{M_{\tau_\gamma}}^2\right]&=E\left[{M_0}^2+\sum_{n<\gamma}\left({M_{\tau_{n+1}}}^2-{M_{\tau_n}}^2\right)\right]\\&=E\left[{M_0}^2+\sum_{n<\gamma}\left(M_{\tau_{n+1}}-M_{\tau_n}\right)^2\right].\end{aligned} \tag{5.3.1}$$

On the other hand, by our definition of $E_{a,b}$ and $(\tau_n)_{n\in{}^*\mathbb{N}}$, we have

$$\left(M_{\tau_{n+1}}-M_{\tau_n}\right)^2\geq|b-a|^2 \text{ on } E_{a,b}$$

for infinitely many $n<\gamma$ (i.e. for all $n\in\mathbb{N}_0$). Equation (5.3.1) hence shows us that $E[{M_R}^2]\geq|b-a|^2L(\mu)[E_{a,b}]\cdot n$ for all $n\in\mathbb{N}$. Thus $E[{M_R}^2]$ becomes infinite by our previous assumption of $E_{a,b}$ having positive probability. On the other hand, $E[{M_t}^2]$ was assumed to be finite for all finite $t\in T$ in the statement of the Lemma! This contradiction completes the proof of the theorem. □

Lemma 5.3.1. *[263, Proposition 6.3] If X is a hyperfinite Lévy process, then almost all paths of X have one-sided limits for all $s\in\mathbb{R}_{\geq 0}$.*

Proof. First assume X to have finite increments. In the proof of Lemma 5.2.5, we have already seen that due to the independence and stationarity of the increments, $E[\Delta X_t - m_X\Delta t|\mathcal{F}_t] = E[\Delta X_t]-m_X\Delta t = E[\Delta X_0]-m_X\Delta t = 0$ for all $t\in T$, hence $(X_t - m_X t)_{t\in T}$ is a martingale. Furthermore, Lemma 5.2.5 makes sure that m_X is finite whence $(X_t-m_Xt)_{t\in T}$ is a hyperfinite Lévy process with finite increments. By means of Theorem 5.2.1 we see that $E[|X_t-m_Xt|^2]$ is finite for all finite $t\in T$.

Thus, Theorem 5.3.1 may be applied to the components of the ${}^*\mathbb{R}^d$-valued martingale $(X_t-m_Xt)_{t\in T}$ in order to see that the components of $(X_t-m_Xt)_{t\in T}$, and therefore $(X_t-m_Xt)_{t\in T}$ itself, have S-one sided limits for all paths in a set of Loeb probability 1 – which completes the proof for the case where X has finite increments.

If X does not have finite increments, we simply refer to Lemma 5.2.3 (stochastic approximation of hyperfinite Lévy processes by hyperfinite Lévy processes with finite increments) and use the conclusion of the present Lemma for hyperfinite Lévy processes with finite increments: Let us choose a $t_0 \in T \setminus \mathrm{st}^{-1}\{0\}$. Then for all $n \in \mathbb{N}$ and $\varepsilon \in \mathbb{R}_{>0}$, Lemma 5.2.3 provides us with a hyperfinite Lévy process with finite increments $\hat{X}^{(n,\varepsilon)}$ such that

$$\mu\left[\bigcap_{t\leq nt_0}\left\{X_t = \hat{X}_t^{(n,\varepsilon)}\right\}\right] > 1-\varepsilon.$$

Hence, $\bigcup_{\varepsilon\in\mathbb{Q}_{>0}}\bigcap_{s\leq nt_0}\{X_s = \hat{X}_s^{(n,\varepsilon)}\}$ has Loeb probability 1 for all $n \in \mathbb{N}$ and therefore,

$$L(\mu)\left[\bigcap_{n\in\mathbb{N}}\bigcup_{\varepsilon\in\mathbb{Q}_{>0}}\bigcap_{t\leq nt_0}\left\{X_t = \hat{X}_t^{(n,\varepsilon)}\right\}\right] = 1.$$

Hence there exists a set $\Omega_1 \subset \Omega$ of Loeb probability 1 such that for all finite $t_1 \in T$ there exists a hyperfinite Lévy process $\hat{X}$ with finite increments satisfying $\hat{X}_t(\omega) = X_t(\omega)$ for all $\omega \in \Omega_1$ and $t \in T \cap [0, t_1]$. Since we have already shown in the first part of this proof that $\hat{X}$ has one-sided limits almost surely, we are done. □

Whenever one-sided limits exist, we can introduce a well-defined standard part:

Definition 5.3.2. Assume $F : T \to {}^*\mathbb{R}^d$ to have one-sided limits for all $s \in \mathbb{R}_{\geq 0}$. Then we can define the *(right) standard part* of F, denoted $\mathrm{st}(F)$, via

$$\forall s \in \mathbb{R}_{\geq 0} \quad \mathrm{st}(F)(s) = S\text{-}\lim_{{}^\circ t\downarrow s} F(t).$$

It is an easy exercise to show the following result:

Lemma 5.3.2. *If $F : T \to {}^*\mathbb{R}^d$ has one-sided limits for all $s \in \mathbb{R}_{\geq 0}$, then $\mathrm{st}(F) : \mathbb{R}_{\geq 0} \to \mathbb{R}^d$ is right-continuous with left limits.*

Now we can define the standard part of a hyperfinite Lévy process.

Definition 5.3.3. If X is a hyperfinite Lévy process, the *standard part* of X, denoted by $x = (x_s)_{s\in\mathbb{R}_{\geq 0}}$ or $\mathrm{st}(X)$, is defined as

$$x_s(\omega) = \mathrm{st}\left(X_\cdot(\omega)\right)(s)$$

for all $\omega \in \Omega$ such that $X_\cdot(\omega)$ has one-sided limits on all of $\mathbb{R}_{\geq 0}$.

Remark 5.3.2. In light of Lemma 5.3.1, $\mathrm{st}(X_\cdot(\omega))$ is well-defined for almost all $\omega \in \Omega$, making the process $x = (x_s)_{s \in \mathbb{R}_{\geq 0}}$ well-defined on a set of probability 1. Lemma 5.3.2, in addition, implies that all the paths of $x = \mathrm{st}(X)$ are right-continuous with left limits.

In order to prove that $\mathrm{st}(X)$ is indeed a Lévy process, we will need the following result, which may be viewed as stating the S-continuity in probability of $t \mapsto X_t$:

Lemma 5.3.3. *[263, Lemma 6.4] If X is a hyperfinite Lévy process, $s \in \mathbb{R}_{\geq 0}$ and $t \in T \cap \mathrm{st}^{-1}\{s\}$, then ${}^\circ X_t = x_s$ holds $L(\mu)$-almost surely.*

Proof. First we prove that it is sufficient to show that

$$\forall t \in T \cap \mathrm{Fin}({}^*\mathbb{R}) \quad \forall \varepsilon \in \mathbb{R}_{>0} \quad \exists \delta_X(\varepsilon, t) \in \mathbb{R}_{>0} \quad \forall u \in T \cap \mathrm{Fin}({}^*\mathbb{R}) \quad \begin{pmatrix} |t-u| < \delta_X(\varepsilon,t) \Rightarrow \\ \mu\{|X_t - X_u| \geq \varepsilon\} < \varepsilon \end{pmatrix}. \tag{5.3.2}$$

In order to convince ourselves of the sufficiency of assertion (5.3.2), note that whenever $(s_n)_n \in T^{\mathbb{N}}$ such that ${}^\circ s_n \downarrow s \in \mathbb{R}_{\geq 0}$ as $n \to \infty$, then one will have ${}^\circ X_{s_n} \longrightarrow x_s$ $L(\mu)$-almost surely as $n \to \infty$ by the definition of S-right limits. Hence also ${}^\circ X_{s_n} \longrightarrow x_s$ in $L(\mu)$-probability as $n \to \infty$. On the other hand, a consequence of formula (5.3.2) is that for all $t \in T \in \mathrm{st}^{-1}\{s\}$, ${}^\circ X_{s_n} \longrightarrow {}^\circ X_t$ in $L(\mu)$-probability as $n \to \infty$. Since limits in probability are almost surely unique, we conclude that ${}^\circ X_t = x_s$ $L(\mu)$-almost surely.

Hence what remains to show is the validity of the formula (5.3.2). Using Lemma 5.2.3, we will first show that if we know (5.3.2) for all hyperfinite Lévy processes X with finite increments, then it will also follow for arbitrary hyperfinite Lévy processes X: Given $t \in T \cap \mathrm{Fin}({}^*\mathbb{R})$ and $\varepsilon \in \mathbb{R}_{>0}$, let us assume $1 \in T$ without loss of generality and consider the hyperfinite Lévy process with finite increments $\hat{X} = \hat{X}^{(\varepsilon/2, t+1)}$ – which was constructed in Lemma 5.2.3 – and find the corresponding $\delta = \delta_{\hat{X}^{(\varepsilon/2,t+1)}}(\frac{\varepsilon}{2}, t)$ as in formula (5.3.2). Then

$$\forall u \in T \cap \mathrm{Fin}({}^*\mathbb{R}) \begin{pmatrix} |t-u| < \delta \Rightarrow \\ \mu\left\{\left|\hat{X}_t - \hat{X}_u\right| \geq \varepsilon/2\right\} < \frac{\varepsilon}{2}, \\ \mu\left[\bigcup_{v \leq t+1}\left\{X_v \neq \hat{X}_v\right\}\right] < \frac{\varepsilon}{2} \end{pmatrix}$$

which implies

$$\forall u \in T \cap \mathrm{Fin}({}^*\mathbb{R}) \begin{pmatrix} |t-u| < \delta \wedge 1 \Rightarrow \\ \mu\{|X_t - X_u| \geq \varepsilon/2\} < \frac{\varepsilon}{2} + \frac{\varepsilon}{2} = \varepsilon \end{pmatrix}.$$

Now all that is left to prove is that (5.3.2) holds for X being a hyperfinite Lévy process with finite increments. Consider $t, u \in T$ with $t < u$. If we combine Remark 5.2.4 with Lemma 5.2.5, we may write

$$E\left[|X_u - X_t|^2\right] = E\left[|X_{u-t}|^2\right] = {\sigma_X}^2(u-t) + |m_X|^2 \cdot (u-t)(u-t-\Delta t),$$

which via the transferred Chebyshev inequality yields

$$\begin{aligned}\mu\left\{|X_t - X_u| \geq \varepsilon\right\} &\leq \frac{E\left[|X_u - X_t|^2\right]}{\varepsilon^2} \\ &\leq \frac{{\sigma_X}^2(u-t) + |m_X|^2 \cdot (u-t)(u-t-\Delta t)}{\varepsilon^2} \\ &\leq \frac{{\sigma_X}^2\delta + |m_X|^2 \cdot \delta^2}{\varepsilon^2}\end{aligned}$$

whenever $|u - t| < \delta$. Since both ${\sigma_X}^2$ and m_X are finite – due to Lemma 5.2.5 and the finiteness of the increments X –, we can choose $\delta \in \mathbb{R}_{>0}$ so small that ${\sigma_X}^2\delta + |m_X|^2 \cdot \delta^2 < \varepsilon^3$ and we are done. □

In order to be able to justify the choice of the term "hyperfinite Lévy process", we should expect the subsequent theorem to hold (which we are now ready to prove, based on Lemma 5.3.3).

Theorem 5.3.2. *[263, Theorem 6.6] If X is a hyperfinite Lévy process, then $x = \mathrm{st}(X)$ is a Lévy process.*

Proof. We simply verify that $\mathrm{st}(X)$ has all the defining features of a Lévy process, as laid down in Definition 5.1.1:

- Almost all the paths of $x = \mathrm{st}(X)$ are right-continuous with left limits by Remark 5.3.2.
- Because of the previous Lemma 5.3.3, $x_0 = 0$ almost surely.
- In order to show the independence of the increments of $x = \mathrm{st}(X)$, we will mimic the proof given by Anderson for the fact that the *-independence of $^*\mathbb{R}$-valued internal random variables implies the S-independence of their standard parts [61, Lemma 20]. (Essentially, the same argument can be found in the volume by Albeverio, Fenstad, Høegh-Krohn, and Lindstrøm [25].) Suppose, for this purpose, that we are given real numbers $s_0 < \cdots < s_n \in \mathbb{R}_{\geq 0}$ and compact sets $A_0, \ldots, A_{n-1} \subset \mathbb{R}^d$. Then for all $t_0, \ldots, t_n \in T$ such that $t_0 \approx s_0, \ldots, t_n \approx s_n$, we will on the one hand have

$$\begin{aligned}B &= \left\{{}^\circ X_{t_n} - {}^\circ X_{t_{n-1}} \in A_{n-1}\right\} \cap \bigcap_{i=1}^{n-2} \left\{{}^\circ X_{t_i} \in A_i\right\} \\ &= \left\{{}^\circ X_{t_n} - {}^\circ X_{t_{n-1}} \in \mathrm{st}\left({}^*A_{n-1}\right)\right\} \cap \bigcap_{i=1}^{n-2} \left\{{}^\circ X_{t_i} \in \mathrm{st}\left({}^*A_i\right)\right\}\end{aligned}$$

$$= \bigcap_{m\in\mathbb{N}} \left(\underbrace{\left\{ {}^*d\left(X_{t_n} - X_{t_{n-1}}, {}^*A_{n-1}\right) \leq \frac{1}{m} \right\}}_{=:C_m} \cap \underbrace{\bigcap_{i=1}^{n-2} \left\{ {}^*d\left(X_{t_i}, {}^*A_i\right) \leq \frac{1}{m} \right\}}_{=:D_m} \right)$$

(wherein d is the function such that $d(x, A)$, for $x \in \mathbb{R}^d$ and compact $A \subset \mathbb{R}^d$ denotes the minimal distance of x from A), since $\mathrm{st}({}^*A) = A$ for all compact $A \subset \mathbb{R}^d$ – cf. Proposition A.1.1 or Albeverio et al. [25, Proposition 2.1.6 (iv)']. Note that due to Remark 5.2.3, we also know that $X_{t_n} - X_{t_{n-1}}$ is independent of $\mathcal{F}_{t_{n-1}}$, hence C_m is independent of D_m for all $m \in \mathbb{N}$ and therefore

$$\forall m \in \mathbb{N} \quad L(\mu)\left[C_m \cap D_m\right] = {}^\circ\mu\left[C_m\right]\mu\left[D_m\right].$$

Using that both $(C_m)_{m\in\mathbb{N}}$ and $(D_m)_{m\in\mathbb{N}}$ are increasing,

$$\begin{aligned} L(\mu)\left[B\right] &= \inf_{m\in\mathbb{N}} L(\mu)\left[C_m \cap D_m\right] = \inf_{m\in\mathbb{N}} L(\mu)\left[C_m\right] \cdot L(\mu)\left[D_m\right] \\ &= \inf_{m\in\mathbb{N}} L(\mu)\left[C_m\right] \cdot \inf_{m\in\mathbb{N}} L(\mu)\left[D_m\right] \\ &= L(\mu)\left[\bigcap_{m\in\mathbb{N}} C_m\right] \cdot L(\mu)\left[\bigcap_{m\in\mathbb{N}} D_m\right] \\ &= L(\mu)\left\{{}^\circ X_{t_n} - {}^\circ X_{t_{n-1}} \in A_{n-1}\right\} L(\mu)\left[\bigcap_{i=1}^{n-2}\left\{{}^\circ X_{t_i} \in A_i\right\}\right] \quad (5.3.3) \end{aligned}$$

(where again we have used that $\mathrm{st}({}^*A) = A$ for all compact $A \subset \mathbb{R}^d$). On the other hand, Lemma 5.3.3 yields

$$L(\mu)\left[B\right] = L(\mu)\left[\left\{x_{s_n} - x_{s_{n-1}} \in A_{n-1}\right\} \cap \bigcap_{i=1}^{n-2}\left\{x_{s_i} \in A_i\right\}\right]$$

as well as

$$L(\mu)\left\{{}^\circ X_{t_n} - {}^\circ X_{t_{n-1}} \in A_{n-1}\right\} = L(\mu)\left\{x_{s_n} - x_{s_{n-1}} \in A_{n-1}\right\}$$

and

$$L(\mu)\left[\bigcap_{i=1}^{n-2}\left\{{}^\circ X_{t_i} \in A_i\right\}\right] = L(\mu)\left[\bigcap_{i=1}^{n-2}\left\{x_{s_i} \in A_i\right\}\right].$$

Via (5.3.3), we conclude that

$$L(\mu)\left[\left\{x_{s_n}-x_{s_{n-1}}\in A_{n-1}\right\}\cap\bigcap_{i=1}^{n-2}\left\{x_{s_i}\in A_i\right\}\right]$$
$$=L(\mu)\left\{x_{s_n}-x_{s_{n-1}}\in A_{n-1}\right\}L(\mu)\left[\bigcap_{i=1}^{n-2}\left\{x_{s_i}\in A_i\right\}\right].$$

Since this last equation holds for arbitrary choice of compact $A_1,\ldots,A_{n-1}$, it must even hold for all Borel sets $A_1,\ldots,A_{n-1}\subset\mathbb{R}^d$, thereby completing the proof for the independence of $x_{s_n}-x_{s_{n-1}}$ from $\mathcal{F}_{s_n}$.

- Using the same notation as in the previous bullet point, what remains to show is that the law of $x_{s_n}-x_{s_{n-1}}$ under $L(\mu)$ equals the law of $x_{s_n-s_{n-1}}-x_0$ under $L(\mu)$.

We have already seen that for all compact $A_{n-1}\subset\mathbb{R}^d$,

$$L(\mu)\left\{x_{s_n}-x_{s_{n-1}}\in A_{n-1}\right\}=\inf_{m\in\mathbb{N}}L(\mu)\left[C_m\right]. \tag{5.3.4}$$

Since this holds for arbitrary $s_n>s_{n-1}$, we may replace s_n by s_n-s_{n-1} and s_{n-1} by 0, in order to get (via $t_n-t_{n-1}\approx s_n-s_{n-1}$) the identity

$$L(\mu)\left\{x_{s_n-s_{n-1}}-x_0\in A_{n-1}\right\}=\inf_{m\in\mathbb{N}}L(\mu)\left[C'_m\right] \tag{5.3.5}$$

wherein $C'_m=\{{}^*d(X_{t_n-t_{n-1}}-X_0,{}^*A_{n-1})\leq\frac{1}{m}\}$ for all $m\in\mathbb{N}$. But since X is a hyperfinite Lévy process, we can simply recall Remark 5.2.4 and obtain $\mu[C_m]=\mu[C'_m]$ for all $m\in\mathbb{N}$. This readily yields, via (5.3.4) and (5.3.5):

$$L(\mu)\left\{x_{s_n}-x_{s_{n-1}}\in A_{n-1}\right\}=L(\mu)\left\{x_{s_n-s_{n-1}}-x_0\in A_{n-1}\right\}$$

for every compact $A_{n-1}\subset\mathbb{R}^d$ and therefore even for all Borel sets $A_{n-1}\subset\mathbb{R}^d$. Hence $x_{s_n}-x_{s_{n-1}}$ and $x_{s_n-s_{n-1}}-x_0$ have the same law under $L(\mu)$ for all $s_{n-1}<s_n\in\mathbb{R}_{>0}$.

□

5.4 Lindstrøm's Hyperfinite Lévy-Khintchine Formula

Definition 5.4.1. Let X be a hyperfinite Lévy process. A positive infinitesimal $\eta\in\mathrm{st}^{-1}\{0\}\cap{}^*\mathbb{R}_{>0}$ is called a *splitting infinitesimal* if and only if

$$\forall\varepsilon\approx 0\quad\frac{1}{\Delta t}\sum_{\substack{a\in A\\|a|\in[\eta,\varepsilon]}}|a|^2p_a\approx 0 \tag{5.4.1}$$

Remark 5.4.1. Underspill and overspill show that the defining formula (5.4.1) for a splitting infinitesimal is equivalent to

$$S\text{-}\lim_{{}^\circ c \downarrow 0} \frac{1}{\Delta t} \sum_{\substack{a \in A \\ |a| \in [\eta, c]}} |a|^2 p_a = 0.$$

Lemma 5.4.1. *[263, p. 528] If X is a hyperfinite Lévy process, then there exists a positive infinitesimal $\eta_0 \approx 0$ such that all positive infinitesimals $\eta \geq \eta_0$ are splitting infinitesimals.*

Proof. Since for all ε the function $\eta \mapsto \frac{1}{\Delta t} \sum_{\substack{a \in A \\ |a| \in [\eta, \varepsilon]}} |a|^2 p_a$ is decreasing, we only have to find one splitting infinitesimal. To construct such an infinitesimal, we observe that by the second condition in Theorem 5.2.2, the quantity $\frac{1}{\Delta t} \sum_{\substack{a \in A \\ |a| \in [0, \eta)}} |a|^2 p_a$ is not only increasing in η, but also finite for all finite η. Now put

$$\beta = \inf \left\{ {}^\circ\left(\frac{1}{\Delta t} \sum_{\substack{a \in A \\ |a| \in [0, \zeta]}} |a|^2 p_a \right) \mid \zeta \in {}^*\mathbb{R}_{>0}, \quad {}^\circ\zeta > 0 \right\}$$

as well as

$$B = \left\{ \eta \in {}^*\mathbb{R}_{>0} \;\middle|\; \frac{1}{\Delta t} \sum_{\substack{a \in A \\ |a| \in [0, \eta)}} |a|^2 p_a > \beta - \eta \right\} \supseteq \mathrm{st}^{-1}\left(\mathbb{R}_{>0}\right).$$

Note that $B \cap {}^*(0,1]$ is internal and contains ${}^*(0,1] \setminus \mathrm{st}^{-1}\{0\}$, therefore there must be an infinitesimal element $\eta \in B \cap {}^*(0,1] \subset B$ (by underspill).

We shall now prove that this η is indeed a splitting infinitesimal.

Since $\varepsilon \mapsto \frac{1}{\Delta t} \sum_{\substack{a \in A \\ |a| \in [0, \varepsilon]}} |a|^2 p_a$ is increasing, we readily know ${}^\circ(\frac{1}{\Delta t} \sum_{\substack{a \in A \\ |a| \in [0, \varepsilon]}} |a|^2 p_a) \leq \beta$ for all infinitesimal ε. Hence we will have for any infinitesimal $\varepsilon \geq \eta$ an infinitely small hyperreal $\delta \in \mathrm{st}^{-1}\{0\}$ such that

$$\frac{1}{\Delta t} \sum_{\substack{a \in A \\ |a| \in [0, \eta)}} |a|^2 p_a > \beta - \eta \geq \frac{1}{\Delta t} \sum_{\substack{a \in A \\ |a| \in [0, \varepsilon]}} |a|^2 p_a - \eta + \delta$$

and thus

$$0 \approx \eta - \delta > \frac{1}{\Delta t} \sum_{\substack{a \in A \\ |a| \in [0, \varepsilon]}} |a|^2 p_a - \frac{1}{\Delta t} \sum_{\substack{a \in A \\ |a| \in [0, \eta)}} |a|^2 p_a = \frac{1}{\Delta t} \sum_{\substack{a \in A \\ |a| \in [\eta, \varepsilon]}} |a|^2 p_a \geq 0.$$

□

The name of a splitting infinitesimal alludes to the rôle it plays in decomposing hyperfinite Lévy processes into a sum of a drift term, a hyperfinite Lévy martingale with infinitesimal increments and a so-called *hyperfinite jump martingale* (i.e. a hyperfinite Lévy martingale where the contribution of the infinitesimal increments is negligible), cf. Lindstrøm [263, Theorem 5.3].

Therefore, we would expect that it has some importance in formulating and proving a hyperfinite analogon to the Lévy-Khintchine formula. The previously defined $\hat{\nu}$ induces a measure ν_x on $\mathbb{R}^d \setminus \{0\}$ which shall play the rôle of the Lévy measure in the hyperfinite Lévy-Khintchine formula. In order to prove the said formula, we need some facts about this measure ν_x, and also some information about candidates for the hyperfinite analogue of the covariance matrix.

Definition 5.4.2. Let X be a hyperfinite Lévy process. Then

$$\nu_x[C] = \lim_{\mathbb{R}\ni\varepsilon\downarrow 0} L(\hat{\nu})\left[\mathrm{st}^{-1}(C) \cap \left\{x \in {}^*\mathbb{R}^d \mid |x| \geq \varepsilon\right\}\right]$$

for all those C for which the right-hand side is well-defined.

Remark 5.4.2. 1. For all compact cubes $C = \bigotimes_{i=1}^{d}[a_i, b_i] \subset \mathbb{R}^d$, $\mathrm{st}^{-1}(C) = \bigcap_{n\in\mathbb{N}} \bigotimes_{i=1}^{d} {}^*(a_i - \frac{1}{n}, b_i + \frac{1}{n}) \in L(2^{\Omega})$, 2^{Ω} denoting the internal power set of Ω. Hence $\nu_x[C]$ is defined whenever C is a compact cube.

2. Thanks to the Carathéodory extension theorem, ν_x can be extended to a completed measure on the Borel σ-algebra of $\mathbb{R}^d$.
3. By Lemma 5.2.7, $L(\hat{\nu})[\mathrm{st}^{-1}(C) \cap \{x \in {}^*\mathbb{R}^d \mid |x| \geq \varepsilon\}]$ is finite for all $\varepsilon \in \mathbb{R}_{>0}$ and all Borel sets $C \subset \mathbb{R}^d$.

Definition 5.4.3. Define $C_X \in {}^*\mathbb{R}^{d\times d}$ via

$$\forall i, j \in \{1, \ldots, d\} \quad (C_X)_{i,j} = \frac{1}{\Delta t} \sum_{a\in A} a_i a_j p_a.$$

Lemma 5.4.2. *[263, Proposition 7.1, Lemma 7.4] Let X be a hyperfinite Lévy process. Then ν_x satisfies*

1. $\int_{\overline{B_1(0)}} |y|^2 \, \nu_x(dy) < +\infty$;
2. $\nu_x[\complement B_1(0)] < +\infty$;
3. $\nu_x\{0\} = 0$,

   and C_X has the following properties:
4. *C_X is symmetric and for all $y \in {}^*\mathbb{R}^d$,*

$$y \cdot (C_X)y = \frac{1}{\Delta t} |a \cdot y|^2 \, p_a \leq {\sigma_X}^2 |y|^2.$$

Thus, C_X is positive semi-definite, which also implies $C \in {{}^*\mathbb{R}_{\geq 0}}^{d\times d}$.

Proof. 1. By construction of ν_x (cf. Remark 5.4.2) and due to the fact that $^\circ \int X \, d\nu = \int X \, dL(\nu)$ for all internal measures ν and S-integrable X (cf. [267, Theorem 3] and [61, Theorem 6]), we have for all $\varepsilon \in \mathbb{R}_{>0}$,

$$\begin{aligned}\int_{\overline{B_1(0)} \setminus B_\varepsilon(0)} |y|^2 \ \nu_x(dy) &\leq \int_{^*\left(\overline{B_{1+\varepsilon}(0)} \setminus B_{\varepsilon/2}(0)\right)} {}^\circ|z|^2 \ L\left(\hat{\nu}\right)(dz) \\ &= {}^\circ\!\!\int_{^*\left(\overline{B_{1+\varepsilon}(0)} \setminus B_{\varepsilon/2}(0)\right)} |a|^2 \hat{\nu}(da) \\ &\leq {}^\circ \sum_{\substack{a \in A \\ |a| \leq 2}} |a|^2 p_a < +\infty,\end{aligned}$$

where in the last step we have used Theorem 5.2.2. Combining this with

$$\int_{\overline{B_1(0)}} |y|^2 \ \nu_x(dy) = \lim_{\varepsilon \downarrow 0} \int_{\overline{B_1(0)} \setminus B_\varepsilon(0)} |y|^2 \ \nu_x(dy)$$

(which follows from monotone convergence since $\nu_x\{0\} = 0$), we arrive at $\int_{\overline{B_1(0)}} |y|^2 \ \nu_x(dy) \leq {}^\circ \sum_{|a| \leq 2} |a|^2 p_a < +\infty$.

2. Similarly, the definition of ν_x implies

$$\nu_x \left[\complement B_1(0) \right] \leq L\left(\hat{\nu}\right) \left\{ y \in {}^*\mathbb{R}^d \ \middle| \ |y| \geq \frac{1}{2} \right\} = \sum_{|a| \geq \frac{1}{2}} |a| p_a$$

which is finite by Theorem 5.2.2.

3. Since $\mathrm{st}^{-1}\{0\} \setminus B_\varepsilon(0) = \emptyset$ for all $\varepsilon \in \mathbb{R}_{>0}$, this part of the Lemma is immediate from the definition of ν_x.

4. Since $a_i a_j = a_j a_i$ for all i, j, the symmetry of C_X is trivial. The formula follows from rearranging terms and using the Cauchy-Bunyakovski-Schwarz inequality:

$$\begin{aligned} y \cdot C_X y &= \sum_{i,j=1}^d y_i \left(C_X\right)_{i,j} y_j = \sum_{i,j=1}^d y_i \frac{1}{\Delta t} \sum_{a \in A} a_i a_j p_a y_j \\ &= \frac{1}{\Delta t} \sum_{a \in A} p_a \underbrace{\sum_{i,j=1}^d y_i a_i a_j y_j}_{=(y \cdot a)(a \cdot y)} = \frac{1}{\Delta t} \sum_{a \in A} p_a \left|a \cdot y\right|^2 \\ &\leq \underbrace{\frac{1}{\Delta t} \sum_{a \in A} p_a |a|^2}_{= \sigma_X{}^2} |y|^2. \end{aligned}$$

□

Lemma 5.4.3. *[263, Proposition 7.2] Let X be a hyperfinite Lévy process. Consider an internal function $F : {}^*\mathbb{R}^d \to \mathbb{R}$, and assume F to be S-continuous in a for all $a \in \mathrm{Fin}({}^*\mathbb{R}^d) \setminus \mathrm{st}^{-1}\{0\}$. Suppose, furthermore, that there exists some $C \in \mathbb{R}_{>0}$ such that $|F(a)| \leq C \cdot (|a|^2 \wedge 1)$ for all $a \in {}^*\mathbb{R}^d$. Then*

$$\int_{\mathbb{R}^d} {}^\circ F(y)\ \nu_x(dy) = \int_{\complement \overline{B_\eta(0)}} {}^\circ F(a)\ L(\hat{\nu})(da) < +\infty$$

for all splitting infinitesimals η.

Proof. First note that both $\int_{\complement \overline{B_\eta(0)}} |a|^2\ \hat{\nu}(da) = \frac{1}{\Delta t} \sum_{|a|>\eta} |a|^2 p_a$ and $\hat{\nu}[\complement \overline{B_1(0)}]$ are finite due to Theorem 5.2.2. Then, by employing the estimates on F and the fact that ${}^\circ \int X\ d\nu = \int X\ dL(\nu)$ for all internal measures ν and S-integrable X (cf. [267, Theorem 3] and [61, Theorem 6]), one gets

$$\begin{aligned} \int_{\complement \overline{B_\eta(0)}} {}^\circ F(a)\ L(\hat{\nu})(da) &= {}^\circ\!\!\int_{\complement \overline{B_\eta(0)}} F(a)\ \hat{\nu}(da) \\ &\leq C \cdot {}^\circ\!\!\int_{\complement \overline{B_\eta(0)}} |a|^2\ \hat{\nu}(da) + C \cdot {}^\circ\hat{\nu}\left[\complement \overline{B_1(0)}\right] \end{aligned}$$

and we have proven $\int_{\complement \overline{B_\eta(0)}} {}^\circ F(a)\ L(\hat{\nu})(da) < +\infty$.

On the other hand, if we assume F to be nonnegative, we see that for all $n \in \mathbb{N}$,

$$\begin{aligned} {}^\circ\!\!\int_{\overline{B_{n-\frac{1}{n}}(0)} \setminus B_{2/n}(0)} F(a)\ \hat{\nu}(da) &\leq \int_{\overline{B_n(0)} \setminus B_{1/n}(0)} {}^\circ F(y)\ \nu_x(dy) \\ &\leq {}^\circ\!\!\int_{\overline{B_{n+\frac{1}{n}}(0)} \setminus B_{1/2n}(0)} F(a)\ \hat{\nu}(da). \quad (5.4.2) \end{aligned}$$

Next, using the estimates for F on $B_1(0)$ and the assumption that η is a splitting infinitesimal, we get

$$\begin{aligned} &\int_{\overline{B_{n+\frac{1}{n}}(0)} \setminus B_{1/2n}(0)} F(a)\ \hat{\nu}(da) - \int_{\overline{B_{n+\frac{1}{n}}(0)} \setminus B_\eta(0)} F(a)\ \hat{\nu}(da) \\ &\leq 2C \int_{B_{1/2n}(0) \setminus B_\eta(0)} |a|^2\ \hat{\nu}(da) \approx 0, \end{aligned}$$

hence $\int_{\overline{B_{n+\frac{1}{n}}(0)} \setminus B_{1/2n}(0)} F(a)\ \hat{\nu}(da) \approx \int_{\overline{B_{n+\frac{1}{n}}(0)} \setminus B_\eta(0)} F(a)\ \hat{\nu}(da)$, and similarly, $\int_{\overline{B_{n-\frac{1}{n}}(0)} \setminus B_{2/n}(0)} F(a)\ \hat{\nu}(da) \approx \int_{\overline{B_{n-\frac{1}{n}}(0)} \setminus B_\eta(0)} F(a)\ \hat{\nu}(da)$ for all $n \in \mathbb{N}$.

If we combine these results with the fact that ${}^{\circ}\!\int X\, d\nu = \int X\, dL(\nu)$ for all internal measures ν and S-integrable X (cf. [267, Theorem 3] and [61, Theorem 6]), the estimates (5.4.2) can be written

$$\int_{\overline{B_{n-\frac{1}{n}}(0)\setminus B_\eta(0)}} {}^{\circ}F(a)\; L(\hat{\nu})\,(da) \leq \int_{\overline{B_n(0)\setminus B_{1/n}(0)}} {}^{\circ}F(y)\; \nu_x(dy)$$
$$\leq \int_{\overline{B_{n+\frac{1}{n}}(0)\setminus B_\eta(0)}} {}^{\circ}F(a)\; L(\hat{\nu})\,(da). \quad (5.4.3)$$

Using the standard monotone convergence theorem for the Loeb measure $L(\hat{\nu})$ and for ν_x, we find that

$$\int_{\complement\overline{B_\eta(0)}} {}^{\circ}F(a)\; L(\hat{\nu})\,(da) \leq \int_{\mathbb{R}^d} {}^{\circ}F(y)\; \nu_x(dy) \leq \int_{\complement\overline{B_\eta(0)}} {}^{\circ}F(a)\; L(\hat{\nu})\,(da),$$

hence

$$\int_{\mathbb{R}^d} {}^{\circ}F(y)\; \nu_x(dy) = \int_{\complement\overline{B_\eta(0)}} {}^{\circ}F(a)\; L(\hat{\nu})\,(da) = {}^{\circ}\!\int_{\complement\overline{B_\eta(0)}} F(a)\; \hat{\nu}(da).$$

Hence we have proven the assertion of the Lemma for nonnegative F. Needless to say, the Lemma's assertion in full generality follows immediately from applying this partial result to the nonnegative functions $F\chi_{\{F>0\}}$ and $(-F)\chi_{\{F<0\}}$, and afterwards exploiting the linearity of the integral. □

We conclude this section by stating and proving the hyperfinite Lévy-Khintchine formula:

Theorem 5.4.1. *[263, Theorem 8.1] If X is a hyperfinite Lévy process, $k \in {}^*\mathbb{R}_{>0}$ finite and noninfinitesimal and η a splitting infinitesimal for X, then for all finite $y \in {}^*\mathbb{R}^d$ and any finite $t \in T$,*

$$E\left[e^{iy\cdot X_t}\right] \approx \exp\left(\begin{array}{c} ity\cdot m_k - \frac{t}{2} y\cdot C_\eta y + \\ \int_{\complement\overline{B_\eta(0)}} \left(e^{iy\cdot a} - 1 - iy\cdot a\chi_{\overline{B_k(0)}}(a)\right)\hat{\nu}(da) \end{array}\right),$$

wherein $m_k = m_{X^{\leq k}}$ and $C_\eta = C_{X^{\leq\eta}}$.

Proof. First note that due to Taylor's Theorem, we can find for all $N \in \mathbb{N}$ and $a \in A \cap \overline{B_\eta(0)}$ an internal S-bounded function $\xi_0^a : {}^*B_N(0) \to {}^*\mathbb{R}$ such that

$$\forall y \in {}^*B_N(0) \quad e^{iy\cdot a} - 1 - iy\cdot a = -\frac{1}{2}(y\cdot a)^2 + \xi_0^a(y)\,|\Delta t|^2.$$

(Since $A \cap \overline{B_\eta(0)}$ is hyperfinite, one can choose the functions ξ_0^a, $a \in A \cap \overline{B_\eta(0)}$, simultaneously whence $\{\xi_0^a\}_{a\in A\cap\overline{B_\eta(0)}}$ is internal.) If we set $\xi_0(y) = \sum_{|a|\le\eta} \xi_0^a(y)$ for all $y \in {}^*\mathbb{R}^d$ and recall that $\sum_{|a|\le\eta} p_a \le 1$, we have thus constructed an internal S-bounded function $\xi_0 : {}^*B_N \to {}^*\mathbb{R}$ such that

$$\forall y \in {}^*B_N(0) \quad \sum_{|a|\le\eta} \left(e^{iy\cdot a} - 1 - iy\cdot a\right) p_a = -\sum_{|a|\le\eta} \frac{1}{2}(y\cdot a)^2 p_a + \xi_0(y)\,|\Delta t|^2 . \tag{5.4.4}$$

Let us now fix a finite $y \in {}^*\mathbb{R}^d$ and put $\xi = \xi_0(y)$. Using the formula for $y\cdot(C_{X^{\le\eta}})y$ of Lemma 5.4.2, (5.4.4) then becomes

$$\sum_{|a|\le\eta} \left(e^{iy\cdot a} - 1 - iy\cdot a\right) p_a = -\frac{1}{2} y\cdot C_\eta y \Delta t + \xi\,|\Delta t|^2 . \tag{5.4.5}$$

On the other hand, by definition of $m_k = m_{X^{\le k}}$, we have $iy\cdot m_k\Delta t = E[iy\cdot\Delta X^{\le k}]$. This allows us to perform the following calculation:

$$\begin{aligned}
E\left[e^{iy\cdot\Delta X_0}\right] &= 1 + E\left[iy\cdot\Delta X^{\le k}\right] + E\left[e^{iy\cdot\Delta X_0} - 1 - iy\cdot\Delta X^{\le k}\right] \\
&= 1 + iy\cdot m_k\Delta t + E\left[e^{iy\cdot\Delta X_0} - 1 - iy\cdot\Delta X^{\le k}\right] \\
&= 1 + iy\cdot m_k\Delta t + \sum_{a\in A}\left(e^{iy\cdot a} - 1 - iy\cdot a\chi_{\overline{B_k(0)}}(a)\right) \\
&= 1 + iy\cdot m_k\Delta t + \Delta t\int_{\complement\overline{B_\eta(0)}}\left(e^{iy\cdot a} - 1 - iy\cdot a\chi_{\overline{B_k(0)}}(a)\right)\hat{\nu}(da) \\
&\quad + \sum_{|a|\le\eta}\left(e^{iy\cdot\Delta X_0} - 1 - iy\cdot a\chi_{\overline{B_k(0)}}(a)\right)
\end{aligned}$$

(where we have exploited that $\hat{\nu}\{a\} = p_a/\Delta t$). On combining this result with identity (5.4.5), we obtain

$$E\left[e^{iy\cdot\Delta X_0}\right] = 1 + R\Delta t + \xi\,|\Delta t|^2 , \tag{5.4.6}$$

where

$$R = iy\cdot m_k + \int_{\complement\overline{B_\eta(0)}}\left(e^{iy\cdot a} - 1 - iy\cdot a\chi_{\overline{B_k(0)}}(a)\right)\hat{\nu}(da) - \frac{1}{2}y\cdot C_\eta y.$$

The *-integral $\int_{\complement\overline{B_\eta(0)}}(e^{iy\cdot a} - 1 - iy\cdot a\chi_{\overline{B_k(0)}}(a))\hat{\nu}(da)$ can be shown to be finite if we observe that the integrand is dominated according to

$$\forall a \in A \quad \left|e^{iy\cdot a} - 1 - iy\cdot a\chi_{\overline{B_k(0)}}(a)\right| \le \underbrace{\left|e^{iy\cdot a}\right|}_{=1} + 1 + \left|y\cdot a\chi_{\overline{B_k(0)}}(a)\right|$$
$$\le 2 + |y|\cdot \underbrace{\left|a\chi_{\overline{B_k(0)}}(a)\right|}_{\le k}$$
$$\le 2 + k|y| \in \mathrm{Fin}\,({}^*\mathbb{R})$$

(where, of course, the Cauchy-Bunyakovski-Schwarz inequality was used) and

$$\forall a \in A\cap B_{k\wedge 1}(0) \quad {}^\circ\left|e^{iy\cdot a} - 1 - iy\cdot a\chi_{\overline{B_k(0)}}(a)\right| = \left|e^{i{}^\circ y\cdot{}^\circ a} - 1 - i{}^\circ y\cdot{}^\circ a\right|$$
$$= \left|\sum_{n=2}^{\infty}\frac{(i{}^\circ y\cdot{}^\circ a)^n}{n!}\right| = \sum_{n=2}^{\infty}\frac{(|{}^\circ y|\,|{}^\circ a|)^n}{n!} \le |{}^\circ a|^2\sum_{n=2}^{\infty}\frac{|{}^\circ y|^n}{n!}$$
$$\le |{}^\circ a|^2\, e^{|{}^\circ y|}$$

which can be combined to get

$$\left|e^{iy\cdot a} - 1 - iy\cdot a\chi_{\overline{B_k(0)}}(a)\right| \le \frac{1}{(k\wedge 1)^2}\left(2e^{|y|}\vee(2+k|y|)\right)\left(|a|^2\wedge 1\right)$$

for all $a \in A$. Then, by Lemma 5.4.3, we conclude that $\int_{\complement\overline{B_\eta(0)}}(e^{iy\cdot a} - 1 - iy\cdot a\chi_{\overline{B_k(0)}}(a))\hat{\nu}(da)$ must be finite, too. Also, C_η is finite by Lemma 5.4.2, therefore we obtain that R must be finite as well.

Next, we recall that since the increments of X are all *-independent and identically distributed, we have

$$E\left[e^{iy\cdot X_t}\right] = E\left[e^{iy\cdot\sum_{s<t}\Delta X_s}\right] = \prod_{s<t}E\left[e^{iy\cdot\Delta X_s}\right] = E\left[e^{iy\cdot\Delta X_0}\right]^{t/\Delta t}$$

which via (5.4.6) becomes

$$\begin{aligned} E\left[e^{iy\cdot X_t}\right] &= \left(1 + R\Delta t + \xi\left|\Delta t\right|^2\right)^{t/\Delta t} \\ &= \left(1 + (tR + t\xi\Delta t)\frac{1}{t/\Delta t}\right)^{t/\Delta t} \\ &= {}^\circ e^{tR + t\xi\Delta t} \end{aligned} \tag{5.4.7}$$

(in the last step we have once again used that $(1 + x/N)^N \longrightarrow e^x$ locally uniformly in x as $N \to \infty$). But since $\xi = \xi_0(y)$ is finite, one also has $t\xi\Delta t \approx 0$ and hence $e^{tR+t\xi\Delta t} \approx e^{tR}$. Inserting this into (5.4.7) gives us ${}^\circ E[e^{iy\cdot X_t}] = e^{tR}$, and resubstituting R completes the proof. □

5.5 Lindstrøm's Hyperfinite Representation Theorem for Lévy Processes

In this section, we shall prove what may be regarded as the central result of this chapter: that for each triple consisting of a drift vector $\gamma \in \mathbb{R}^d$, a symmetric matrix $C \in {\mathbb{R}_{\geq 0}}^{d\times d}$ and a Lévy measure ν, there is a corresponding hyperfinite Lévy process X such that its standard part $x = \mathrm{st}(X)$ has covariance matrix C, drift γ and Lévy measure $\nu_x = \nu$.

In light of the Lévy-Khintchine formula, this means that the notion of a Lévy process and a hyperfinite Lévy process are essentially equivalent. In the proof of the subsequent result, the hyperfinite Lévy-Khintchine formula of Theorem 5.4.1 will play a crucial rôle.

Theorem 5.5.1. *[263, Theorem 9.1] Given any generating triplet (γ, C, ν), there is a hyperfinite Lévy process X such that γ, C, ν are the drift vector, covariance matrix and Lévy measure of X's standard part $x = \mathrm{st}(X)$ (cf.* (5.1.1)*), respectively.*

Proof. It is enough to merely construct a hyperfinite Lévy process whose standard part has covariance matrix C and Lévy measure ν. We will pay no attention to the drift part, since after having constructed X with a finite drift vector $\gamma_0 \in \mathrm{Fin}({}^*\mathbb{R}^d)$, we can always translate the set of increments of X by $(\gamma - \gamma_0)\Delta t$ to ensure that $\mathrm{st}(X)$ has drift vector γ.

Furthermore, we observe that by Lemma 5.3.3, $E[e^{iy\cdot X_t}] = E[e^{i^\circ y\cdot x_{^\circ t}}]$ for all finite $y \in \mathrm{Fin}({}^*\mathbb{R}^d)$ and finite $t \in T$. Therefore and owing to the fact that we may ignore the drift part (as remarked previously), comparing the standard Lévy-Khintchine formula (Theorem 5.1.1) with the hyperfinite Lévy-Khintchine formula (Theorem 5.4.1) yields that for any hyperfinite Lévy process X, ${}^\circ C_{X^{\leq\eta}}$ is the covariance matrix of $\mathrm{st}(X)$ for any splitting infinitesimal. If we use Lemma 5.4.3 in addition, we see that ν_x will always be the Lévy measure of $\mathrm{st}(X)$.

So, we are left with the task of constructing a hyperfinite Lévy process X such that:

1. $C_{X^{\leq\eta}} \approx C$ for some splitting infinitesimal η, and
2. $\lim_{\mathbb{R}\ni\varepsilon\downarrow 0} L(\hat{\nu})[\mathrm{st}^{-1}(D) \setminus B_\varepsilon(0)] = \nu(D)$ for all compact cubes D in $\mathbb{R}^d$ (cf. Remark 5.4.2).

First, we shall construct the part of X corresponding to ν (also known as the *internal jump part*). That is, we will look for an internal hyperfinite set $A_1 \subset {}^*\mathbb{R}^d$ and a family $(p_a)_{a\in A_1} \in (0,1)^{A_1}$ with $\sum_{a\in A_1} p_a < 1$ For this purpose, set $B_N = \{y \in {}^*\mathbb{R}^d \mid \frac{1}{N} \leq |y| \leq N\}$ and note that for all $N \in \mathbb{N}$, ${}^*\nu[B_N] = \nu[\overline{B_N(0)} \setminus B_{1/N}(0)]$ is finite as ν is a Lévy measure. Since $|y| \leq N$

for all $y \in B_N$, this also implies the finiteness of $\int_{B_N} |y| \, {}^*\nu(dy)$ for all $N \in \mathbb{N}$. Therefore, the set

$$A_n = \left\{ N \in {}^*\mathbb{N} \;\middle|\; \begin{array}{c} N \geq n, \quad \sqrt{\Delta t} \cdot \int_{B_N} |y| \, {}^*\nu(dy) \leq \frac{1}{N}, \\ \Delta t \cdot {}^*\nu\left[B_N\right] \leq \frac{1}{N}, \quad \sqrt{\Delta t} \cdot N \leq \frac{1}{N} \end{array} \right\}$$

is not only internal, but also nonempty, for all n. Furthermore, the sequence $(A_n)_{n \in \mathbb{N}}$ is decreasing. The $\aleph_1$-saturation of our nonstandard universe now yields that $\bigcap_{n \in \mathbb{N}} A_n \neq \emptyset$ whence we have found an $N \in {}^*\mathbb{N} \setminus \mathbb{N}$ such that

$$\sqrt{\Delta t} \cdot \int_{B_N} |y| \, {}^*\nu(dy) \approx \Delta t \cdot {}^*\nu\left[B_N\right] \approx \sqrt{\Delta t} \cdot N \approx 0. \tag{5.5.1}$$

Putting

$$\eta = 1/N,$$

we will show later on that η is a splitting infinitesimal. First, we have to construct the set A_1, and for this sake we choose an internal lattice $\mathbb{L}$ of the shape $\zeta \mathbb{Z}^d$ with infinitesimal spacing ζ and in such a way that $\underline{\mathbb{L} \cap B_N \subset}$ $\overline{\complement B_\eta(0)}$ (rather than merely $\subset \complement B_\eta(0)$). Furthermore, we denote by $\rho = \rho_{\mathbb{L}}$ the corresponding rounding operation $\rho_{\mathbb{L}} : \mathbb{R}^d \to \mathbb{L}$, that is

$$\rho_{\mathbb{L}} : y \mapsto \left(\underbrace{\max\left(\zeta\mathbb{Z} \cap \left(y_i + \frac{\varepsilon}{2} + {}^*\mathbb{R}_{\leq 0} \right) \right)}_{=(\zeta\mathbb{Z})_{\leq y_i + \varepsilon/2}} \right)_{i=1}^{d}.$$

Then, all preimages of singletons in $\mathbb{L}$ under $\rho_{\mathbb{L}}$ are *-Borel measurable, and so are hence all preimages of hyperfinite subsets of $\mathbb{L}$ under $\rho_{\mathbb{L}}$.

Using this and the fact that $\mathbb{L} \cap [-N, N]^d$ is hyperfinite, we can define an internal measure $\hat{\nu}$ on $\mathbb{L}$ via

$$\hat{\nu} : B \mapsto {}^*\nu\left[{\rho_{\mathbb{L}}}^{-1}(B) \cap B_N \right],$$

and we may set

$$A_1 = \left\{ a \in \mathbb{L} \mid \hat{\nu}\{a\} > 0 \right\}.$$

Also note that

$$\hat{\nu}\left[A_1\right] = \sum_{a \in A_1} \hat{\nu}\{a\} = \sum_{\substack{a \in \mathbb{L} \\ \hat{\nu}\{a\} > 0}} \hat{\nu}\{a\} = \sum_{a \in \mathbb{L}} \hat{\nu}\{a\} = \hat{\nu}\left[\mathbb{L}\right] = {}^*\nu\left[B_N\right]$$

is finite. Therefore, if we set

$$\forall a \in A_1 \quad p_a = \Delta t \hat{\nu}\{a\},$$

we get

$$\sum_{a \in A_1} p_a = \Delta t \cdot \hat{\nu}\,[A_1] \approx 0.$$

The complementary probability $q = 1 - \sum_{a \in A_1} p_a \approx 1$ will be the weight that we will assign to the part of X that corresponds to C (also referred to as the *internal diffusion part* of X). We shall use the abbreviation

$$\delta = \frac{1}{\Delta t} \sum_{\substack{a \in A_1 \\ |a| \leq 1}} a p_a$$

in the following.

In order to define A_2, choose $C^{1/2} \in \mathbb{R}^{d \times d}$ such that $C = C^{1/2} \cdot {}^t C^{1/2}$ (${}^t D$ denoting the transpose of any $D \in \mathbb{R}^{d \times d}$), and let $\{e_1, \ldots, e_d\}$ be the standard basis of $\mathbb{R}^d$. We will set

$$\begin{aligned}
\forall k \in \{1, \ldots, d\} \;\; a_{+k} &= C^{1/2} e_k \sqrt{d \frac{\Delta t}{q}} - \delta \frac{\Delta t}{q} \\
\forall k \in \{1, \ldots, d\} \;\; a_{-k} &= -C^{1/2} e_k \sqrt{d \frac{\Delta t}{q}} - \delta \frac{\Delta t}{q} \\
A_2 &= \{a_{+1}, a_{-1}, \ldots, a_{+d}, a_{-d}\} \\
\forall a \in A_2 \;\; p_a &= \frac{q}{2d} \\
A &= A_1 \cup A_2
\end{aligned}$$

Thus, A_2 only contains $2d$ points, viz. just the $a_{\pm k}$, $k \in \{1, \ldots, d\}$, which are all assigned equal weight $\approx \frac{1}{2d}$.

Note that by construction $\sum_{a \in A} p_a = \sum_{a \in A_1} p_a + \sum_{a \in A_2} p_a = (1 - q) + 2d \frac{q}{2d} = 1$, so we can indeed define a hyperfinite random walk with increments $A = A_1 \cup A_2$ and transition probabilities $(p_a)_{a \in A}$. For the rest of this proof, X will always denote this hyperfinite random walk.

As we shall see in a minute, subtracting $\delta \frac{\Delta t}{q} = \frac{1}{q} \sum_{\substack{a \in A_1 \\ |a| \leq 1}} a p_a$ in the definition of the $a_{\pm k}$ will ensure that the first condition of Theorem 5.2.2 is satisfied, and if we can show that the second and third conditions also hold for X, the said theorem may be applied to prove X to be a hyperfinite Lévy process (rather than merely a hyperfinite random walk).

By the choice of $\mathbb{L}$ and $\rho_{\mathbb{L}}$, one has $|b_i| \geq \frac{|\rho_{\mathbb{L}}(b_i)|}{2}$ for all $i \in \{1, \ldots, d\}$ and $b \in {}^*\mathbb{R}^d$. This implies $|b| \geq |\rho_{\mathbb{L}}(b)|/2$ for arbitrary $b \in {}^*\mathbb{R}^d$. On the other hand, the facts that ν is a Radon measure on $\mathbb{R}^d \setminus \{0\}$ and $\int_{\overline{B_M(0)} \setminus B_\varepsilon(0)} |y| \ \nu(dy) < +\infty$ for all $M, \varepsilon \in \mathbb{R}_{>0}$ mean that $|y| \ \nu(dy)$ must be a Radon measure on the Polish space $\overline{B_M(0)} \setminus B_\varepsilon(0)$ for all $M, \varepsilon \in \mathbb{R}_{>0}$. Therefore, $|y| \ \nu(dy)$ is also outer-regular (cf. e.g. [76, Korollar 26.4]). This can be used, together with the nonstandard characterization of limits to get $\int_{\overline{B_{1+\varsigma}(0)} \setminus B_{1/N}(0)} |y| \ {}^*\nu(dy) \approx 2 \int_{\overline{B_1(0)} \setminus B_{1/N}(0)} |y| \ {}^*\nu(dy)$. So,

$$\begin{aligned}
\frac{1}{\Delta t} \sum_{\substack{a \in A_1 \\ |a| \leq 1}} |a| p_a &= \int_{\overline{B_1(0)} \setminus \overline{B_{1/N}(0)}} |a| \ \hat{\nu}(da) \\
&= \int_{\rho^{-1}\left(\overline{B_1(0)}\right) \setminus \rho^{-1}\left(\overline{B_{1/N}(0)}\right)} |\rho(y)| \ {}^*\nu(dy) \\
&\leq 2 \int_{\overline{B_{1+\varsigma}(0)} \setminus B_{1/N}(0)} |y| \ {}^*\nu(dy) \approx 2 \int_{\overline{B_1(0)} \setminus B_{1/N}(0)} |y| \ {}^*\nu(dy).
\end{aligned}$$

Therefore and in light of (5.5.1) as well as the definition of δ, we may conclude that

$$|\delta| \sqrt{\Delta t} \leq 2 \int_{\overline{B_1(0)} \setminus B_{1/N}(0)} |y| \ {}^*\nu(dy) \approx 0. \tag{5.5.2}$$

Hence all elements of A_2 are infinitely small.

Combining this with the fact that $\frac{1}{\Delta t} \sum_{a \in A_2} a p_a = -\delta$ by construction, we arrive at

$$\begin{aligned}
\frac{1}{\Delta t} \sum_{\substack{a \in A \\ |a| \leq k}} a p_a &= \underbrace{\frac{1}{\Delta t} \sum_{\substack{a \in A_1 \\ |a| \leq 1}} a p_a}_{=\delta} + \underbrace{\frac{1}{\Delta t} \sum_{a \in A_2} a p_a}_{=-\delta} + \frac{1}{\Delta t} \sum_{\substack{a \in A_1 \\ 1 < |a| \leq k}} a p_a \\
&= \int_{\overline{B_k(0)} \setminus \overline{B_1(0)}} y \ \hat{\nu}(dy)
\end{aligned} \tag{5.5.3}$$

for all noninfinitesimal k. But, if we exploit once again that $|b| \geq |\rho_{\mathbb{L}}(b)|/2$ for all $b \in {}^*\mathbb{R}^d$ and that $\hat{\nu}$ is the image of ${}^*\nu$ under ρ when restricted to B_N, we obtain

$$\begin{aligned}
\left| \int_{\overline{B_k(0)} \setminus \overline{B_1(0)}} a \ \hat{\nu}(da) \right| &\leq \int_{\overline{B_k(0)} \setminus \overline{B_1(0)}} |a| \ \hat{\nu}(da) \leq \int_{\overline{B_{k+\varsigma}(0)} \setminus \overline{B_{1-\varsigma}(0)}} 2|y| \ {}^*\nu(dy) \\
&\leq \int_{\overline{B_{2k}(0)} \setminus \overline{B_{1/2}(0)}} 2|y| \ {}^*\nu(dy)
\end{aligned}$$

which is finite for all finite and noninfinitesimal k as ν is a Lévy measure. Thereby (5.5.3) actually implies that $\frac{1}{\Delta t}\sum_{\substack{a\in A\\|a|\leq k}} ap_a$ is finite for all finite and noninfinitesimal k. Hence we have proven that the first condition of Theorem 5.2.2 holds for X.

In order to prove the second one, we employ the triangle inequality as well as (5.5.2) and the fact that $q \approx 1$ to see that

$$\begin{aligned}\forall a\in A_2 \quad |a| &\leq \sqrt{d}\max_{i,j}\left|\left(C^{1/2}\right)_{i,j}\right|\cdot\sqrt{\frac{d\Delta t}{q}+\frac{\delta\Delta t}{q}}\\ &\approx d\max_{i,j}\left|\left(C^{1/2}\right)_{i,j}\right|\cdot\sqrt{\Delta t}\approx 0,\end{aligned} \tag{5.5.4}$$

hence (exploiting that A_2 is finite and $\sum_{a\in A_2} p_a \leq 1$), we get $\sum_{a\in A_2}|a|^2 p_a \approx 0$. Therefore we have for all $k \in \mathbb{R}$

$$\begin{aligned}\frac{1}{\Delta t}\sum_{\substack{a\in A\\|a|\leq k}}|a|^2 p_a \approx \frac{1}{\Delta t}\sum_{\substack{a\in A_1\\|a|\leq k}}|a|^2 p_a &\leq \int_{\overline{B_k(0)}}|a|^2\,\hat{\nu}(da) \leq \int_{B_{k+1}(0)}|y|^2\,{}^*\nu(dy)\\ &= \int_{B_{k+1}(0)}|y|^2\,\nu(dy)\end{aligned}$$

which is finite since ν is a Lévy measure. And because the expression on the left is increasing in k, the finiteness of $\frac{1}{\Delta t}\sum_{|a|\leq k}|a|^2 p_a$ also follows for all finite $k \in {}^*\mathbb{R}$.

The third condition of Theorem 5.2.2 also holds for X: For all finite and noninfinitesimal $k \in {}^*\mathbb{R}_{>0}$,

$$\frac{1}{\Delta t}\sum_{\substack{a\in A\\|a|>k}} p_a = \hat{\nu}\left[\complement\overline{B_k(0)}\right] \leq {}^*\nu\left[\complement\overline{B_{k-\varsigma}(0)}\right] \leq \nu\left[\complement\overline{B_{{}^\circ 2k}(0)}\right],$$

which converges to zero as ${}^\circ k \to \infty$ since ν is a Radon (in particular, inner-regular) measure on $\mathbb{R}^d \setminus \{0\}$ satisfying $\nu[\complement\overline{B_1(0)}] < +\infty$.

So we have verified all three conditions stated in Theorem 5.2.2, hence X is a hyperfinite Lévy process.

Moreover, in a similar way we can prove η to be a splitting infinitesimal:

$$\begin{aligned}\frac{1}{\Delta t}\sum_{\substack{a\in A\\|a|\leq c}}|a^2|p_a &\leq \int_{\overline{B_c(0)}}|a|^2\,\hat{\nu}(da) \leq \int_{B_{2c}(0)}|y|^2\,{}^*\nu(dy)\\ &= \int_{B_{2c}(0)}|y|^2\,\nu(dy)\end{aligned}$$

which converges to zero as $c \downarrow 0$, since $\int_{B_1(0)} |y|^2 \, \nu(dy) < +\infty$ implies that $|y|^2\nu(dy)$ is a Radon measure (even a finite one) on the Polish space $B_1(0) \subset \mathbb{R}^d$ and therefore $|y|^2\nu(dy)$ is outer-regular (cf. e.g. [76, Korollar 26.4]).

What remains to prove is that $C_{X^{\leq\eta}} \approx C$ and that $\nu_x[D] = \lim_{\varepsilon\downarrow 0} L(\hat{\nu})[\mathrm{st}^{-1}(D) \setminus B_\varepsilon(0)]$ equals $\nu[D]$ for all compact cubes D in $\mathbb{R}^d$.

For any such cube $D = \bigotimes_{i=1}^d [a_i, b_i]$, $\mathrm{st}^{-1}(D) = \bigcap_{n\in\mathbb{N}} \bigotimes_{i=1}^d {}^*(a_i - \frac{1}{n}, b_i + \frac{1}{n})$ and one also has, since the internal measure $\hat{\nu}$ is the image of ${}^*\nu$ under ρ when restricted to B_N,

$$
\begin{aligned}
&{}^*\nu\left[\bigotimes_{i=1}^d {}^*\left(a_i - \frac{1}{n+1}, b_i + \frac{1}{n+1}\right) \setminus {}^*B_{2\varepsilon}(0)\right] \\
&\leq \hat{\nu}\left[\bigotimes_{i=1}^d {}^*\left(a_i - \frac{1}{n}, b_i + \frac{1}{n}\right) \setminus {}^*B_{\varepsilon}(0)\right] \\
&\leq {}^*\nu\left[\bigotimes_{i=1}^d {}^*\left(a_i - \frac{1}{n-1}, b_i + \frac{1}{n-1}\right) \setminus {}^*B_{\varepsilon/2}(0)\right],
\end{aligned}
$$

that is

$$
\begin{aligned}
&\nu\left[\bigotimes_{i=1}^d \left(a_i - \frac{1}{n+1}, b_i + \frac{1}{n+1}\right) \setminus B_{2\varepsilon}(0)\right] \\
&\leq \hat{\nu}\left[\bigotimes_{i=1}^d {}^*\left(a_i - \frac{1}{n}, b_i + \frac{1}{n}\right) \setminus {}^*B_{\varepsilon}(0)\right] \\
&\leq \nu\left[\bigotimes_{i=1}^d \left(a_i - \frac{1}{n-1}, b_i + \frac{1}{n-1}\right) \setminus B_{\varepsilon/2}(0)\right]
\end{aligned}
$$

for all $n \geq 2$. Hence

$$
\begin{aligned}
&\inf_{n\in\mathbb{N}} \nu\left[\bigotimes_{i=1}^d \left(a_i - \frac{1}{n}, b_i + \frac{1}{n}\right) \setminus B_{2\varepsilon}(0)\right] \\
&\leq \inf_{n\in\mathbb{N}} {}^\circ\hat{\nu}\left[\bigotimes_{i=1}^d {}^*\left(a_i - \frac{1}{n}, b_i + \frac{1}{n}\right) \setminus {}^*B_{\varepsilon}(0)\right] \\
&\leq \inf_{n\in\mathbb{N}} \nu\left[\bigotimes_{i=1}^d \left(a_i - \frac{1}{n}, b_i + \frac{1}{n}\right) \setminus B_{\varepsilon/2}(0)\right].
\end{aligned}
$$

But by the definition of the Loeb measure and the outer regularity of the Radon measure ν on the Polish space $\complement B_\varepsilon(0)$ (cf. again [76, Korollar 26.4]), this can be written as

$$\nu\left[\bigotimes_{i=1}^{d}[a_i,b_i]\setminus B_{2\varepsilon}\right]\le L(\hat\nu)\left[\bigcap_{n\in\mathbb{N}}\bigotimes_{i=1}^{d}{}^*\left(a_i-\frac{1}{n},b_i+\frac{1}{n}\right)\setminus{}^*B_\varepsilon(0)\right]$$
$$\le\nu\left[\bigotimes_{i=1}^{d}[a_i,b_i]\setminus B_{\varepsilon/2}(0)\right],$$

hence

$$\nu\left[D\setminus B_{2\varepsilon}(0)\right]\le L(\hat\nu)\left[\mathrm{st}^{-1}(D)\setminus{}^*B_\varepsilon(0)\right]\le\nu\left[D\setminus B_{\varepsilon/2}(0)\right]$$

and therefore

$$\lim_{\varepsilon\downarrow 0}L(\hat\nu)\left[\mathrm{st}^{-1}(D)\setminus{}^*B_\varepsilon(0)\right]=\lim_{\varepsilon\downarrow 0}\nu\left[D\setminus B_\varepsilon(0)\right].$$

If $0\notin D$, then the compactness of D entails that we will find some $\varepsilon\in\mathbb{R}_{>0}$ such that $D\setminus B_\varepsilon(0)=D$ and $\mathrm{st}^{-1}(D)\setminus{}^*B_\varepsilon(0)=\mathrm{st}^{-1}(D)$, which readily yields $\lim_{\varepsilon\downarrow 0}L(\hat\nu)[\mathrm{st}^{-1}(D)\setminus B_\varepsilon(0)]=\nu[D]$. In case $0\in D$ and ν has finite mass, then ν can be extended to a Radon measure on $\mathbb{R}^d$, which again will be outer-regular and therefore $\lim_{\varepsilon\downarrow 0}\nu[D\setminus B_\varepsilon(0)]=\nu[D]$. If $0\in D$ and ν has infinite mass, then both $\lim_{\varepsilon\downarrow 0}\nu[D\setminus B_\varepsilon(0)]$ and $\nu[D]$ will be infinite, hence also $\lim_{\varepsilon\downarrow 0}\nu[D\setminus B_\varepsilon(0)]=\nu[D]$. Thus, even when $0\in D$, $\lim_{\varepsilon\downarrow 0}L(\hat\nu)[\mathrm{st}^{-1}(D)\setminus{}^*B_\varepsilon(0)]=\lim_{\varepsilon\downarrow 0}\nu[D\setminus B_\varepsilon(0)]=\nu[D]$ which completes the proof of $\nu_x=\nu$.

The last step is to convince ourselves that $C_{X^{\le\eta}}\approx C$. Recall that $\mathbb{L}$ was chosen such that $\mathbb{L}\cap B_N\subset\complement\overline{B_\eta(0)}$, hence $|a|>\eta$ for all $a\in A_1$. On the other hand, by virtue of relation (5.5.1), we have $\sqrt{\Delta t}/\eta\approx 0$, and due to estimate (5.5.4), also that $|a|/\sqrt{\Delta t}$ is finite for all $a\in A_2$. Therefore, $|a|/\eta=\frac{|a|}{\sqrt{\Delta t}}\frac{\sqrt{\Delta t}}{\eta}\approx 0$ for all $a\in A_2$, in particular $A_2\subset B_\eta(0)$. Summarizing this, $A\cap\overline{B}_\eta(0)=A_2$ and therefore for all $i,j\in\{1,\dots,d\}$,

$$\begin{aligned}
(C_{X^{\le\eta}})_{i,j}&=\frac{1}{\Delta t}\sum_{|a|\le\eta}a_ia_jp_a=\frac{1}{\Delta t}\sum_{a\in A_2}a_ia_jp_a\\
&=\frac{1}{\Delta t}\frac{q}{2d}\sum_{k=1}^{d}\left(\left(C^{1/2}\right)_{i,k}\sqrt{d\frac{\Delta t}{q}}-\delta_i\frac{\Delta t}{q}\right)\left(\left(C^{1/2}\right)_{j,k}\sqrt{d\frac{\Delta t}{q}}-\delta_j\frac{\Delta t}{q}\right)\\
&\quad+\frac{1}{\Delta t}\frac{q}{2d}\sum_{k=1}^{d}\left(-\left(C^{1/2}\right)_{i,k}\sqrt{d\frac{\Delta t}{q}}-\delta_i\frac{\Delta t}{q}\right)\left(-\left(C^{1/2}\right)_{j,k}\sqrt{d\frac{\Delta t}{q}}-\delta_j\frac{\Delta t}{q}\right)\\
&=\frac{1}{\Delta t}\frac{q}{2d}\sum_{k=1}^{d}\left(\left(C^{1/2}\right)_{i,k}\left(C^{1/2}\right)_{j,k}2d\frac{\Delta t}{q}+\delta_i\delta_j\cdot 2d\frac{(\Delta t)^2}{q^2}\right)\\
&=\delta_i\delta_k\frac{\Delta t}{q}+\underbrace{\sum_{k=1}^{d}\left(C^{1/2}\right)_{i,k}\left(C^{1/2}\right)_{j,k}}_{=\left(\left(C^{1/2}\right)\cdot{}^t\left(C^{1/2}\right)\right)_{i,k}}
\end{aligned}$$

However, $(C^{1/2}) \cdot {}^t(C^{1/2}) = C$ by definition of $C^{1/2}$, and $|\delta_i \delta_j|^2 \leq |\delta|^2 \Delta t = (|\delta|\sqrt{\Delta t})^2 \approx 0$ by relation (5.5.2); furthermore $q \approx 1$. Hence we conclude

$$\forall i, j \in \{1, \ldots, d\} \quad (C_{X^{\leq \eta}})_{i,j} \approx C_{i,j}$$

and we are done. □

5.6 Hyperfinite Lévy Processes: Extensions and Applications

Theorem 5.5.1 has been extended substantially by Albeverio and Herzberg [14]: They prove that the jump part of a Lévy process is essentially a convolution of Poisson processes in the following sense:

Theorem 5.6.1. *[14, Theorem 4.4] Given any generating triplet (γ, C, ν), there is a hyperfinite Lévy process X, with $A_1 \subset \mathbb{L}$ for some lattice $\mathbb{L}$ such that*

1. *γ, C, ν are the drift vector, covariance matrix and Lévy measure of X's standard part $x = \mathrm{st}(X)$ (cf. (5.1.1)).*
2. *$\hat{\nu} = \ast_{\underline{\alpha} \in \mathbb{L} \setminus \{0\}} (\lambda_{\underline{\alpha}} \delta_{\underline{\alpha}} + (1 - \lambda_{\underline{\alpha}}) \delta_0)$, where the set $\{\lambda_{\underline{\alpha}} \mid \underline{\alpha} \in \mathbb{L}\}$ can be derived from ν via*
$$\lambda_{\underline{\alpha}} = \left({}^*\nu \circ (\rho_{\mathbb{L}})^{-1} \right) \{\underline{\alpha}\}.$$

This result has a financial interpretation in that it allows to regard a one-dimensional Lévy market model as essentially a binomial model which is unique modulo the order in which the jumps occur. In this sense, there is a very weak notion of completeness (viz. completeness up to the order in which the jumps occur) with respect to which any one-dimensional Lévy market model can be considered internally complete [14, Sects. 5, 6]; generalizations to arbitrary finite dimensions are conceivable, using the results on dynamically complete markets of Anderson and Raimondo [64].

However, the order in which the jumps occur can be neglected, if A_1, the set of jumps, is sufficiently small in cardinality when compared to the inverse time scale and some regularity parameters of the Lévy measure ν. This is basically the reason why they have to use the following very technically looking bound on the cardinality of $\mathbb{L} \supset A_1$ to prove Theorem 5.6.1:

Theorem 5.6.2. *[14, Theorem 4.7] Let (γ, C, ν) be a generating triplet and use the notation of the proof of Theorem 5.5.1. Then there is a hyperfinite Lévy process X, with $A_1 \subset \mathbb{L} \setminus \{0\}$ for some lattice $\mathbb{L} = \varepsilon \mathbb{Z}^d \cap B_N$ such that*

1. *γ, C, ν are the drift vector, covariance matrix and Lévy measure of X's standard part $x = \mathrm{st}(X)$ (cf. (5.1.1)).*

2. $N \in {}^*\mathbb{N} \setminus \mathbb{N}_0$ *is chosen so small that*

$$\begin{pmatrix} 4{D_{0,N}}^2 + 4D_{0,N}D_{2,N} + ND_{0,N}D_{3,N} + \\ D_{0,N}D_1 + D_{2,N}D_1 \end{pmatrix} \cdot \Delta t \cdot 3^{2^d \cdot N^{2d}} \cdot ((D_{0,N} + D_{2,N}) \vee 1) \approx 0$$

where $D_{0,N}, \ldots, D_{3,N}$ *are defined as* $D_{0,N} = {}^*\nu[\overline{B_N(0)} \setminus B_1(0)]$, $D_1 = \int_{\overline{B_1(0)}} |x|^2 \; \nu(dx)$, $D_{2,N} = {}^*\nu[\overline{B_1(0)} \setminus B_{1/N}(0)]$, *and* $D_{3,N} = {}^*\int_{\overline{B_1(0)} \setminus B_{1/N}(0)} |x| \; {}^*\nu(dx)$.

Proof. Most of the theorem is a restatement of Lindstrøm's representation theorem, Theorem 5.5.1. The fact that we can choose N sufficiently small follows from the $\aleph_1$-saturation principle since for all finite N, the hyperreals $D_0, \ldots, D_3$ are standard real numbers: Setting

$$f(N) := 3^{2^d \cdot N^{2d}} \cdot \begin{pmatrix} 4{D_{0,N}}^2 + 4D_{0,N}D_{2,N} + ND_{0,N}D_{3,N} + \\ D_{0,N}D_1 + D_{2,N}D_1 \end{pmatrix} ((D_{0,N} + D_{2,N}) \vee 1),$$

one gets that the set

$$A_n := \left\{ N \geq n \;\middle|\; \begin{matrix} \Delta t \cdot f(N) \leq \frac{1}{n}, \\ \sqrt{\Delta t} \cdot D_{3,N} \leq \frac{1}{n}, \quad \sqrt{\Delta t} \cdot N \leq \frac{1}{n} \end{matrix} \right\}$$

is not only internal, but also nonempty, for all n. Given that $(A_n)_{n \in \mathbb{N}}$ is also decreasing, $\aleph_1$-saturation yields an element $N \in \bigcap_{n \in \mathbb{N}} A_n$. By definition of A_n, the reciprocal of this N, $\frac{1}{N}$, will on the one hand serve as the splitting infinitesimal η which we introduced in the proof of Theorem 5.5.1. On the other hand, N is going to satisfy $\Delta t \cdot f(N) \approx 0$. This completes the proof of Theorem 5.6.2. □

Now we can outline the proof of Theorem 5.6.1 as well:

Proof (Proof sketch for Theorem 5.6.1). Thanks to Lindstrøm's representation theorem for standard Lévy processes [263, Theorem 9.1], there is a hyperfinite adapted probability space Ω on which one can define a hyperfinite Lévy process $Y_\cdot = (Y_t)_{t \in T}$ whose standard part $y_\cdot$ has drift vector γ, covariance matrix C and Lévy measure ν, respectively, and which takes values in some lattice $\mathbb{L}$.

In the proof of Theorem 5.5.1, however, the choice of the lattice spacing has been left open. On the other hand, (the second part of) the previous Theorem 5.6.2 asserts that we may assume

$$\Delta t \cdot 3^{2^d \cdot N^{2d}} \begin{pmatrix} 4D_{0,N}{}^2 + 4D_{0,N}D_{2,N} + ND_{0,N}D_{3,N} + \\ D_{0,N}D_1 + D_{2,N}D_1 \end{pmatrix} ((D_{0,N} + D_{2,N}) \vee 1)$$
$$\approx 0.$$

In light of the trivial estimate $\text{card}\,\mathbb{L} \leq (2 \cdot N/\varepsilon)^d$ that holds for any lattice of mesh size ε and radius N, we may simply put $\varepsilon = \frac{1}{N}$ and thereby obtain an infinitesimal spacing for $\mathbb{L}$ which ensures that

$$3^{\text{card}\,\mathbb{L}} \Delta t \left(4C_0{}^2 + 4C_0C_2 + NC_0C_3 + C_0C_1 + 4C_2C_1\right)((C_0 + C_2) \vee 1) \approx 0$$

in the notation of Theorem 5.6.2.

One can prove (cf. [14, Theorems 3.4 and 4.3]) that for all hyperfinite pure-jump Lévy processes Y with infinitesimal generator $L_{\Delta t} : f \mapsto \sum_{\underline{\alpha} \in \mathbb{L}} \lambda_{\underline{\alpha}} \cdot (f(\cdot + \underline{\alpha}) - f)$ and which satisfy $3^{\text{card}\,\mathbb{L}} \Delta t (4C_0{}^2 + 4C_0C_2 + NC_0C_3 + C_0C_1 + 4C_2C_1)((C_0 + C_2) \vee 1) \approx 0$, there is a hyperfinite Lévy process X with internal infinitesimal generator

$$K_{\Delta t} : f \mapsto \frac{\left(\ast_{\underline{\alpha} \in \mathbb{L}} \left(\lambda_{\underline{\alpha}} \delta_{\underline{\alpha}} + \left(1 - \lambda_{\underline{\alpha}}\right) \delta_0\right)\right) \ast f - f}{\Delta t}$$

which has the same standard part (pointwise for all smooth functions with compact support and all elements of $\mathbb{L}$) as the internal infinitesimal generator $L_{\Delta t}$.

However, one can also establish (cf. [14, Lemma 4.5]) that two hyperfinite pure-jump Lévy processes whose internal infinitesimal generators have the same standard part (pointwise for all smooth functions with compact support and all elements of $\mathbb{L}$), will have standard parts with the same infinitesimal generator, thus standard parts with identical finite-dimensional distributions. Hence the standard part $x. = \text{st}(X)$ coincides with the standard part $y. = \text{st}(Y)$. This brings the proof to a close. □

As another application, Lindstrøm himself [264] has used hyperfinite Lévy processes to develop a theory of nonlinear stochastic integration with respect to Lévy processes and to study certain minimal martingale measure problems for stochastic integrals with respect to a Lévy process. Current research on hyperfinite Lévy processes includes a pathwise theory of stochastic integration with respect to (smooth functions of) Lévy processes with jump part of bounded variation [205] and applications in equilibrium asset pricing with stochasticity derived from Lévy processes [204].

Chapter 6
Epilogue: Genericity of Hyperfinite Loeb Path Spaces

An important part of this book has been concerned with the investigation of hyperfinite Markov chains $X = (X_t)_{t \in T}$ on an internal probability space Ω. If we confine ourselves to a finite time horizon – say, 1 – and if we assume Δt to be a reciprocal hypernatural number, then $H = 1/\Delta t \in {}^*\mathbb{N} \setminus \mathbb{N}$ will be the number of possible transitions occurring between $0 \in T$ and $1 \in T$. Because X was assumed to be a hyperfinite Markov chain starting at a given point x_0 in the state space, the number of possible paths between 0 and 1 will be hyperfinite. Thus, provided we are only interested in studying events that lie in the filtration generated by X, we may assume Ω to be hyperfinite.

Already very early in the history of nonstandard probability theory, important Markov processes, such as the Brownian motion or the Poisson process, have been found to have simple internal analogues (by [61,267], respectively), and in Chap. 5, we have seen how to construct an internal analogue of an arbitrary Lévy process.

However, the analogy between the original Lévy process (on an arbitrary probability space) and the standard part of its internal analogue (on an internal probability pace) is limited to the equality of the corresponding Markovian semigroups, and hence the finite-dimensional distributions. It cannot be assumed *a priori* that the original process – defined on an arbitrary given probability space – and the standard part of its internal analogue on some hyperfinite probability space will be indistinguishable by propositions in a suitably formalized language of adapted probability theory with a finite time horizon.

We will see in this final chapter that this is indeed the case – even for arbitrary, not necessarily Markovian stochastic processes –, and that in the special case of Markov processes, on which this book is putting particular emphasis, indistinguishability with respect to the language of adapted probability theory with a finite time horizon is the same as equality of the respective processes' finite-dimensional distributions.

S. Albeverio et al., *Hyperfinite Dirichlet Forms and Stochastic Processes*,
Lecture Notes of the Unione Matematica Italiana 10,
DOI 10.1007/978-3-642-19659-1_6, © Springer-Verlag Berlin Heidelberg 2011

In this short chapter, we shall not give any proofs and formal references. Rather, we will largely follow Fajardo and Keisler's comprehensive, however dense, monograph [156] and refer the interested reader to their work as well as to the original articles by Hoover and Keisler [210] and Keisler [238].

6.1 Adapted Probability Logic

Definition 6.1.1. A triple $(\Omega, (\mathcal{F}_t)_{t\in[0,1]}, P)$ consisting of a set Ω, a filtration $(\mathcal{F}_t)_{t\in[0,1]}$ over Ω and a probability measure $P : \mathcal{F}_1 \to [0,1]$ is called an *adapted probability space* (for short: *adapted space*).

An adapted function can be viewed as a scheme to construct a real-valued random variable from a stochastic process on a given Polish space – provided that this random variable only depends on the process at finitely many times and that one confines oneself to the use of bounded continuous functions and conditional expectations.

Definition 6.1.2 (Adapted functions and their values). Let M be a Polish space. The set of *adapted functions* is defined inductively as the smallest set of expressions f such: that

1. *Base Step.* For all $n \in \mathbb{N}$ and any n-tuple $(t_1, \ldots, t_n)$ of mutually distinct elements of $[0,1]$ and for all bounded continuous functions $\phi : M^n \to \mathbb{R}$, $\phi_{t_1,\ldots,t_n}$ is an adapted function. If $(x_t)_{t\in[0,1]}$ is an M-valued stochastic process, then the $\mathbb{R}$-valued random variable $\phi_{t_1,\ldots,t_n}(x.) = \phi(x_{t_1}, \ldots, x_{t_n})$ is the value of $\phi_{t_1,\ldots,t_n}$ for $x.$.
2. *Composition Step.* Let $m \in \mathbb{N}$, assume $\psi : M^m \to \mathbb{R}$ is bounded and continuous, and let $f_1, \ldots, f_m$ be adapted functions. Then $\psi(f_1, \ldots, f_m)$ is an adapted function as well, and if $(x_t)_{t\in[0,1]}$ is an M-valued stochastic process, then the real-valued random variable $\psi(f_1, \ldots, f_m)(x.) = \psi(f_1(x.), \ldots, f_m(x.))$ is the value of $\psi(f_1, \ldots, f_m)$ for $x.$.
3. *Conditional Expectation Step.* Let $t \in [0,1]$ and let f be an adapted function. Then, $E[f|t]$ also is an adapted function, and if $(x_t)_{t\in[0,1]}$ is an M-valued stochastic process on an adapted probability space $(\Omega, (\mathcal{F}_t)_{t\in[0,1]}, P)$, then the real-valued random variable $E[f|t](x.) = E_P[f(x.)|\mathcal{F}_t]$ is the value of $E[f|t]$ for $x.$.

This gives rise to the following notion for the equivalence of two processes:

Definition 6.1.3. Let M be a Polish space, let $(x_t)_{t\in[0,1]}$ be an M-valued stochastic process on an adapted probability space $(\Omega, (\mathcal{F}_t)_{t\in[0,1]}, P)$ and let $(y_t)_{t\in[0,1]}$ be an M-valued stochastic process on an adapted probability space $(\Gamma, (\mathcal{G}_t)_{t\in[0,1]}, Q)$. The processes $(x_t)_{t\in[0,1]}$ and $(y_t)_{t\in[0,1]}$ shall be called *adapted-equivalent* (in symbols: $x. \equiv y.$) if and only if for all adapted functions f, one has $E_P[f(x.)] = E_Q[f(y.)]$.

This notion is at least as strong as equality of finite-dimensional distributions:

Lemma 6.1.1. *Let $(x_t)_{t\in[0,1]}$ be an M-valued stochastic process on an adapted probability space $(\Omega, (\mathcal{F}_t)_{t\in[0,1]}, P)$ and let $(y_t)_{t\in[0,1]}$ be an M-valued stochastic process on an adapted probability space $(\Gamma, (\mathcal{G}_t)_{t\in[0,1]}, Q)$. If $x_\cdot \equiv y_\cdot$ and $t_1, \ldots, t_n \in [0,1]$ are mutually distinct, then $P_{(x_{t_1},\ldots,x_{t_n})} = Q_{(y_{t_1},\ldots,y_{t_n})}$ as probability measures on M^n.*

We shall see later on, in Theorem 6.4.1, that the converse holds for Markov processes (due to a theorem of [210]).

6.2 Universality, Saturation, Homogeneity

For the rest of this chapter, we assume all processes to have Polish spaces as state spaces.

First, the – fairly self-explanatory – concept of universality is introduced:

Definition 6.2.1. Let $\Omega = (\Omega, (\mathcal{F}_t)_{t\in[0,1]}, P)$ be an adapted probability space. The adapted space Ω is *adapted-universal* if and only if for all stochastic processes $y_\cdot$ on any other adapted probability space, there is a stochastic process $x_\cdot$ on $(\Omega, (\mathcal{F}_t)_{t\in[0,1]}, P)$ such that $x_\cdot \equiv y_\cdot$.

For the next Definition, note that whenever M is a Polish space, so will be M^2 (and, in fact, any M^n for $n \in \mathbb{N}$). Hence, adapted functions (as well as the notion of adapted equivalence) are also defined for M^2-valued stochastic processes.

If an adapted probability space Ω has the following property, many probabilistic arguments on any other adapted probability space can be translated into arguments on Ω:

Definition 6.2.2. Let $\Omega = (\Omega, (\mathcal{F}_t)_{t\in[0,1]}, P)$ be an adapted probability space. The adapted space Ω shall be called *adapted-saturated* if and only if for all $\Gamma = (\Gamma, (\mathcal{G}_t)_{t\in[0,1]}, Q)$ and for all stochastic processes $x_\cdot$ on Ω and $y_\cdot, z_\cdot$ on Γ satisfying $x_\cdot \equiv y_\cdot$, there exists another stochastic process $w_\cdot$ on Ω such that $(x_\cdot, w_\cdot) \equiv (y_\cdot, z_\cdot)$.

An adapted probability space can be regarded as uniquely determined by the expected values of adapted functions whenever it satisfies the following condition:

Definition 6.2.3. If $\Omega = (\Omega, (\mathcal{F}_t)_{t\in[0,1]}, P)$ is an adapted probability space, then Ω shall be called *adapted-homogeneous* if and only if for all stochastic processes $x_\cdot, y_\cdot$ on Ω satisfying $x_\cdot \equiv y_\cdot$, there exists a bijection $h : \Omega_0 \leftrightarrow \Omega_1$

such that:

1. $P[\Omega_0] = P[\Omega_1] = 1$,
2. for all $t \in [0,1]$, $h[\overline{\mathcal{F}_t}] = \overline{\mathcal{F}_t}$ (where $\overline{\mathcal{F}_t}$ denotes the completion of the σ-algebra $\mathcal{F}_t$ with respect to P),
3. for all $A \in \mathcal{F}_1$, $P[h(A)] = P[A]$, and
4. for all $t \in [0,1]$, one has $x_t = y_t \circ h$ P-almost surely.

(Such a map h is called *adapted automorphism.*)

Finally, these three features of an adapted probability space are not unrelated – saturation implies universality, and homogeneity together with universality implies saturation:

Theorem 6.2.1. *Let $\Omega = (\Omega, (\mathcal{F}_t)_{t\in[0,1]}, P)$ be an adapted probability space. Then:*

- *If Ω is both adapted-universal and adapted-homogeneous, then it is adapted-saturated.*
- *If Ω is adapted-saturated, then it is adapted-universal.*

6.3 Hyperfinite Adapted Spaces

Definition 6.3.1. A *hyperfinite adapted probability space* is a triple $(\Omega, (\mathcal{F}_t)_{t\in[0,1]}, L(\mu))$ such that:

- $\Omega = {\Omega_0}^T$, where Ω_0 is a hyperfinite set with at least two elements and $T = \{0, \frac{1}{H!}, \dots, \frac{H!-1}{H!}, 1\}$ (for $H \in {}^*\mathbb{N} \setminus \mathbb{N}$) is the *hyperfinite time line.*
- The external filtration $(\mathcal{F}_t)_{t\in[0,1]}$ is derived from some internal filtration $(\mathcal{A}_s)_{s\in T}$ over Ω via $\mathcal{F}_t = \sigma(\bigcap_{{}^\circ s=t} \mathcal{A}_s)$ for all $t \in [0,1]$, and

$$\mathcal{A}_s = \left\{ A \in 2^\Omega \;\middle|\; \begin{array}{c} (\forall r \leq s \quad \forall \omega \in \Omega \quad \forall \omega' \in \Omega \quad \omega(r) = \omega'(r)) \\ \Rightarrow (\omega \in A \Leftrightarrow \omega' \in A) \end{array} \right\}.$$

- $L(\mu)$ is the Loeb extension of the internal normalized counting measure $\mu : A \mapsto |A|/|\Omega|$ on Ω.

Example 6.3.1. For instance, the probability space on which Anderson's lifting $(B_t)_{t\in T}$ of the Brownian motion [61] is defined, is $\Omega = \{-1,1\}^T$. Let $(\mathcal{A}_s)_{s\in T}$ be the internal filtration generated by this lifting and derive an external filtration $\mathcal{F}_\cdot$ from $\mathcal{A}_\cdot$ in the manner of the previous Definition 6.3.1. Also, let μ be the joint internal distribution of all the B_t, $t \in T$ – which is a finitely additive probability measure on $\mathcal{A}_1$. Then $(\Omega, (\mathcal{F}_t)_{t\in[0,1]}, L(\mu))$ is a hyperfinite adapted probability space.

Theorem 6.3.1. *Every hyperfinite adapted probability space* $(\Omega, (\mathcal{F}_t)_{t\in[0,1]}, L(\mu))$ *is adapted-universal, adapted-homogeneous, and thus also adapted-saturated. The witness for the adapted homogeneity can even be chosen as an internal bijection* $h : \Omega \leftrightarrow \Omega$.

The special importance of hyperfinite adapted probability spaces is emphasized by the fact that no Polish adapted probability space is universal, let alone saturated:

Theorem 6.3.2. *Let* Ω *be a Polish space, let* Q *be the completion of a probability measure on the Borel* σ*-algebra of* Ω, *and let* $(\mathcal{F}_t)_{t\in[0,1]}$ *be a filtration such that each element of* $\mathcal{F}_t$ *is* Q*-measurable, for all* $t \in [0,1]$. *Then* $(\Omega, (\mathcal{F}_t)_{t\in[0,1]}, Q)$ *cannot be universal.*

6.4 Adapted Probability Logic and Markov Processes

For Markov processes – and hardly any other stochastic processes have been treated in this book –, conditioning with respect to the information up to a certain time actually can be reduced to simply looking at the transition kernel at this given time, since Markov processes have no memory. In this sense, we should not be too surprised by the following result (the converse to Lemma 6.1.1):

Theorem 6.4.1. *Any two Markov processes* $x.$ *on* Ω *and* $y.$ *on* Γ *are adapted-equivalent if and only if their finite-dimensional distributions are equal (i.e. they are equivalent in the sense of, e.g., [77]).*

Summarizing this short chapter, we have seen that hyperfinite adapted probability spaces can be regarded as the generic choice of an adapted probability space, in a rigorous sense (referring to adapted universality, saturation and homogeneity). Furthermore, as Theorem 6.4.1 states, any two Markov processes $x.$ and $y.$ with the same finite-dimensional distributions cannot be discerned by adapted functions in expectation: There is no adapted function which would attain a different expected value when applied to $x.$ instead of $y.$.

We close this book by addressing two possible foundational objections against nonstandard analysis. First, it has to be acknowledged that many foundational questions about nonstandard analysis (particularly with regard to its set-theoretic strength) can be resolved by adopting a syntactic approach to nonstandard analysis, as developed by Nelson [284] (Internal Set Theory, IST), Hrbáček [211] and Kanovei and Reeken [228] (Bounded Set Theory, BST). As part of this well-developed and still-growing literature, several authors have proposed nonstandard set theories, i.e. axiomatisations of

Robinson's nonstandard analysis (and fragments thereof), and proven that they are conservative extensions of subsystems of ZFC.

On the other hand, a folklore objection to the application of nonstandard methods in analysis used to be the well-known fact that the existence of nonstandard enlargements is dependent on the choice of a non-principal ultrafilter and is thus somewhat "non-constructive". (Indeed, the Ultrafilter Existence Theorem does not follow from Zermelo-Fraenkel set theory only.) However, this objection against the use of nonstandard analysis can be refuted in at least two ways. Firstly, the Ultrafilter Existence Theorem – which is equivalent to the Boolean Prime Ideal Theorem – is not as strong as the Axiom of Choice (cf. [200] as well as [71]). Secondly, recent studies in model theory have revealed that there exist nonstandard universes which are definable through a formula of set theory: Kanovei and Shelah [230] were the first to prove the existence of a definable nonstandard model of the reals; Kanovei and Reeken [229] even established the existence of definable nonstandard set universes, and Herzberg [202, 203] showed that there are nonstandard universes in the usual sense (as enlargements of the superstructure over the reals). Admittedly, the Axiom of Choice is still employed to prove that the structures thus defined have the properties of nonstandard models – in particular, κ-saturation for uncountable cardinals κ. Nevertheless, we now know that nonstandard universes are far more tractable objects than popular opinion expected only one decade ago.

Appendix

We summarize here some basic facts of nonstandard analysis. For details and proofs, see e.g. Albeverio et al. [25], Courant and Robbins [123], Courant et al. [124], Cutland [125], Dauben [130], Davis [135], Stroyan and Bayod [341], and Stroyan and Luxemburg [342].

A.1 General Topology

Let Y be a Hausdorff topological space. First of all, let us construct ${}^*Y$. Let η be a finitely additive measure on the set $\mathbb{N}$ of positive integers such that

(a) For all $A \subset \mathbb{N}$, $\eta(A)$ is defined and either 0 or 1.
(b) $\eta(\mathbb{N}) = 1$, and $\eta(A) = 0$ for all finite $A \subset \mathbb{N}$.

It is well-known that such a measure exists as a consequence of, e.g., the axiom of choice.

Given two sequences $y = \{y_n\}_{n\in\mathbb{N}}$ and $z = \{z_n\}_{n\in\mathbb{N}}$, $y_n, z_n \in Y$, we define the following equivalence relation

$$y \sim z \text{ if and only if } \eta(\{n \in \mathbb{N} \mid y_n = z_n\}) = 1.$$

Having defined this relation, we introduce

$${}^*Y = Y^{\mathbb{N}}/\sim .$$

We call ${}^*Y$ the *nonstandard version* or *extension* of Y. In particular, we call ${}^*\mathbb{R} = \mathbb{R}^{\mathbb{N}}/\sim$ the *nonstandard reals* or *hyperreals*. In the same manner, we can define ${}^*\mathbb{N}$, etc.

Fixed an element $y \in Y$, the *monad* of y is the subset of ${}^*Y$ defined by

$$\mu(y) = \cap\{{}^*O \mid y \in O \text{ and } O \text{ is open}\}.$$

S. Albeverio et al., *Hyperfinite Dirichlet Forms and Stochastic Processes*, Lecture Notes of the Unione Matematica Italiana 10,
DOI 10.1007/978-3-642-19659-1,

An element $a \in {}^*Y$ is a *nearstandard point* if it belongs to $\mu(y)$ for some $y \in Y$. The set of all nearstandard points is denoted by $Ns({}^*Y)$. Since Y is Hausdorff, each element $a \in Ns({}^*Y)$ is nearstandard to exactly one element y of Y. Let us call y the *standard part of point* a and denote it by ${}^\circ a$ or $\mathrm{st}(a)$. Moreover, we have the following characterizations of open, closed and compact sets. If A is a subset of ${}^*Y$, the standard part of set A is defined by

$$^\circ A = \mathrm{st}(A) = \{y \in Y \mid \exists s \in A(\mathrm{st}(s) = y)\}.$$

Proposition A.1.1. *Let A be subset of Y. Then*

(i) A is open if and only if $\mathrm{st}^{-1}(A) \subset {}^*A$.
(ii) A is closed if and only if ${}^*A \cap Ns({}^*Y) \subset \mathrm{st}^{-1}(A)$.
(iii) A is compact if and only if ${}^*A \subset \mathrm{st}^{-1}(A)$.

A subset A of ${}^*Y$ is called *S-dense* in ${}^*Y$ if for all $y \in Y$, there is an element $a \in A$ such that $a \in \mu(y)$. A subset A of ${}^*Y$ is called *S-rich* in ${}^*Y$ if it is hyperfinite and S-dense in ${}^*Y$.

A.2 Structure of ${}^*\mathbb{R}$

For two sequences $a = \{a_n \mid n \in \mathbb{N}\}$ and $b = \{b_n \mid n \in \mathbb{N}\}$ in $\mathbb{R}^{\mathbb{N}}$, let us define addition and multiplication operations by

$$a + b = \{a_n + b_n \mid n \in \mathbb{N}\} \text{ and } ab = \{a_n b_n \mid n \in \mathbb{N}\}.$$

The zero for addition, denoted by 0, is the sequence with only zeros, while the unit of multiplication, denoted by 1, is the sequence with only ones.

For any $a = \{a_n \mid n \in \mathbb{N}\}$ in $\mathbb{R}^{\mathbb{N}}$, we denote by $\langle a \rangle$ its equivalence class with respect to $\sim$. It is easy to verify

$$\langle a \rangle + \langle b \rangle = \langle a + b \rangle, \quad \langle a \rangle \langle b \rangle = \langle ab \rangle.$$

If $\langle a \rangle \neq 0$ we can define the inverse $\langle a \rangle^{-1}$ by $(a^{-1})_n = (a_n)^{-1}$ whenever $a_n \neq 0$. The above definitions make ${}^*\mathbb{R}$ into a field of numbers. ${}^*\mathbb{R}$ can be linearly ordered by defining

$$\langle a \rangle < \langle b \rangle \text{ if and only if } \eta(\{n \in \mathbb{N} \mid a_n < b_n\}) = 1.$$

We can consider $\mathbb{R}$ as embedded in ${}^*\mathbb{R}$ by identifying any $r \in \mathbb{R}$ with ${}^*r \in {}^*\mathbb{R}$ where all the elements equal r. From now on, we will write r for the element ${}^*r \in {}^*\mathbb{R}$.

Let $a \in {}^*\mathbb{R}$, we call a *infinitesimal* if and only if for all $r \in \mathbb{R}, r > 0$

$$-r < a < r.$$

We call $a \in {}^*\mathbb{R}$ infinite if and only if for all $r \in \mathbb{R}, r > 0$

$$|a| > r,$$

where $|a| = \langle \{|a_n|\}_{n\in\mathbb{N}} \rangle$ for $a = \langle \{a_n\}_{n\in\mathbb{N}} \rangle$, and $a \in {}^*\mathbb{R}$ is called *finite* if and only if there exists an $r \in \mathbb{R}, r > 0$ such that

$$|a| < r.$$

Proposition A.2.1. *Any finite $a \in {}^*\mathbb{R}$ can be written uniquely as*

$$a = r + \varepsilon \tag{A.2.1}$$

with $r \in \mathbb{R}$ and ε infinitesimal.

We call r in the relation (A.2.1) the *standard part* of a and denote it by $\mathrm{st}(a)$ or ${}^\circ a$. Conversely, for each $r \in \mathbb{R}$, the set of all $a \in {}^*\mathbb{R}$ such that $a \approx r$ is called the *monad* of r.

A.3 Internal Sets and Saturation

A sequence $\{A_n\}_{n\in\mathbb{N}}$ of subsets of $\mathbb{R}$ defines a subset $\langle \{A_n\}_{n\in\mathbb{N}} \rangle$ of ${}^*\mathbb{R}$ by

$$\langle \{x_n\}_{n\in\mathbb{N}} \rangle \in \langle \{A_n\}_{n\in\mathbb{N}} \rangle \text{if and only if } \eta(\{n \mid x_n \in A_n\}) = 1.$$

A subset of ${}^*\mathbb{R}$ which can be obtained in this way is called *internal.* A subset of ${}^*\mathbb{R}$ which is not internal is called *external.*

Proposition A.3.1. *Let A be an internal subset of ${}^*\mathbb{R}$.*

(i) If A contains arbitrarily large elements, then A contains an infinite element.
(ii) If A contains arbitrarily small positive infinite elements, then A contains a finite element.

Proposition A.3.2. *Let $\{A^i\}_{i\in\mathbb{N}}$ be a sequence of internal sets such that $\bigcap_{i\leq I} A^i \neq \emptyset$ for all $I \in \mathbb{N}$, then $\bigcap_{i\in\mathbb{N}} A^i \neq \emptyset$.*

Proposition A.3.2 is known as the *Principle of Countable Saturation* or $\aleph_1$-*Saturation.* (It is possible to construct nonstandard universes which satisfy even higher saturation principles.)

Proposition A.3.3. *Let $\{A_n\}_{n\in\mathbb{N}}$ be a sequence of internal sets. The union $\cup_{n\in\mathbb{N}} A_n$ is internal if and only if it equals $\cup_{n\leq N} A_n$ for some $N \in \mathbb{N}$.*

A.4 Loeb Measure

For any set S, we will write $\mathcal{P}(S)$ for the set of all subsets of S. Let us define a sequence $\{V_n(S)\}_{n\in\mathbb{N}_0}$ of sets recursively by

$$V_0(S) = S, \quad V_{n+1}(S) = V_n(S) \cup \mathcal{P}(V_n(S)).$$

The *superstructure* over S is the union $V(S) = \cup_{n\in\mathbb{N}_0} V_n(S)$. In the same way as in A.1, we can construct ${}^*S$ by

$${}^*S = S^{\mathbb{N}}/\sim .$$

Let Ω be an internal set in some superstructure $V({}^*S)$. An *internal algebra* on Ω is an internal set $\mathcal{F}$ of subsets of Ω which contains $\emptyset$ and Ω, and is closed under complements and finite unions. Since $\mathcal{F}$ is internal, it automatically satisfies a stronger property: if $\mathcal{F}_1$ is a hyperfinite subset of $\mathcal{F}$, the union $\cup\{A \mid A \in \mathcal{F}_1\}$ is an element of $\mathcal{F}$.

Let ${}^*\overline{\mathbb{R}}_+ = {}^*\mathbb{R}_+ \cup \{0, \infty\}$ be the set of extended, nonnegative hyperreals. An *internal, finitely additive measure* on the internal algebra $\mathcal{F}$ is an internal function $\mu : \mathcal{F} \longrightarrow {}^*\overline{\mathbb{R}}_+$ such that $\mu(\emptyset) = 0$ and

$$\mu(A \cup B) = \mu(A) + \mu(B)$$

for all disjoint $A, B \in \mathcal{F}$. Since μ is internal, the additivity extends to the following hyperfinite unions:

$$\mu(\cup_{A\in\mathcal{F}_1} A) = \sum_{A\in\mathcal{F}_1} \mu(A)$$

for all disjoint, hyperfinite subsets $\mathcal{F}_1$ of $\mathcal{F}$.

Definition A.4.1. (i) A subset B of Ω is μ-approximately if for each $\varepsilon \in \mathbb{R}_+$, there are sets $A, C \in \mathcal{F}$ such that $A \subset B \subset C$ and $\mu(C) - \mu(A) < \varepsilon$.

(ii) The *Loeb algebra* $L(\mathcal{F})$ consists of those $B \subset \Omega$ such that $B \cap F$ is μ-approximable for all $F \in \mathcal{F}$ with finite μ-measure.

(iii) The *Loeb measure* of μ is the map $L(\mu) : L(\mathcal{F}) \longrightarrow \overline{\mathbb{R}}_+$ defined by

$$L(\mu)(B) = \inf\{{}^\circ\mu(C) \mid C \in \mathcal{F} \text{ and } C \supset B\}.$$

Proposition A.4.1. *$(\Omega, L(\mathcal{F}), L(\mu))$ is a complete measure space.*

A.5 Linear Spaces

Let $(E, ||\cdot||)$ be a linear normal space. An element $x \in {}^*E$ is called *finite* if $||x||$ is a finite hyperreal. Let us denote by $\mathrm{Fin}({}^*E)$ the set of all finite elements of ${}^*E$. We call $x \in {}^*E$ *infinitesimal* if $||x|| \approx 0$. We write $x \approx y$ if $||x - y|| \approx 0$. An element $x \in {}^*E$ is called *nearstandard* if $||x - y|| \approx 0$ for some $y \in E$. An element $x \in {}^*E$ is called *pre-nearstandard* if for all $\varepsilon \in \mathbb{R}_+$ there is some $y \in E$ such that $||x - y|| < \varepsilon$. We denote by $\mathrm{Pns}({}^*E)$ the set of all pre-nearstandard points of ${}^*E$.

Proposition A.5.1. *The space* $\mathrm{Pns}({}^*E)/\approx$ *is the completion of the given space* E.

Proposition A.5.2. *Let* $(F, ||\cdot||)$ *be an internal normed linear space. Then* $\mathrm{Fin}(F)/\sim$ *is a Banach space with the norm* $|||{}^\circ x||| = {}^\circ||x||$*, where* ${}^\circ x$ *denotes the equivalence class of* $x \in \mathrm{Fin}(F)$.

For $N \in {}^*\mathbb{N}$, an internal space F is called *N-dimensional* if there is an internal set $\{e_1, e_2, \cdots, e_N\}$ such that each element $x \in F$ can be expressed by $x = \sum_{i=1}^{N} x_i e_i$ uniquely.

Proposition A.5.3. *Let* E *be an infinite-dimensional linear space over* $\mathbb{R}$. *Then, there exists a hyperfinite-dimensional space* X *over* ${}^*\mathbb{R}$ *such that* $E \subset X \subset {}^*E$.

Historical Notes, Bibliographical Complements

Chapter 1

The theory of Dirichlet forms has its origins in seminal work by Beurling and Deny [87,88]. It relies essentially on the use of Hilbert space methods in potential theory, which in turn originate from "energy methods" and "variational methods". Such methods of potential theory were developed particularly during the period 1935–1950 by analysts such as C.J. De la Vallée Poussin, M. Riesz, O. Frostman and H. Cartan, see, e.g., Cartan [106]. The essential new idea in Beurling and Deny's fundamental work consists in the systematic use of the Dirichlet norm and the corresponding contraction properties of Dirichlet forms.

In the theory of Dirichlet forms, the case of discreteness (in time and/or space) has played, at least in the beginning, an important role, as it facilitated the description of the underlying ideas and methods without having to dwell on the technical problems encountered in implementing the basic ideas of the theory in the continuum case.

Discrete-time potential theory was developed further by J. Deny (e.g. [142], see also [143]), and J.L. Doob [145,146], see also, e.g., Revuz [306], Williams [360,361], Norris [291], Dellacherie and Meyer [138]. The hyperfinite approach to Dirichlet forms originated in work by Albeverio, Fenstad and Høegh-Krohn [10], as well as Albeverio, Fenstad, Høegh-Krohn and Lindstrøm [25]. One of its aims was to preserve as much as possible the spirit of the discrete-time and discrete-space approach, by considering a hyperfinite model both in time and space which by its very essence is discrete but at the same time yields an alternative view of continuum realizations.

Symmetric, nonnegative (or lower semi-bounded) quadratic forms and their association with nonnegative (or lower semi-bounded) self-adjoint operators is a well-developed topic of analysis, occurring, e.g. in the relation with the calculus of variations or problems of spectral theory and quantum mechanics, see e.g. Rellich [303], Reed and Simon [300–302], Faris [168],

Weidmann [356–358], Bratteli and Robinson [101], Kato [235], Arendt and Thomaschewski [67], Zhang [365], Takeda [348], Lenz, Stollmann and Veselić [260].

Forms which are unbounded from below are much less studied, see, however, e.g., McIntosh [279, 280] or Berrahmoune [85].

Nonsymmetric (lower semi-bounded) bilinear forms occur in the theory of dissipative processes, see, e.g., Pazy [299] and Arendt, Batty, Hieber and Neubrander [66], Kuwae [252], Hu, Ma and Sun [214].

The relation with corresponding operators is studied, e.g. in Albeverio and Ma [41, 42], Albeverio, Ma and Röckner [43–45], and Pazy [299].

For applications of nonsymmetric Dirichlet forms, see, e.g., Hu, Ma and Sun [213, 214].

For questions of self-adjoint (maximal dissipative, respectively) extensions, see, e.g., the references in Remark 1.2.1.

For the origins of the theory of nonsymmetric Dirichlet forms, see Albeverio [2].

For applications of the theory of Dirichlet forms to manifolds, see, e.g., Ouhabaz [297], Davies [131, 132], Cycon, Froese, Kirsch and Simon [127], Biroli and Vernole [93], and Bonciocat [97].

For extensions of the theory of Dirichlet forms to the non-commutative case, see Albeverio and Høegh-Krohn [18], Davies and Lindsay [133], Cipriani [120] and for the supersymmetric case Albeverio and Kondratiev [33].

For the case of p-adic state spaces, see Albeverio, Karwowski and Yasuda [29], Albeverio, Karwowski and Zhao [30], Kochubei [243, 244], Kaneko and Yasuda [227], Kaneko [226], and for more general trees, see Evans [153] or Albeverio and Karwoski [28].

For the case of forms related to jump processes, see, e.g., Sato [323], Applebaum [65], Fukushima and Tanaka [186], Uemura [351], Barlow, Bass, Chen and Kassmann [72], Foondun [173], and Chen and Kumagai [115].

For the study of Dirichlet forms on graphs, see, e.g., Bonciocat [97], von Below and Lubary [79, 80], von Below and Nicaise [81], Lumer [269], and Kant, Klauss, Voigt and Weber [231]. For applications of Dirichlet forms to (quantum) graphs and discrete spaces, see, e.g., Bonciocat [97].

For the problem of tubular neighbourhoods of graphs and diffusion operators associated to them and their graph limits, see Albeverio, Cacciapuoti and Finco [7] and also the appendix by P. Exner in Albeverio, Gesztesy, Høegh-Krohn, and Holden [12].

For Dirichlet forms in infinite-dimensional analysis, see, e.g., Albeverio [2], Chen, Ma, and Röckner [116], Eberle [149], Liskevich and Röckner [265], Da Prato [128], Huang and Yan [215], Wang and Wu [354,355], Ma and Röckner [270], Denis, Grorud and Pontier [141], Albeverio, Kondratiev, Kozitsky, and Röckner [31] (and Carmona and Tehranchi [104] for applications to interest-rate models in mathematical finance).

For applications of Dirichlet forms to Schrödinger operators, see, e.g., Albeverio [1], Albeverio, Høegh-Krohn, and Streit [26], Albeverio and Ma [40, 41], Demuth and Krishna [140], Wu [363], Robinson [312], Davies and Simon [134], Kuwae [250], and Laptev and Weidl [256] (see also, e.g., Exner [154] and Davies [132] for relations to dissipative systems).

For relations to the theory of metric measure spaces and spectral theory, see Bonciocat [97], Sturm [343], Paulik [298], Kuwae and Shioya [253, 254], and Grigoryan and Hu [195].

For convergence of families of Dirichlet forms, see, e.g., Albeverio, Kusuoka and Streit [36], Stollmann [339], Kuwae and Shioya [253, 254], Mosco [283], Ben Amor and Brasche [82], Kolesnikov [245], Biroli [91], and Biroli and Tchou [92].

For the construction of Markov processes from a (nonsymmetric) L^p-semigroup, see, e.g., Jacob [220].

For relations between Dirichlet forms, semigroups, and processes, see, e.g., Feller [170], Ma and Röckner [270], Fuhrman [174], Albeverio [2], Taira [344], Fukushima [179, 180], and Chen, Fitzsimmons, Kuwae, and Zhang [111], LeJan [258], and Ôkura [292] (see also Oshima [294]) and Stroock [340].

For the relation of Dirichlet forms with general potential theory, see Beznea and Boboc [89], Beznea, Boboc and Röckner [90], and Bliedtner and Hansen [95]. In particular, quasi-regular semi-Dirichlet forms and their relations with right processes are studied in the former paper.

For relations to non-equilibrium particle systems, random walks on fractals and statistical physics, see, e.g., Chen [109].

For relations of discrete Dirichlet forms to fractals, see, e.g., Chen [109], Kigami [239], Hino [209], and Hu and Zähle [212] (Kigami [239] also treats approximations of Dirichlet forms).

Higher-order operators in relation to Dirichlet forms are discussed, e.g., in Noll [290].

For the theory of generalized Dirichlet forms, see, e.g., Stannat [336,337], Trutnau [349,350], and Conrad and Grothaus [122].

For an analytic treatment of generators of Markov semigroups with Hölder-continuous coefficients, see Lorenzi and Bertoldi [268].

For nonlinear Dirichlet problems and their relations to the theory of harmonic mappings, see, e.g., Hesse [207] and Hesse, Rumpf and Sturm [208].

Canonical diffusion processes on metric measure spaces have been constructed by von Renesse [304] and von Renesse and Sturm [305]. For relations of Dirichlet forms with gradient-flows in metric spaces and the theory of transport in spaces of probability measures, see, e.g., Jordan, Kinderlehrer, and Otto [224], Ambrosio, Gigli, and Savaré [58]. For Dirichlet forms associated with log-concave measures and Fokker-Planck equations, see Albeverio and Kusuoka [34], Albeverio, Kusuoka, and Röckner [35], Ambrosio, Savaré, and Zambotti [59]. For recent work on Wasserstein diffusions, see also Andres and von Renesse [60].

Gauging of Dirichlet forms is discussed in von Renesse [304], Takeda [347], Zhao [366], and Chen and Shiozawa [117].

Applications of Dirichlet forms to statistics (sampling perfect matchings of a graph) are given in Jerrum [223].

Applications to boundary value problems can be found in Bass and Hsu [74], Krall [246], Bass and Kassmann [75], Fang, Fukushima, and Ying [167], Chen, Fukushima, and Ying [114], Chen and Fukushima [110], and applications to semilinear differential equations have been studied by Stollmann and [339], Vogt and Voigt [352], Chen and Zhang [118], and Grossinho and Tersian [197].

Sobolev-type inequalities for Dirichlet forms and corresponding hypercontractivity estimates are discussed, e.g., in Gross [196], Fabes, Fukushima, Gross, Kenig, Röckner, and Stroock [155], Wang [353], Mendez-Hernandez [281], Saloff-Coste [322], Kassmann [233], Del Moral, Ledoux, and Miclo [137], Cattiaux [107], Dai Pra, and Posta [129], and Cattiaux, Gentil, and Guillin [108], Grunewald, Otto, Villani, and Westdickenberg [199], Otto and Reznikoff [296] (see also, e.g., Gestesy, Holden, Jost, Paycha, and Röckner and Scarlatti [182, 183]), Bakry [69], Bakry, Concordet, and Ledoux [70], Albeverio, Kozitsky, Kondratiev, and Röckner [30].

For Dirichlet forms related to the Lévy Laplacian, see M.N. Feller [169] as well as Albeverio, Belopolskaya, and M.N. Feller [3,4] and references therein.

For problems of pseudodifferential operators and Dirichlet forms, see, e.g., Bass and Kassmann [75], Schilling and Uemura [324], or Kassmann [234].

For random Dirichlet forms, see, e.g., Fukushima, Nakao, and Takeda [182], Albeverio and Bernabei [5], and Rhodes [308].

Chapter 2

The relation between probabilistic potential theory and the theory of Dirichlet forms goes back to the origins of the theory of Dirichlet forms, see the notes in Chap. 1.

Fukushima's decomposition theorem has been extended to the infinite-dimensional setting in Albeverio, Ma, and Röckner [38] and in the book by Ma and Röckner [270].

For the theory of Feynman-Kac functionals, see, e.g., Albeverio and Ma [39–42], Albeverio, Blanchard, and Ma [6], Brasche [100], De Leva, Kim, and Kuwae [136], Eberle and Marinelli [150], and Demuth and Van Casteren [139], as well as Albeverio and Fan [8].

For the theory of subordination, see, e.g., Albeverio and Rüdiger [51,53], Jacob [220], Jacob and Moroz [221], Chen, Fukushima, and Ying [113], Fitzsimmons [171], Fukushima and Uemura [187], Fukushima, He, and Ying [181], and Kim, Song, and Vondraček [242]. For a recent application, see Albeverio and Gordina [13].

Chapter 3

The standard representation theory has its origins in Albeverio, Fenstad, Høegh-Krohn and Lindstrøm [25]. It uses heavily results of Lindstrøm [261] and Fukushima [175],

Chapter 4

The association of "good" Markov processes to Dirichlet forms goes back to work by Silverstein [333,334], Fukushima [175], Fukushima, Ôshima and Takeda [183] in the symmetric case, and to Carrillo-Menendez [105], and Albeverio, Ma, and Röckner [44] in the nonsymmetric case. The nonstandard analytic framework is treated in Albeverio, Fenstad, Høegh-Krohn, and Lindstrøm [25], and Fan [165]. See also Albeverio [2] and references therein. Quasi-regularity was found and elaborated in Albeverio and Ma [41] and Ma and Röckner [270].

Chapter 5

The theory of Lévy processes goes back to P. Lévy. For systematic accounts, see, e.g., Bertoin [86], Sato [323], Rüdiger [320], and Applebaum [65]. For analytic aspects, see also Jacob [218–220] and Jacob and Schilling [222].

The nonstandard theory goes back to Lindstrøm [263] and Albeverio and Herzberg [14].

For further applications, see, e.g., Albeverio [2], Albeverio, Rüdiger, and Wu [54, 55] and Garroni and Menaldi [190].

Chapter 6

This chapter has a foundational purpose and is based on the monograph by Fajardo and Keisler [156], where also original work, in particular due to Hoover and Keisler [210], is discussed.

Appendix

Nonstandard analysis originated with the seminal work of Robinson [310,311] (with an early precursor by Schmieden and Laugwitz [325]). Further systematic expositions of nonstandard analysis include Machover and Hirschfeld [276], Bell and Machover [78], Albeverio, Fenstad, Høegh-Krohn, and Lindstrøm [25], Cutland [125], Davis [135], Stroyan and Bayod [341], Stroyan and Luxemburg [342] (for separate articles, see e.g., Albeverio, Luxemburg, and Wolff [37], Arkeryd, Cutland, and Henson [68]). For more recent textbooks and monographs on nonstandard analysis, see e.g. Capiński and Cutland [102], Landers and Rogge [255], Goldblatt [194], Cutland [126], Loeb and Wolff [362], Ng [287], Sousa Pinto [335], and again Ng [289]. Fundamental contributions to nonstandard measure-theoretic probability theory and nonstandard stochastic analysis were made, among others, by Loeb [267] and Anderson [61,62]. Other, hitherto unmentioned, examples of work on stochastic nonstandard analysis can be found in Albeverio and Fan [8], Albeverio, Luxemburg, and Wolff [37], Anderson [62], Arkeryd, Cutland, and Henson [68], Capinski and Cutland [102], to mention but a few; see, however, also the references in Albeverio et al. [25].

For expositions of Edward Nelson's approach to nonstandard analysis (Internal Set Theory and an elementary subsystem thereof, called Radically Elementary Mathematics), we refer to Nelson [284–286], the introductory contributions in Diener and Diener [144], and Robert [309]. Yet another approach to nonstandard analysis which promises to be of significant pedagogical value has been suggested by Benci and Di Nasso [83]. A comparison of different approaches to nonstandard analysis can be found in Benci, Forti and Di Nasso [84].

Foundational questions of nonstandard analysis have been systematically studied in a monumental treatise by Kanovei and Reeken [229]. Early contributions to this field of research include Hrbáček [211]; more recent advances are due to, e.g., Kanovei and Shelah [230].

References

1. Albeverio, S.: Wiener and Feynman - path integrals and their applications. In: Proceedings of the Norbert Wiener Centenary Congress (East Lansing, MI, 1994), Proc. Sympos. Appl. Math., vol. 52, pp. 163–194. Amer. Math. Soc., Providence, RI (1997)
2. Albeverio, S.: Theory of Dirichlet forms and applications. In: P. Bernard (ed.) Lectures on probability theory and statistics (Saint-Flour, 2000), *Lecture Notes in Math.*, vol. 1816, pp. 1–106. Springer, Berlin (2003)
3. Albeverio, S., Belopolskaya, Y., Feller, M.: Lévy-Dirichlet forms. Infin. Dimens. Anal. Quantum Probab. Relat. Top. **9**(3), 435–449 (2006). DOI 10.1142/S0219025706002470
4. Albeverio, S., Belopolskaya, Y., Feller, M.: Lévy-Dirichlet forms. II. Methods Funct. Anal. Topol. **12**(4), 302–314 (2006)
5. Albeverio, S., Bernabei, M.S.: Homogenization in random Dirichlet forms. Stoch. Anal. Appl. **23**(2), 341–364 (2005)
6. Albeverio, S., Blanchard, P., Ma, Z.: Feynman-Kac semigroups in terms of signed smooth measures. Random partial differential equations, Proc. Conf., Oberwolfach/Ger. 1989, ISNM 102, 1-33 (1991). (1991)
7. Albeverio, S., Cacciapuoti, C., Finco, D.: Coupling in the singular limit of thin quantum waveguides. J. Math. Phys. **48**(3), 032,103, 21 (2007). DOI 10.1063/1.2710197
8. Albeverio, S., Fan, R.Z.: Representation of martingale additive functionals and absolute continuity of infinite-dimensional symmetric diffusions. In: Z.M. Ma, M. Röckner (eds.) Dirichlet forms and stochastic processes (Beijing, 1993), pp. 25–45. de Gruyter, Berlin (1995)
9. Albeverio, S., Fan, R.Z., Röckner, M., Stannat, W.: A remark on coercive forms and associated semigroups. In: M. Demuth, B. Schulze (eds.) Partial differential operators and mathematical physics (Holzhau, 1994), *Oper. Theory Adv. Appl.*, vol. 78, pp. 1–8. Birkhäuser, Basel (1995)
10. Albeverio, S., Fenstad, J.E., Høegh-Krohn, R.: Singular perturbations and nonstandard analysis. Trans. Amer. Math. Soc. **252**, 275–295 (1979)
11. Albeverio, S., Fukushima, M., Karwowski, W., Streit, L.: Capacity and quantum mechanical tunneling. Comm. Math. Phys. **81**(4), 501–513 (1981)
12. Albeverio, S., Gesztesy, F., Høegh-Krohn, R., Holden, H.: Solvable models in quantum mechanics. With an appendix by Pavel Exner, second edn. AMS Chelsea Publishing, Providence, RI (2005)
13. Albeverio, S., Gordina, M.: Lévy processes and their subordination in matrix Lie groups. Bull. Sci. Math. **131**(8), 738–760 (2007)
14. Albeverio, S., Herzberg, F.S.: A combinatorial infinitesimal representation of Lévy processes and an application to incomplete markets. Stochastics **78**(5), 301–325 (2006). DOI 10.1080/17442500600900707
15. Albeverio, S., Herzberg, F.S.: Lifting Lévy processes to hyperfinite random walks. Bull. Sci. Math. **130**(8), 697–706 (2006). DOI 10.1016/S0007-4497(06)00019-4

16. Albeverio, S., Høegh-Krohn, R.: Quasi invariant measures, symmetric diffusion processes and quantum fields. In: Les méthodes mathématiques de la théorie quantique des champs (Colloq. Internat. CNRS, No. 248, Marseille, 1975), pp. 11–59. Éditions Centre Nat. Recherche Sci., Paris (1976)
17. Albeverio, S., Høegh-Krohn, R.: Dirichlet forms and diffusion processes on rigged Hilbert spaces. Z. Wahrscheinlichkeitstheorie und verw. Gebiete **40**(1), 1–57 (1977)
18. Albeverio, S., Høegh-Krohn, R.: Dirichlet forms and Markov semigroups on C^*-algebras. Comm. Math. Phys. **56**(2), 173–187 (1977)
19. Albeverio, S., Høegh-Krohn, R.: Hunt processes and analytic potential theory on rigged Hilbert spaces. Ann. Inst. H. Poincaré Sect. B (N.S.) **13**(3), 269–291 (1977)
20. Albeverio, S., Høegh-Krohn, R.: Some Markov processes and Markov fields in quantum theory, group theory hydrodynamics and C^*-algebras. In: D. Williams (ed.) Stochastic integrals (Proc. Sympos., Univ. Durham, Durham, 1980), *Lecture Notes in Math.*, vol. 851, pp. 497–540. Springer, Berlin (1981)
21. Albeverio, S., Høegh-Krohn, R.: Stochastic methods in quantum field theory and hydrodynamics. New stochastic methods in physics. Phys. Rep. **77**(3), 193–214 (1981). DOI 10.1016/0370-1573(81)90071-5
22. Albeverio, S., Høegh-Krohn, R.: Diffusions, quantum fields and groups of mappings. In: M. Fukushima (ed.) Functional analysis in Markov processes (Katata/Kyoto, 1981), *Lecture Notes in Math.*, vol. 923, pp. 133–145. Springer, Berlin (1982)
23. Albeverio, S., Høegh-Krohn, R.: Some remarks on Dirichlet forms and their applications to quantum mechanics and statistical mechanics. In: M. Fukushima (ed.) Functional analysis in Markov processes (Katata/Kyoto, 1981), *Lecture Notes in Math.*, vol. 923, pp. 120–132. Springer, Berlin (1982)
24. Albeverio, S., Høegh-Krohn, R.: Diffusion fields, quantum fields, and fields with values in Lie groups. In: M. Pinsky (ed.) Stochastic analysis and applications, *Adv. Probab. Related Topics*, vol. 7, pp. 1–98. Dekker, New York (1984)
25. Albeverio, S., Fenstad, J.E., Høegh-Krohn, R., Lindstrøm, T.: Nonstandard methods in stochastic analysis and mathematical physics, *Pure and Applied Mathematics*, vol. 122. Academic Press Inc., Orlando, FL (1986). Reprint: Dover Publications, Mineola, NY (2009)
26. Albeverio, S., Høegh-Krohn, R., Streit, L.: Energy forms, Hamiltonians, and distorted Brownian paths. J. Mathematical Phys. **18**(5), 907–917 (1977)
27. Albeverio, S., Høegh-Krohn, R., Streit, L.: Regularization of Hamiltonians and processes. J. Math. Phys. **21**(7), 1636–1642 (1980). DOI 10.1063/1.524649
28. Albeverio, S., Karwowski, W.: Jump processes on leaves of multibranching trees. J. Math. Phys. **49**(9), 093,503, 20 (2008)
29. Albeverio, S., Karwowski, W., Yasuda, K.: Trace formula for p-adics. Acta Appl. Math. **71**(1), 31–48 (2002)
30. Albeverio, S., Karwowski, W., Zhao, X.: Asymptotics and spectral results for random walks on p-adics. Stochastic Process. Appl. **83**(1), 39–59 (1999)
31. Albeverio, S., Kondratiev, Y., Kozitsky, Y., Röckner, M.: The statistical mechanics of quantum lattice systems. A path integral approach, *EMS Tracts in Mathematics*, vol. 8. European Mathematical Society (EMS), Zürich (2009)
32. Albeverio, S., Kondratiev, Y., Röckner, M.: Strong Feller properties for distorted Brownian motion and applications to finite particle systems with singular interactions. In: Finite and infinite dimensional analysis in honor of Leonard Gross (New Orleans, LA, 2001), *Contemp. Math.*, vol. 317, pp. 15–35. Amer. Math. Soc., Providence, RI (2003)
33. Albeverio, S., Kondratiev, Y.G.: Supersymmetric Dirichlet operators. Ukraïn. Mat. Zh. **47**(5), 583–592 (1995)
34. Albeverio, S., Kusuoka, S.: Maximality of infinite-dimensional Dirichlet forms and Høegh-Krohn's model of quantum fields. In: S. Albeverio, J.E. Fenstad, H. Holden, T. Lindstrøm (eds.) Ideas and methods in quantum and statistical physics (Oslo, 1988), pp. 301–330. Cambridge University Press, Cambridge (1992)

35. Albeverio, S., Kusuoka, S., Röckner, M.: On partial integration in infinite-dimensional space and applications to Dirichlet forms. J. London Math. Soc. (2) **42**(1), 122–136 (1990)
36. Albeverio, S., Kusuoka, S., Streit, L.: Convergence of Dirichlet forms and associated Schrödinger operators. J. Funct. Anal. **68**(2), 130–148 (1986)
37. Albeverio, S., Luxemburg, W.A.J., Wolff, M.P.H. (eds.): Advances in analysis, probability and mathematical physics. Contributions of nonstandard analysis, *Mathematics and its Applications*, vol. 314. Kluwer Academic Publishers Group, Dordrecht (1995)
38. Albeverio, S., Ma, Z., Röckner, M.: Characterization of (non-symmetric) semi-Dirichlet forms associated with Hunt processes. Random Oper. Stoch. Equ. **3**(2), 161–179 (1995). DOI 10.1515/rose.1995.3.2.161
39. Albeverio, S., Ma, Z.M.: Nowhere Radon smooth measures, perturbations of Dirichlet forms and singular quadratic forms. Stochastic differential systems, Proc. 4th Conf., Bad Honnef/FRG 1988, Lect. Notes Control Inf. Sci. 126, 32–45 (1989). (1989)
40. Albeverio, S., Ma, Z.M.: Diffusion processes with singular Dirichlet forms. Stochastic analysis and applications, Proc. 1st Lisbon Conf., Lisbon/Port. 1989, Prog. Probab. 26, 11–28 (1991). (1991)
41. Albeverio, S., Ma, Z.M.: Necessary and sufficient conditions for the existence of m-perfect processes associated with Dirichlet forms. In: Séminaire de Probabilités, XXV, *Lecture Notes in Math.*, vol. 1485, pp. 374–406. Springer, Berlin (1991). DOI 10.1007/BFb0100871
42. Albeverio, S., Ma, Z.M.: Additive functionals, nowhere Radon and Kato class smooth measures associated with Dirichlet forms. Osaka J. Math. **29**(2), 247–265 (1992)
43. Albeverio, S., Ma, Z.M., Röckner, M.: A Beurling-Deny type structure theorem for Dirichlet forms on general state spaces. In: Ideas and methods in mathematical analysis, stochastics, and applications (Oslo, 1988), pp. 115–123. Cambridge University Press, Cambridge (1992)
44. Albeverio, S., Ma, Z.M., Röckner, M.: Nonsymmetric Dirichlet forms and Markov processes on general state space. C. R. Acad. Sci. Paris Sér. I Math. **314**(1), 77–82 (1992)
45. Albeverio, S., Ma, Z.M., Röckner, M.: Regularization of Dirichlet spaces and applications. C. R. Acad. Sci. Paris Sér. I Math. **314**(11), 859–864 (1992)
46. Albeverio, S.; Mandrekar, V.; Rüdiger, B.: Existence of mild solutions for stochastic differential equations and semilinear equations with non-Gaussian Lévy noise. Stochastic Process. Appl. **119**(3), 835–863 (2009)
47. Albeverio, S., Morato, L.M., Ugolini, S.: Non-symmetric diffusions and related Hamiltonians. Potential Anal. **8**(2), 195–204 (1998)
48. Albeverio, S., Röckner, M.: Classical Dirichlet forms on topological vector spaces—the construction of the associated diffusion process. Probab. Theory Related Fields **83**(3), 405–434 (1989)
49. Albeverio, S., Röckner, M.: Classical Dirichlet forms on topological vector spaces—closability and a Cameron-Martin formula. J. Funct. Anal. **88**(2), 395–436 (1990)
50. Albeverio, S., Röckner, M.: Stochastic differential equations in infinite dimensions: solutions via Dirichlet forms. Probab. Theory Related Fields **89**(3), 347–386 (1991)
51. Albeverio, S., Rüdiger, B.: Infinite-dimensional stochastic differential equations obtained by subordination and related Dirichlet forms. J. Funct. Anal. **204**(1), 122–156 (2003)
52. Albeverio, S., Rüdiger, B.: Stochastic integrals and the Lévy-Ito decomposition theorem on separable Banach spaces. Stoch. Anal. Appl. **23**(2), 217–253 (2005). DOI 10.1081/SAP-200026429
53. Albeverio, S., Rüdiger, B.: Subordination of symmetric quasi-regular Dirichlet forms. Random Oper. Stochastic Equations **13**(1), 17–38 (2005)
54. Albeverio, S., Rüdiger, B., Wu, J.L.: Invariant measures and symmetry property of Lévy type operators. Potential Anal. **13**(2), 147–168 (2000)

55. Albeverio, S., Rüdiger, B., Wu, J.L.: Analytic and probabilistic aspects of Lévy processes and fields in quantum theory. In: O. Barndorff-Nielsen, T. Mikosch, S. Resnick (eds.) Lévy processes, pp. 187–224. Birkhäuser Boston, Boston, MA (2001)
56. Albeverio, S., Song, S.: Closability and resolvent of Dirichlet forms perturbed by jumps. Potential Anal. **2**(2), 115–130 (1993). DOI 10.1007/BF01049296
57. Albeverio, S., Ugolini, S.: Complex Dirichlet forms: nonsymmetric diffusion processes and Schrödinger operators. Potential Anal. **12**(4), 403–417 (2000)
58. Ambrosio, L., Gigli, N., Savaré, G.: Gradient flows in metric spaces and in the space of probability measures. Second edition. Lectures in Mathematics ETH Zürich. Birkhäuser Verlag, Basel (2008)
59. Ambrosio, L., Savaré, G., Zambotti, L.: Existence and stability for Fokker-Planck equations with log-concave reference measure. Probab. Theory Related Fields **145**(3-4), 517–564 (2009)
60. Andres, S., von Renesse, M.K.: Particle approximation of the Wasserstein diffusion. J. Funct. Anal. **258**(11), 3879–3905 (2010)
61. Anderson, R.M.: A non-standard representation for Brownian motion and Itô integration. Israel J. Math. **25**(1-2), 15–46 (1976)
62. Anderson, R.M.: Star-finite representations of measure spaces. Trans. Amer. Math. Soc. **271**(2), 667–687 (1982)
63. Anderson, R.M.: Nonstandard analysis with applications to economics. In: W. Hildenbrand, H. Sonnenschein (eds.) Handbook of mathematical economics, Vol. IV, *Handbooks in Econom.*, vol. 1, pp. 2145–2208. North-Holland, Amsterdam (1991)
64. Anderson, R.M., Raimondo, R.C.: Equilibrium in continuous-time financial markets: endogenously dynamically complete markets. Econometrica **76**(4), 841–907 (2008). DOI 10.1111/j.1468-0262.2008.00861.x
65. Applebaum, D.: Lévy processes and stochastic calculus, *Cambridge Studies in Advanced Mathematics*, vol. 116, second edn. Cambridge University Press, Cambridge (2009)
66. Arendt, W., Batty, C.J.K., Hieber, M., Neubrander, F.: Vector-valued Laplace transforms and Cauchy problems, *Monographs in Mathematics*, vol. 96. Birkhäuser Verlag, Basel (2001)
67. Arendt, W., Thomaschewski, S.: Local operators and forms. Positivity **9**(3), 357–367 (2005)
68. Arkeryd, L.O., Cutland, N.J., Henson, C.W. (eds.): Nonstandard analysis, *NATO Advanced Science Institutes Series C: Mathematical and Physical Sciences*, vol. 493. Kluwer Academic Publishers Group, Dordrecht (1997)
69. Bakry, D.: On Sobolev and logarithmic Sobolev inequalities for Markov semigroups. In: New trends in stochastic analysis (Charingworth, 1994), pp. 43–75, World Scientific Publishers, River Edge, NJ (1997)
70. Bakry, D., Concordet, D., Ledoux, M.: Optimal heat kernel bounds under logarithmic Sobolev inequalities. ESAIM Probab. Statist. **1**, 391–407 (1997)
71. Banaschewski, B.: The power of the ultrafilter theorem. J. London Math. Soc. (2) **27**(2), 193–202 (1983)
72. Barlow, M.T., Bass, R.F., Chen, Z.Q., Kassmann, M.: Non-local Dirichlet forms and symmetric jump processes. Trans. Amer. Math. Soc. **361**(4), 1963–1999 (2009). DOI 10.1090/S0002-9947-08-04544-3. URL http://dx.doi.org/10.1090/S0002-9947-08-04544-3
73. Barndorff-Nielsen, O.E., Mikosch, T., Resnick, S.I. (eds.): Lévy processes. Birkhäuser Boston Inc., Boston, MA (2001)
74. Bass, R.F., Hsu, P.: Some potential theory for reflecting Brownian motion in Hölder and Lipschitz domains. Ann. Probab. **19**(2), 486–508 (1991)
75. Bass, R.F., Kassmann, M.: Hölder continuity of harmonic functions with respect to operators of variable order. Comm. Partial Differential Equations **30**(7-9), 1249–1259 (2005)

76. Bauer, H.: Maß- und Integrationstheorie, second edn. de Gruyter Lehrbuch. Walter de Gruyter & Co., Berlin (1992)
77. Bauer, H.: Wahrscheinlichkeitstheorie, fifth edn. de Gruyter Lehrbuch. Walter de Gruyter & Co., Berlin (2002)
78. Bell, J.L., Machover, M.: A course in mathematical logic. North-Holland Publishing Co., Amsterdam (1977)
79. von Below, J., Lubary, J.A.: The eigenvalues of the Laplacian on locally finite networks. Results Math. **47**(3-4), 199–225 (2005)
80. von Below, J., Lubary, J.A.: The eigenvalues of the Laplacian on locally finite networks under generalized node transition. Results Math. **54**(1-2), 15–39 (2009). DOI 10.1007/s00025-009-0376-y
81. von Below, J., Nicaise, S.: Dynamical interface transition in ramified media with diffusion. Comm. Partial Differential Equations **21**(1-2), 255–279 (1996). DOI 10.1080/03605309608821184
82. Ben Amor, A., Brasche, J.F.: Sharp estimates for large coupling convergence with applications to Dirichlet operators. J. Funct. Anal. **254**(2), 454–475 (2008)
83. Benci, V., Di Nasso, M.: Alpha-theory: an elementary axiomatics for nonstandard analysis. Expo. Math. **21**(4), 355–386 (2003)
84. Benci, V., Forti, M., Di Nasso, M.: The eightfold path to nonstandard analysis. In: Nonstandard methods and applications in mathematics, *Lect. Notes Log.*, vol. 25, pp. 3–44. Assoc. Symbol. Logic, La Jolla, CA (2006)
85. Berrahmoune, L.: A note on admissibility for unbounded bilinear control systems. Bull. Belg. Math. Soc. Simon Stevin **16**(2), 193–204 (2009)
86. Bertoin, J.: Lévy processes, *Cambridge Tracts in Mathematics*, vol. 121. Cambridge University Press, Cambridge (1996)
87. Beurling, A., Deny, J.: Espaces de Dirichlet. I. Le cas élémentaire. Acta Math. **99**, 203–224 (1958)
88. Beurling, A., Deny, J.: Dirichlet spaces. Proc. Nat. Acad. Sci. U.S.A. **45**, 208–215 (1959)
89. Beznea, L., Boboc, N.: Potential theory and right processes. Mathematics and its Applications (Dordrecht) 572. Dordrecht: Kluwer Academic Publishers. vi, 370 pp. (2004)
90. Beznea, L., Boboc, N., Röckner, M.: Markov processes associated with L^p-resolvents, applications to quasi-regular Dirichlet forms and stochastic differential equations. C. R. Math. Acad. Sci. Paris **346**(5-6), 323–328 (2008)
91. Biroli, M.: Γ-convergence for strongly local Dirichlet forms in open sets with holes. Tr. Mat. Inst. Steklova **250**(Differ. Uravn. i Din. Sist.), 262–271 (2005)
92. Biroli, M., Tchou, N.: Γ-convergence for strongly local Dirichlet forms in perforated domains with homogeneous Neumann boundary conditions. Adv. Math. Sci. Appl. **17**(1), 149–179 (2007)
93. Biroli, M., Vernole, P.: Harnack inequality for harmonic functions relative to a nonlinear p-homogeneous Riemannian Dirichlet form. Nonlinear Anal. **64**(1), 51–68 (2006)
94. Bliedtner, J.: Dirichlet forms on regular functional spaces. In: Seminar on Potential Theory, II, pp. 15–62. Lecture Notes in Math., Vol. 226. Springer, Berlin (1971)
95. Bliedtner, J., Hansen, W.: Potential theory. An analytic and probabilistic approach to balayage. Universitext. Springer-Verlag, Berlin (1986)
96. Blumenthal, R.M., Getoor, R.K.: Markov processes and potential theory. Pure and Applied Mathematics, Vol. 29. Academic Press, New York (1968)
97. Bonciocat, A.I.: Curvature bounds and heat kernels. Discrete versus continuous spaces. Bonn: Univ. Bonn, Mathematisch-Naturwissenschaftliche Fakultät (Diss.). 86 p. (2008)
98. Bouleau, N.: Error calculus for finance and physics: the language of Dirichlet forms, *de Gruyter Expositions in Mathematics*, vol. 37. Walter de Gruyter & Co., Berlin (2003)

99. Bouleau, N., Hirsch, F.: Dirichlet forms and analysis on Wiener space, *de Gruyter Studies in Mathematics*, vol. 14. Walter de Gruyter & Co., Berlin (1991)
100. Brasche, J.: On eigenvalues and eigensolutions of the Schrödinger equation on the complement of a set with classical capacity zero. Methods Funct. Anal. Topol. **9**(3), 189–206 (2003)
101. Bratteli, O., Robinson, D.W.: Operator algebras and quantum statistical mechanics. 1, second edn. Texts and Monographs in Physics. Springer-Verlag, New York (1987)
102. Capiński, M., Cutland, N.J.: Nonstandard methods for stochastic fluid mechanics, *Series on Advances in Mathematics for Applied Sciences*, vol. 27. World Scientific Publishing Co. Inc., River Edge, NJ (1995)
103. Carmona, R.: Regularity properties of Schrödinger and Dirichlet semigroups. J. Funct. Anal. **33**(3), 259–296 (1979). DOI 10.1016/0022-1236(79)90068-5
104. Carmona, R.A., Tehranchi, M.R.: Interest rate models: an infinite dimensional stochastic analysis perspective. Springer Finance. Berlin: Springer. xiv, 235 pp. (2006)
105. Carrillo-Menendez, S.: Processus de Markov associé à une forme de Dirichlet non symétrique. Z. Wahrscheinlichkeitstheorie und Verw. Gebiete **33**(2), 139–154 (1975/76)
106. Cartan, H.: Théorie du potentiel newtonien: énergie, capacité, suites de potentiels. Bull. Soc. Math. France **73**, 74–106 (1945)
107. Cattiaux, P.: Hypercontractivity for perturbed diffusion semigroups. Ann. Fac. Sci. Toulouse Math. (6) **14**(4), 609–628 (2005)
108. Cattiaux, P., Gentil, I., Guillin, A.: Weak logarithmic Sobolev inequalities and entropic convergence. Probab. Theory Related Fields **139**(3-4), 563–603 (2007)
109. Chen, M.F.: From Markov chains to non-equilibrium particle systems. 2nd ed. River Edge, NJ: World Scientific. xii, 597 pp. (2004)
110. Chen, Z., Fukushima, M.: Symmetric Markov processes, time change and boundary theory. Princeton University Press, Princeton, NJ (forthcoming)
111. Chen, Z.Q., Fitzsimmons, P.J., Kuwae, K., Zhang, T.S.: Perturbation of symmetric Markov processes. Probab. Theory Related Fields **140**(1-2), 239–275 (2008)
112. Chen, Z.Q., Fitzsimmons, P.J., Takeda, M., Ying, J., Zhang, T.S.: Absolute continuity of symmetric Markov processes. Ann. Probab. **32**(3A), 2067–2098 (2004)
113. Chen, Z.Q., Fukushima, M., Ying, J.: Entrance law, exit system and Lévy system of time changed processes. Illinois J. Math. **50**(1-4), 269–312 (electronic) (2006)
114. Chen, Z.Q., Fukushima, M., Ying, J.: Traces of symmetric Markov processes and their characterizations. Ann. Probab. **34**(3), 1052–1102 (2006)
115. Chen, Z.Q., Kumagai, T.: A priori Hölder estimate, parabolic Harnack principle and heat kernel estimates for diffusions with jumps. Rev. Mat. Iberoam. **26**(2), 551–589 (2010)
116. Chen, Z.Q., Ma, Z.M., Röckner, M.: Quasi-homeomorphisms of Dirichlet forms. Nagoya Math. J. **136**, 1–15 (1994)
117. Chen, Z.Q., Shiozawa, Y.: Limit theorems for branching Markov processes. J. Funct. Anal. **250**(2), 374–399 (2007)
118. Chen, Z.Q., Zhang, T.: Time-reversal and elliptic boundary value problems. Ann. Probab. **37**(3), 1008–1043 (2009)
119. Chung, K.L.: Markov chains with stationary transition probabilities, *Grundlehren der mathematischen Wissenschaften*, vol. 104. Springer-Verlag, Berlin (1960)
120. Cipriani, F.: Dirichlet forms on noncommutative spaces. In: Quantum potential theory, *Lecture Notes in Math.*, vol. 1954, pp. 161–276. Springer, Berlin (2008). DOI 10.1007/978-3-540-69365-9_5
121. Cohn, D.L.: Measure theory. Birkhäuser Boston, Mass. (1980)
122. Conrad, F., Grothaus, M.: Construction of N-particle Langevin dynamics for $H^{1,\infty}$-potentials via generalized Dirichlet forms. Potential Anal. **28**(3), 261–282 (2008)
123. Courant, R., Robbins, H.: What is mathematics? Oxford University Press, New York (1979)

124. Courant, R., Robbins, H., Stewart, I.: What is Mathematics? An Elementary Approach to Ideas and Methods, second edn. Oxford University Press, New York (1996)
125. Cutland, N.J. (ed.): Nonstandard analysis and its applications. Papers from a conference held at the University of Hull, Hull, 1986, *London Mathematical Society Student Texts*, vol. 10. Cambridge University Press, Cambridge (1988)
126. Cutland, N.J.: Loeb measures in practice: recent advances, *Lecture Notes in Mathematics*, vol. 1751. Springer-Verlag, Berlin (2000)
127. Cycon, H.L., Froese, R.G., Kirsch, W., Simon, B.: Schrödinger operators with application to quantum mechanics and global geometry, study edn. Texts and Monographs in Physics. Springer-Verlag, Berlin (1987)
128. Da Prato, G.: An introduction to infinite-dimensional analysis. Universitext. Springer-Verlag, Berlin (2006)
129. Dai Pra, P., Posta, G.: Logarithmic Sobolev inequality for zero-range dynamics. Ann. Probab. **33**(6), 2355–2401 (2005)
130. Dauben, J.W.: Abraham Robinson. With a foreword by Benoit B. Mandelbrot. Princeton University Press, Princeton, NJ (1995)
131. Davies, E.B.: Quantum theory of open systems. Academic Press [Harcourt Brace Jovanovich Publishers], London (1976)
132. Davies, E.B.: Heat kernels and spectral theory, *Cambridge Tracts in Mathematics*, vol. 92. Cambridge University Press, Cambridge (1989). DOI 10.1017/CBO9780511566158
133. Davies, E.B., Lindsay, J.M.: Noncommutative symmetric Markov semigroups. Math. Z. **210**(3), 379–411 (1992)
134. Davies, E.B., Simon, B.: Ultracontractivity and the heat kernel for Schrödinger operators and Dirichlet Laplacians. J. Funct. Anal. **59**(2), 335–395 (1984)
135. Davis, M.: Applied nonstandard analysis. Wiley-Interscience [John Wiley & Sons], New York (1977)
136. De Leva, G., Kim, D., Kuwae, K.: L^p-independence of spectral bounds of Feynman-Kac semigroups by continuous additive functionals. J. Funct. Anal. **259**(3), 690–730 (2010)
137. Del Moral, P., Ledoux, M., Miclo, L.: On contraction properties of Markov kernels. Probab. Theory Related Fields **126**(3), 395–420 (2003)
138. Dellacherie, C., Meyer, P.A.: Probabilités et potentiel. Chapitres IX à XI, revised edn. Publications de l'Institut de Mathématiques de l'Université de Strasbourg, XVIII. Hermann, Paris (1983)
139. Demuth, M., van Casteren, J.A.: Stochastic spectral theory for selfadjoint Feller operators. A functional integration approach. Probability and its Applications. Birkhäuser Verlag, Basel (2000)
140. Demuth, M., Krishna, M.: Determining spectra in quantum theory. Progress in Mathematical Physics 44. Boston, MA: Birkhäuser. x, 219 pp. (2005)
141. Denis, L., Grorud, A., Pontier, M.: Formes de Dirichlet sur un espace de Wiener-Poisson. Application au grossissement de filtration. In: Séminaire de Probabilités, XXXIV, *Lecture Notes in Math.*, vol. 1729, pp. 198–217. Springer, Berlin (2000)
142. Deny, J.: Méthodes hilbertiennes en théorie du potentiel. In: Potential Theory (C.I.M.E., I Ciclo, Stresa, 1969), pp. 121–201. Edizioni Cremonese, Rome (1970)
143. Deny, J., Lions, J.L.: Les espaces du type de Beppo Levi. Ann. Inst. Fourier, Grenoble **5**, 305–370 (1955) (1953–54)
144. Diener, F., Diener, M. (eds.): Nonstandard analysis in practice. Universitext. Springer-Verlag, Berlin (1995)
145. Doob, J.L.: Discrete potential theory and boundaries. J. Math. Mech. **8**, 433–458; erratum 993 (1959)
146. Doob, J.L.: Boundary properties for functions with finite Dirichlet integrals. Ann. Inst. Fourier (Grenoble) **12**, 573–621 (1962)

147. Dynkin, E.B.: Markov processes. Vols. I, II. Translated with the authorization and assistance of the author by J. Fabius, V. Greenberg, A. Maitra, G. Majone, *Grundlehren der Mathematischen Wissenschaften*, vol. 121,122. Academic Press Inc., New York (1965)
148. Dynkin, E.B., Yushkevich, A.A.: Markov processes: Theorems and problems. Translated from the Russian by James S. Wood. Plenum Press, New York (1969)
149. Eberle, A.: Uniqueness and non-uniqueness of semigroups generated by singular diffusion operators, *Lecture Notes in Mathematics*, vol. 1718. Springer-Verlag, Berlin (1999)
150. Eberle, A., Marinelli, C.: L^p estimates for Feynman-Kac propagators with time-dependent reference measures. J. Math. Anal. Appl. **365**(1), 120–134 (2010)
151. ter Elst, A.F.M., Robinson, D.W.: Invariant subspaces of submarkovian semigroups. J. Evol. Equ. **8**(4), 661–671 (2008)
152. Ethier, S.N., Kurtz, T.G.: Markov processes. Wiley Series in Probability and Mathematical Statistics: Probability and Mathematical Statistics. John Wiley & Sons Inc., New York (1986)
153. Evans, S.N.: Probability and real trees. Lectures from the 35th Summer School on Probability Theory held in Saint-Flour, July 6–23, 2005, *Lecture Notes in Mathematics*, vol. 1920. Springer, Berlin (2008)
154. Exner, P.: Open quantum systems and Feynman integrals. Fundamental Theories of Physics. D. Reidel Publishing Co., Dordrecht (1985)
155. Fabes, E., Fukushima, M., Gross, L., Kenig, C., Röckner, M., Stroock, D.W.: Dirichlet forms. Lectures given at the First C.I.M.E. Session held in Varenna, June 8–19, 1992, *Lecture Notes in Mathematics*, vol. 1563. Springer-Verlag, Berlin (1993)
156. Fajardo, S., Keisler, H.J.: Model theory of stochastic processes, *Lecture Notes in Logic*, vol. 14. Association for Symbolic Logic, Urbana, IL (2002)
157. Fan, R.Z.: A class of stochastic differential equations with local time and skew Brownian motion with two boundaries. Chinese J. Appl. Probab. Statist. **3**(2), 130–136 (1987)
158. Fan, R.Z.: An extension of Dynkin formula and probabilistic solutions of some second-order partial differential equations. J. Yunnan Univ., Nat. Sci. Ed. **11**(1), 10–14 (1989)
159. Fan, R.Z.: Representation of martingale additive functionals on Banach spaces. Acta Math. Appl. Sinica (English Ser.) **6**(1), 74–80 (1990)
160. Fan, R.Z.: Closability of a certain symmetric form on Banach spaces. Chinese Ann. Math. Ser. A **12**(2), 202–209 (1991)
161. Fan, R.Z.: Decomposition of a class of functionals. Acta Math. Sinica (N.S.) **7**(3), 224–240 (1991)
162. Fan, R.Z.: Beurling-Deny formulae on Banach spaces. Acta Math. Sci. (English Ed.) **12**(1), 79–84 (1992)
163. Fan, R.Z.: Decomposition of a class of functionals and the predictable representation theorem on Banach spaces. Acta Math. Appl. Sinica (English Ser.) **8**(2), 153–167 (1992)
164. Fan, R.Z.: Some diffusion processes on a half-space and Dirichlet forms related to them. Acta Math. Sinica **35**(3), 418–430 (1992)
165. Fan, R.Z.: Nonstandard construction of symmetric strong Markov processes associated with Dirichlet forms. Albeverio, Sergio (ed.) et al., Stochastic processes, physics and geometry II. Proceedings of the 3rd international conference held in Locarno, Switzerland, 24-29 June 1991. Singapore: World Scientific. 247-277 (1995).
166. Fan, R.Z.: Potential theory of hyperfinite Dirichlet forms. Potential Anal. **5**(5), 417–462 (1996)
167. Fang, X., Fukushima, M., Ying, J.: On regular Dirichlet subspaces of $H^1(I)$ and associated linear diffusions. Osaka J. Math. **42**(1), 27–41 (2005)
168. Faris, W.G.: Self-adjoint operators. Lecture Notes in Mathematics, Vol. 433. Springer-Verlag, Berlin (1975)

169. Feller, M.N.: The Lévy Laplacian. Cambridge Tracts in Mathematics 166. Cambridge: Cambridge University Press. vi, 153 pp. (2005)
170. Feller, W.: The parabolic differential equations and the associated semi-groups of transformations. Ann. of Math. (2) **55**, 468–519 (1952)
171. Fitzsimmons, P.J.: Time changes of symmetric Markov processes and a Feynman-Kac formula. J. Theoret. Probab. **2**(4), 487–501 (1989)
172. Fitzsimmons, P.J., Kuwae, K.: Non-symmetric perturbations of symmetric Dirichlet forms. J. Funct. Anal. **208**(1), 140–162 (2004). DOI 10.1016/j.jfa.2003.10.005
173. Foondun, M.: Heat kernel estimates and Harnack inequalities for some Dirichlet forms with non-local part. Electron. J. Probab. **14**, no. 11, 314–340 (2009)
174. Fuhrman, M.: Analyticity of transition semigroups and closability of bilinear forms in Hilbert spaces. Studia Math. **115**(1), 53–71 (1995)
175. Fukushima, M.: Dirichlet forms and Markov processes, *North-Holland Mathematical Library*, vol. 23. North-Holland Publishing Co., Amsterdam (1980)
176. Fukushima, M.: Energy forms and diffusion processes. In: L. Streit (ed.) Mathematics + physics. Vol. 1, pp. 65–97. World Sci. Publishing, Singapore (1985)
177. Fukushima, M.: *BV* functions and distorted Ornstein Uhlenbeck processes over the abstract Wiener space. J. Funct. Anal. **174**(1), 227–249 (2000)
178. Fukushima, M.: Decompositions of symmetric diffusion processes and related topics in analysis. Sugaku Expositions **14**(1), 1–13 (2001)
179. Fukushima, M.: On the works of Kiyosi Itô and stochastic analysis. Jpn. J. Math. **2**(1), 45–53 (2007)
180. Fukushima, M.: From one dimensional diffusions to symmetric Markov processes. Stochastic Process. Appl. **120**(5), 590–604 (2010)
181. Fukushima, M., He, P., Ying, J.: Time changes of symmetric diffusions and Feller measures. Ann. Probab. **32**(4), 3138–3166 (2004)
182. Fukushima, M., Nakao, S., Takeda, M.: On Dirichlet forms with random data—recurrence and homogenization. In: Stochastic processes—mathematics and physics, II (Bielefeld, 1985), *Lecture Notes in Math.*, vol. 1250, pp. 87–97. Springer, Berlin (1987). DOI 10.1007/BFb0077349. URL http://dx.doi.org/10.1007/BFb0077349
183. Fukushima, M., Ōshima, Y., Takeda, M.: Dirichlet forms and symmetric Markov processes. 2nd revised and extended *ed., de Gruyter Studies in Mathematics*, vol. 19. Walter de Gruyter & Co., Berlin (2011).
184. Fukushima, M., Takeda, M.: A transformation of a symmetric Markov process and the Donsker-Varadhan theory. Osaka J. Math. **21**(2), 311–326 (1984)
185. Fukushima, M., Tanaka, H.: Poisson point processes attached to symmetric diffusions. Ann. Inst. H. Poincaré Probab. Statist. **41**(3), 419–459 (2005)
186. Fukushima, M., Tanaka, H.: Poisson point processes attached to symmetric diffusions. Ann. Inst. H. Poincaré Probab. Statist. **41**(3), 419–459 (2005)
187. Fukushima, M., Uemura, T.: Capacitary bounds of measures and ultracontractivity of time changed processes. J. Math. Pures Appl. (9) **82**(5), 553–572 (2003)
188. Fukushima, M., Ying, J.: A note on regular Dirichlet subspaces. Proc. Amer. Math. Soc. **131**(5), 1607–1610 (electronic) (2003)
189. Fukushima, M., Ying, J.: Erratum to: "A note on regular Dirichlet subspaces" [Proc. Amer. Math. Soc. **131** (2003), no. 5, 1607–1610. Proc. Amer. Math. Soc. **132**(5), 1559 (electronic) (2004)
190. Garroni, M.G., Menaldi, J.L.: Second order elliptic integro-differential problems, *Chapman & Hall/CRC Research Notes in Mathematics*, vol. 430. Chapman & Hall/CRC, Boca Raton, FL (2002)
191. Gesztesy, F., Holden, H., Jost, J., Paycha, S., Röckner, M., Scarlatti, S. (eds.): Stochastic processes, physics and geometry: new interplays. I. A volume in honor of Sergio Albeverio, *CMS Conference Proceedings*, vol. 28. Published by the American Mathematical Society, Providence, RI (2000)

192. Gesztesy, F., Holden, H., Jost, J., Paycha, S., Röckner, M., Scarlatti, S. (eds.): Stochastic processes, physics and geometry: new interplays. II. A volume in honor of Sergio Albeverio, *CMS Conference Proceedings*, vol. 29. American Mathematical Society, Providence, RI (2000)
193. Getoor, R.K.: Markov processes: Ray processes and right processes, *Lecture Notes in Mathematics*, vol. 440. Springer-Verlag, Berlin (1975)
194. Goldblatt, R.: Lectures on the hyperreals. An introduction to nonstandard analysis, *Graduate Texts in Mathematics*, vol. 188. Springer-Verlag, New York (1998)
195. Grigor′yan, A., Hu, J.: Off-diagonal upper estimates for the heat kernel of the Dirichlet forms on metric spaces. Invent. Math. **174**(1), 81–126 (2008)
196. Gross, L.: Logarithmic Sobolev inequalities. Amer. J. Math. **97**(4), 1061–1083 (1975)
197. Grossinho, M.d.R., Tersian, S.A.: An introduction to minimax theorems and their applications to differential equations. Nonconvex Optimization and Its Applications. 52. Dordrecht: Kluwer Academic Publishers. xii, 269 pp. (2001)
198. Grothaus, M., Kondratiev, Y.G., Lytvynov, E., Röckner, M.: Scaling limit of stochastic dynamics in classical continuous systems. Ann. Probab. **31**(3), 1494–1532 (2003)
199. Grunewald, N., Otto, F., Villani, C., Westdickenberg, M.G.: A two-scale approach to logarithmic Sobolev inequalities and the hydrodynamic limit. Ann. Inst. Henri Poincaré Probab. Stat. **45**(2), 302–351 (2009)
200. Halpern, J.D., Lévy, A.: The Boolean prime ideal theorem does not imply the axiom of choice. In: Axiomatic Set Theory (Proc. Sympos. Pure Math., Vol. XIII, Part I, Univ. California, Los Angeles, Calif., 1967), pp. 83–134. Amer. Math. Soc., Providence, R.I. (1971)
201. Henson, C.W.: Analytic sets, Baire sets and the standard part map. Canad. J. Math. **31**(3), 663–672 (1979)
202. Herzberg, F.: Addendum to: "A definable nonstandard enlargement" [MLQ Math. Log. Q. **54** (2008), no. 2, 167–175]. MLQ Math. Log. Q. **54**(6), 666–667 (2008)
203. Herzberg, F.: A definable nonstandard enlargement. MLQ Math. Log. Q. **54**(2), 167–175 (2008)
204. Herzberg, F.: On the foundations of Lévy finance: Equilibrium for a single-agent financial market with jumps. IMW Working Paper 406, Institut für Mathematische Wirtschaftsforschung, Universität Bielefeld (2008)
205. Herzberg, F.: Hyperfinite stochastic integration for Lévy processes with finite-variation jump part. Bull. Sci. Math. **134**(4), 423–445 (2010)
206. Herzberg, F., Lindstrøm, T.: Corrigendum and addendum to 'Hyperfinite Lévy processes' by Tom Lindstrøm (Stochastics 76(6):517–548, 2004). Stochastics **81**(6), 567–570 (2009)
207. Hesse, M.: Harmonic maps into trees and graphs. Analytical and numerical aspects. Ph.D. thesis, Bonner Mathematische Schriften 370. Bonn: Univ. Bonn, Mathematisches Institut (Dissertation 2004). 100 p. (2005)
208. Hesse, M., Rumpf, M., Sturm, K.T.: Discretization and convergence for harmonic maps into trees. Calculus of Variations and Partial Differential Equations **21**(2), 113–136 (2004)
209. Hino, M.: Energy measures and indices of Dirichlet forms, with applications to derivatives on some fractals. Proc. Lond. Math. Soc. (3) **100**(1), 269–302 (2010). DOI 10.1112/plms/pdp032. URL http://dx.doi.org/10.1112/plms/pdp032
210. Hoover, D.N., Keisler, H.J.: Adapted probability distributions. Trans. Amer. Math. Soc. **286**(1), 159–201 (1984)
211. Hrbáček, K.: Axiomatic foundations for nonstandard analysis. Fund. Math. **98**(1), 1–19 (1978)
212. Hu, J., Zähle, M.: Potential spaces on fractals. Studia Math. **170**(3), 259–281 (2005)
213. Hu, Z.C., Ma, Z.M., Sun, W.: Nonlinear filtering of semi-Dirichlet processes. Stochastic Process. Appl. **119**(11), 3890–3913 (2009)

214. Hu, Z.C., Ma, Z.M., Sun, W.: On representations of non-symmetric Dirichlet forms. Potential Anal. **32**(2), 101–131 (2010)
215. Huang, Z.Y., Yan, J.A.: Introduction to infinite dimensional stochastic analysis, *Mathematics and its Applications*, vol. 502, chinese edn. Kluwer Academic Publishers, Dordrecht (2000)
216. Hurd, A.E. (ed.): Nonstandard analysis—recent developments. Including papers presented at the Second Victoria Symposium on Nonstandard Analysis held at the University of Victoria, Victoria, B.C., 1980, *Lecture Notes in Mathematics*, vol. 983. Springer-Verlag, Berlin (1983)
217. Hurd, A.E., Loeb, P.A.: An introduction to nonstandard real analysis, *Pure and Applied Mathematics*, vol. 118. Academic Press Inc., Orlando, FL (1985)
218. Jacob, N.: Pseudo differential operators and Markov processes. Vol. I. Fourier analysis and semigroups. Imperial College Press, London (2001)
219. Jacob, N.: Pseudo differential operators & Markov processes. Vol. II. Generators and their potential theory. Imperial College Press, London (2002)
220. Jacob, N.: Pseudo differential operators and Markov processes. Vol. III. Markov processes and applications. Imperial College Press, London (2005)
221. Jacob, N., Moroz, V.: On the semilinear Dirichlet problem for a class of nonlocal operators generating Dirichlet forms. In: J. Appell (ed.) Recent trends in nonlinear analysis, *Progr. Nonlinear Differential Equations Appl.*, vol. 40, pp. 191–204. Birkhäuser, Basel (2000)
222. Jacob, N., Schilling, R.L.: Fractional derivatives, non-symmetric and time-dependent Dirichlet forms and the drift form. Z. Anal. Anwendungen **19**(3), 801–830 (2000)
223. Jerrum, M.: Counting, sampling and integrating: algorithms and complexity. Lectures in Mathematics, ETH Zürich. Basel: Birkhäuser. xi, 112 pp. (2003)
224. Jordan, R., Kinderlehrer, D., Otto, F.: The variational formulation of the Fokker-Planck equation. SIAM J. Math. Anal. **29**(1), 1–17 (1998).
225. Jost, J., Kendall, W., Mosco, U., Röckner, M., Sturm, K.T.: New directions in Dirichlet forms, *AMS/IP Studies in Advanced Mathematics*, vol. 8. American Mathematical Society, Providence, RI (1998)
226. Kaneko, H.: Capacities and function spaces on the local field. In: p-adic mathematical physics, *AIP Conf. Proc.*, vol. 826, pp. 91–104. Amer. Inst. Phys., Melville, NY (2006)
227. Kaneko, H., Yasuda, K.: Capacities associated with Dirichlet space on an infinite extension of a local field. Forum Math. **17**(6), 1011–1032 (2005)
228. Kanovei, V., Reeken, M.: Internal approach to external sets and universes. I. Bounded set theory. Studia Logica **55**(2), 229–257 (1995)
229. Kanovei, V., Reeken, M.: Nonstandard analysis, axiomatically. Springer Monographs in Mathematics. Springer-Verlag, Berlin (2004)
230. Kanovei, V., Shelah, S.: A definable nonstandard model of the reals. J. Symbolic Logic **69**(1), 159–164 (2004)
231. Kant, U., Klauss, T., Voigt, J., Weber, M.: Dirichlet forms for singular one-dimensional operators and on graphs. J. Evol. Equ. **9**(4), 637–659 (2009)
232. Kaßmann, M.: Harnack-Ungleichungen für nichtlokale Differentialoperatoren und Dirichlet-Formen, *Bonner Mathematische Schriften*, vol. 336. Universität Bonn, Mathematisches Institut, Bonn (2001). Dissertation, Rheinische Friedrich-Wilhelms-Universität Bonn, 2000
233. Kassmann, M.: On regularity for Beurling–Deny type Dirichlet forms. Potential Analysis **19**(1), 69–87 (2003). DOI 10.1023/A:1022486631020
234. Kassmann, M.: The theory of De Giorgi for non-local operators. C. R. Math. Acad. Sci. Paris **345**(11), 621–624 (2007). DOI 10.1016/j.crma.2007.10.007. URL http://dx.doi.org/10.1016/j.crma.2007.10.007
235. Kato, T.: Perturbation theory for linear operators. Reprint of the 1980 edition. Classics in Mathematics. Springer-Verlag, Berlin (1995)

236. Kawabata, T., Takeda, M.: On uniqueness problem for local Dirichlet forms. Osaka J. Math. **33**(4), 881–893 (1996). URL http://projecteuclid.org/getRecord?id=euclid.ojm/1200787222
237. Keisler, H.J.: An infinitesimal approach to stochastic analysis. Mem. Amer. Math. Soc. **48**(297), x+184 (1984)
238. Keisler, H.J.: Infinitesimals in probability theory. In: N.J. Cutland (ed.) Nonstandard analysis and its applications (Hull, 1986), *London Math. Soc. Stud. Texts*, vol. 10, pp. 106–139. Cambridge University Press, Cambridge (1988)
239. Kigami, J.: Analysis on fractals. Cambridge Tracts in Mathematics. 143. Cambridge: Cambridge University Press. viii, 226 pp. (2001)
240. Kim, D., Takeda, M., Ying, J.: Some variational formulas on additive functionals of symmetric Markov chains. Proc. Amer. Math. Soc. **130**(7), 2115–2123 (electronic) (2002)
241. Kim, J.H.: Stochastic calculus related to nonsymmetric Dirichlet forms. Osaka J. Math. **24**(2), 331–371 (1987)
242. Kim, P., Song, R., Vondraček, Z.: Boundary Harnack principle for subordinate Brownian motions. Stochastic Process. Appl. **119**(5), 1601–1631 (2009)
243. Kochubei, A.N.: Pseudo-differential equations and stochastics over non-Archimedean fields, *Monographs and Textbooks in Pure and Applied Mathematics*, vol. 244. Marcel Dekker Inc., New York (2001)
244. Kochubei, A.N.: Analysis in positive characteristic, *Cambridge Tracts in Mathematics*, vol. 178. Cambridge University Press, Cambridge (2009). DOI 10.1017/CBO9780511575624
245. Kolesnikov, A.V.: Mosco convergence of Dirichlet forms in infinite dimensions with changing reference measures. J. Funct. Anal. **230**(2), 382–418 (2006)
246. Krall, A.M.: Hilbert space, boundary value problems and orthogonal polynomials. Operator Theory: Advances and Applications 133. Basel: Birkhäuser. xiv, 352 pp. (2002)
247. Kulik, A.M.: Markov uniqueness and Rademacher theorem for smooth measures on infinite-dimensional space under successful filtration condition. Ukraïn. Mat. Zh. **57**(2), 170–186 (2005)
248. Kumagai, T., Sturm, K.T.: Construction of diffusion processes on fractals, d-sets, and general metric measure spaces. J. Math. Kyoto Univ. **45**(2), 307–327 (2005)
249. Kusuoka, S.: Dirichlet forms and diffusion processes on Banach spaces. J. Fac. Sci. Univ. Tokyo Sect. IA Math. **29**(1), 79–95 (1982)
250. Kuwae, K.: On a strong maximum principle for Dirichlet forms. In: Stochastic processes, physics and geometry: new interplays, II (Leipzig, 1999), *CMS Conf. Proc.*, vol. 29, pp. 423–429. Amer. Math. Soc., Providence, RI (2000)
251. Kuwae, K.: Reflected Dirichlet forms and the uniqueness of Silverstein's extension. Potential Anal. **16**(3), 221–247 (2002)
252. Kuwae, K.: Maximum principles for subharmonic functions via local semi-Dirichlet forms. Canad. J. Math. **60**(4), 822–874 (2008)
253. Kuwae, K., Shioya, T.: Convergence of spectral structures: a functional analytic theory and its applications to spectral geometry. Comm. Anal. Geom. **11**(4), 599–673 (2003)
254. Kuwae, K., Shioya, T.: Sobolev and Dirichlet spaces over maps between metric spaces. J. Reine Angew. Math. **555**, 39–75 (2003). DOI 10.1515/crll.2003.014
255. Landers, D., Rogge, L.: Nichtstandard Analysis. Springer-Lehrbuch. [Springer Textbook]. Springer-Verlag, Berlin (1994)
256. Laptev, A., Weidl, T.: Hardy inequalities for magnetic Dirichlet forms. In: Mathematical results in quantum mechanics (Prague, 1998), *Oper. Theory Adv. Appl.*, vol. 108, pp. 299–305. Birkhäuser, Basel (1999)
257. Le Jan, Y.: Balayage et formes de Dirichlet. Z. Wahrsch. Verw. Gebiete **37**(4), 297–319 (1976/77). DOI 10.1007/BF00533422
258. Le Jan, Y.: Mesures associées à une forme de Dirichlet. Applications. Bull. Soc. Math. France **106**(1), 61–112 (1978)

259. Lemle, L.D.: $L^1(\mathbb{R}^d, dx)$-uniqueness of weak solutions for the Fokker-Planck equation associated with a class of Dirichlet operators. Electron. Res. Announc. Math. Sci. **15**, 65–70 (2008)
260. Lenz, D., Stollmann, P., Veselić, I.: The Allegretto-Piepenbrink theorem for strongly local Dirichlet forms. Doc. Math. **14**, 167–189 (2009)
261. Lindstrøm, T.: Nonstandard energy forms and diffusions on manifolds and fractals. In: Stochastic processes in classical and quantum systems (Ascona, 1985), *Lecture Notes in Phys.*, vol. 262, pp. 363–380. Springer, Berlin (1986)
262. Lindstrøm, T.: An invitation to nonstandard analysis. In: N.J. Cutland (ed.) Nonstandard analysis and its applications (Hull, 1986), *London Math. Soc. Stud. Texts*, vol. 10, pp. 1–105. Cambridge University Press, Cambridge (1988)
263. Lindstrøm, T.: Hyperfinite Lévy processes. Stoch. Stoch. Rep. **76**(6), 517–548 (2004). DOI 10.1080/10451120412331315797
264. Lindstrøm, T.: Nonlinear stochastic integrals for hyperfinite Lévy processes. Log. Anal. **1**(2), 91–129 (2008)
265. Liskevich, V., Röckner, M.: Strong uniqueness for certain infinite-dimensional Dirichlet operators and applications to stochastic quantization. Ann. Scuola Norm. Sup. Pisa Cl. Sci. (4) **27**(1), 69–91 (1998)
266. Loeb, P.A.: A non-standard representation of measurable spaces, L_∞, and L_∞^*. In: Contributions to non-standard analysis (Sympos., Oberwolfach, 1970), pp. 65–80. Studies in Logic and Found. Math., Vol. 69. North-Holland, Amsterdam (1972)
267. Loeb, P.A.: Conversion from nonstandard to standard measure spaces and applications in probability theory. Trans. Amer. Math. Soc. **211**, 113–122 (1975)
268. Lorenzi, L., Bertoldi, M.: Analytical methods for Markov semigroups. Pure and Applied Mathematics (Boca Raton) 283. Boca Raton, FL: Chapman & Hall/CRC. xxxi, 526 pp. (2007)
269. Lumer, G.: Équations de diffusion générales sur des réseaux infinis. In: Seminar on potential theory, Paris, No. 7, *Lecture Notes in Math.*, vol. 1061, pp. 230–243. Springer, Berlin (1984). DOI 10.1007/BFb0099027
270. Ma, Z.M., Röckner, M.: Introduction to the theory of (nonsymmetric) Dirichlet forms. Universitext. Springer-Verlag, Berlin (1992)
271. Ma, Z.M., Röckner, M.: Markov processes associated with positivity preserving coercive forms. Canad. J. Math. **47**(4), 817–840 (1995)
272. Ma, Z.M., Röckner, M.: Construction of diffusions on configuration spaces. Osaka J. Math. **37**(2), 273–314 (2000)
273. Ma, Z.M., Röckner, M.: Diffusions on "simple" configuration spaces. In: F. Gesztesy, H. Holden, J. Jost, S. Paycha, M. Röckner, S. Scarlatti (eds.) Stochastic processes, physics and geometry: new interplays, I (Leipzig, 1999), *CMS Conf. Proc.*, vol. 28, pp. 255–269. Amer. Math. Soc., Providence, RI (2000)
274. Ma, Z.M., Röckner, M., Sun, W.: Approximation of Hunt processes by multivariate Poisson processes. Acta Appl. Math. **63**(1-3), 233–243 (2000)
275. Ma, Z.M., Röckner, M., Yan, J.A. (eds.): Dirichlet forms and stochastic processes. Papers from the International Conference held in Beijing, October 25–31, 1993, and the School on Dirichlet Forms, held in Beijing, October 18–24, 1993. Walter de Gruyter & Co., Berlin (1995)
276. Machover, M., Hirschfeld, J.: Lectures on non-standard analysis. Lecture Notes in Mathematics, Vol. 94. Springer-Verlag, Berlin (1969)
277. Marinelli, C., Prévôt, C., Röckner, M.: Regular dependence on initial data for stochastic evolution equations with multiplicative Poisson noise. J. Funct. Anal. **258**(2), 616–649 (2010)
278. Masamune, J.: Essential self-adjointness of a sublaplacian via heat equation. Comm. Partial Differential Equations **30**(10-12), 1595–1609 (2005)
279. McIntosh, A.: Representation of bilinear forms in Hilbert space by linear operators. Trans. Amer. Math. Soc. **131**, 365–377 (1968)

280. McIntosh, A.: Hermitian bilinear forms which are not semibounded. Bull. Amer. Math. Soc. **76**, 732–737 (1970)
281. Mendez Hernandez, P.J.: Sharp estimates for Dirichlet heat kernels of the Laplacian, fractional Laplacian and applications. Ph.D. Thesis, Purdue University (2001)
282. Meyer, P.A.: Probability and potentials. Blaisdell Ginn, Waltham, Mass. etc. (1966)
283. Mosco, U.: Composite media and asymptotic Dirichlet forms. J. Funct. Anal. **123**(2), 368–421 (1994)
284. Nelson, E.: Internal set theory: a new approach to nonstandard analysis. Bull. Amer. Math. Soc. **83**(6), 1165–1198 (1977)
285. Nelson, E.: Radically elementary probability theory, *Annals of Mathematics Studies*, vol. 117. Princeton University Press, Princeton, NJ (1987)
286. Nelson, E.: The virtue of simplicity. In: The strength of nonstandard analysis, pp. 27–32. SpringerWienNewYork, Vienna (2007)
287. Ng, S.A.: Hypermodels in mathematical finance. Modelling via infinitesimal analysis. World Scientific Publishing Co. Inc., River Edge, NJ (2003)
288. Ng, S.A.: A nonstandard Lévy-Khintchine formula and Lévy processes. Acta Math. Sin. (Engl. Ser.) **24**(2), 241–252 (2008)
289. Ng, S.A.: Nonstandard methods in functional analysis. Lectures and notes. Hackensack, NJ: World Scientific. 300 pp. (2010)
290. Noll, A.: Domain perturbations and capacity in general Hilbert spaces and applications to spectral theory. Ph.D. thesis, Clausthal: TU Clausthal, Mathematisch-Naturwissenschaftliche Fakultät, 111 p. (2000)
291. Norris, J.R.: Markov chains. Reprint of the 1997 original, *Cambridge Series in Statistical and Probabilistic Mathematics*, vol. 2. Cambridge University Press, Cambridge (1998)
292. Ôkura, H.: Recurrence and transience criteria for subordinated symmetric Markov processes. Forum Math. **14**(1), 121–146 (2002)
293. Orey, S.: Conditions for the absolute continuity of two diffusions. Trans. Amer. Math. Soc. **193**, 413–426 (1974)
294. Ōshima, Y.: On absolute continuity of two symmetric diffusion processes. In: S. Albeverio, P. Blanchard, L. Streit (eds.) Stochastic processes—mathematics and physics, II (Bielefeld, 1985), *Lecture Notes in Math.*, vol. 1250, pp. 184–194. Springer, Berlin (1987)
295. Oshima, Y.: Time-dependent Dirichlet forms and related stochastic calculus. Infin. Dimens. Anal. Quantum Probab. Relat. Top. **7**(2), 281–316 (2004)
296. Otto, F., Reznikoff, M.G.: A new criterion for the logarithmic Sobolev inequality and two applications. J. Funct. Anal. **243**(1), 121–157 (2007)
297. Ouhabaz, E.M.: Analysis of heat equations on domains, *London Mathematical Society Monographs Series*, vol. 31. Princeton University Press, Princeton, NJ (2005)
298. Paulik, G.: Gluing spaces and analysis. Ph.D. thesis, Bonner Mathematische Schriften 372. Bonn: Univ. Bonn, Mathematisches Institut (Dissertation 2004). 102 p. (2005)
299. Pazy, A.: Semigroups of linear operators and applications to partial differential equations, *Applied Mathematical Sciences*, vol. 44. Springer-Verlag, New York (1983)
300. Reed, M., Simon, B.: Methods of modern mathematical physics. I. Functional analysis. Academic Press, New York (1972)
301. Reed, M., Simon, B.: Methods of modern mathematical physics. II. Fourier analysis, self-adjointness. Academic Press [Harcourt Brace Jovanovich Publishers], New York (1975)
302. Reed, M., Simon, B.: Methods of modern mathematical physics. III. Academic Press [Harcourt Brace Jovanovich Publishers], New York (1979)
303. Rellich, F.: Perturbation theory of eigenvalue problems. Assisted by J. Berkowitz. With a preface by Jacob T. Schwartz. Gordon and Breach Science Publishers, New York (1969)
304. von Renesse, M.K.: Comparison properties of diffusion semigroups on spaces with lower curvature bounds. Ph.D. thesis, Bonn: Univ. Bonn, Math.-Nat. Fakultät (2003)

305. von Renesse, M.K., Sturm, K.T.: Entropic measure and Wasserstein diffusion. Ann. Probab. **37**(3), 1114–1191 (2009)
306. Revuz, D.: Markov chains, *North-Holland Mathematical Library*, vol. 11, second edn. North-Holland Publishing Co., Amsterdam (1984)
307. Revuz, D., Yor, M.: Continuous martingales and Brownian motion, *Grundlehren der Mathematischen Wissenschaften*, vol. 293, third edn. Springer-Verlag, Berlin (1999)
308. Rhodes, R.: Diffusion in a locally stationary random environment. Probab. Theory Related Fields **143**(3-4), 545–568 (2009)
309. Robert, A.M.: Nonstandard analysis. Dover Publications Inc., Mineola, NY (2003)
310. Robinson, A.: Non-standard analysis. North-Holland Publishing Co., Amsterdam (1966)
311. Robinson, A.: Non-standard analysis. Princeton Landmarks in Mathematics. Princeton University Press, Princeton, NJ (1996)
312. Robinson, D.W.: The thermodynamic pressure in quantum statistical mechanics, *Lecture Notes in Physics*, vol. 9. Springer-Verlag, Berlin (1971)
313. Röckner, M.: Generalized Markov fields and Dirichlet forms. Acta Appl. Math. **3**(3), 285–311 (1985)
314. Röckner, M.: Dirichlet forms on infinite-dimensional state space and applications. In: H. Körezlioğlu, A.S. Üstünel (eds.) Stochastic analysis and related topics (Silivri, 1990), *Progr. Probab.*, vol. 31, pp. 131–185. Birkhäuser Boston, Boston, MA (1992)
315. Röckner, M.: Dirichlet forms on infinite-dimensional "manifold-like" state spaces: a survey of recent results and some prospects for the future. In: Probability towards 2000 (New York, 1995), *Lecture Notes in Statist.*, vol. 128, pp. 287–306. Springer, New York (1998)
316. Röckner, M.: Stochastic analysis on configuration spaces: basic ideas and recent results. In: New directions in Dirichlet forms, *AMS/IP Stud. Adv. Math.*, vol. 8, pp. 157–231. Amer. Math. Soc., Providence, RI (1998)
317. Röckner, M., Wang, F.Y.: On the spectrum of a class of non-sectorial diffusion operators. Bull. London Math. Soc. **36**(1), 95–104 (2004)
318. Röckner, M., Wielens, N.: Dirichlet forms—closability and change of speed measure. In: S. Albeverio (ed.) Infinite-dimensional analysis and stochastic processes (Bielefeld, 1983), *Res. Notes in Math.*, vol. 124, pp. 119–144. Pitman, Boston, MA (1985)
319. Röckner, M., Zhang, T.S.: Lower order perturbations of Dirichlet processes. Forum Math. **15**(2), 285–297 (2003)
320. Rüdiger, B.: Stochastic integration for compensated Poisson measures and the Lévy-Itô formula. In: Proceedings of the International Conference on Stochastic Analysis and Applications, pp. 145–167. Kluwer Acad. Publ., Dordrecht (2004)
321. Saks, S.: Theory of the integral. 2nd revised edition, English translation by L. C. Young. With two additional notes by Stefan Banach. New York: G. E. Stechert & Co. VI, 347 pp. (1937)
322. Saloff-Coste, L.: Aspects of Sobolev-type inequalities. London Mathematical Society Lecture Note Series. 289. Cambridge: Cambridge University Press. x, 190 pp. (2002)
323. Sato, K.I.: Lévy processes and infinitely divisible distributions. Translated from the 1990 Japanese original, revised by the author, *Cambridge Studies in Advanced Mathematics*, vol. 68. Cambridge University Press, Cambridge (1999)
324. Schilling, R.L., Uemura, T.: On the Feller property of Dirichlet forms generated by pseudo differential operators. Tohoku Math. J. (2) **59**(3), 401–422 (2007)
325. Schmieden, C., Laugwitz, D.: Eine Erweiterung der Infinitesimalrechnung. Math. Z. **69**, 1–39 (1958)
326. Schmuland, B.: Some regularity results on infinite-dimensional diffusions via Dirichlet forms. Stochastic Anal. Appl. **6**(3), 327–348 (1988)
327. Schmuland, B.: An alternative compactification for classical Dirichlet forms on topological vector spaces. Stochastics Stochastics Rep. **33**(1-2), 75–90 (1990)

328. Schmuland, B.: A Dirichlet form primer. In: Measure-valued processes, stochastic partial differential equations, and interacting systems (Montreal, PQ, 1992), *CRM Proc. Lecture Notes*, vol. 5, pp. 187–198. Amer. Math. Soc., Providence, RI (1994)
329. Schmuland, B., Sun, W.: The maximum Markovian self-adjoint extensions of Dirichlet operators for interacting particle systems. Forum Math. **15**(4), 615–638 (2003)
330. Sharpe, M.: General theory of Markov processes, *Pure and Applied Mathematics*, vol. 133. Academic Press Inc., Boston, MA (1988)
331. Shiozawa, Y., Takeda, M.: Variational formula for Dirichlet forms and estimates of principal eigenvalues for symmetric α-stable processes. Potential Anal. **23**(2), 135–151 (2005)
332. da Silva, J., Kondratiev, Y., Röckner, M.: On a relation between intrinsic and extrinsic Dirichlet forms for interacting particle systems. Math. Nachr. **222**, 141–157 (2001)
333. Silverstein, M.L.: Symmetric Markov processes. Lecture Notes in Mathematics, Vol. 426. Springer-Verlag, Berlin (1974)
334. Silverstein, M.L.: Boundary theory for symmetric Markov processes. Lecture Notes in Mathematics, Vol. 516. Springer-Verlag, Berlin (1976)
335. Sousa Pinto, J.: Infinitesimal methods of mathematical analysis. Horwood Publishing Limited, Chichester (2004)
336. Stannat, W.: The theory of generalized Dirichlet forms and its applications in analysis and stochastics. Ph.D. thesis, Bielefeld: Univ. Bielefeld, Fak. Math. 129 p. (1996)
337. Stannat, W.: The theory of generalized Dirichlet forms and its applications in analysis and stochastics. Mem. Amer. Math. Soc. **142**(678), viii+101 (1999)
338. Stannat, W.: On the validity of the log-Sobolev inequality for symmetric Fleming-Viot operators. Ann. Probab. **28**(2), 667–684 (2000)
339. Stollmann, P.: A convergence theorem for Dirichlet forms with applications to boundary value problems with varying domains. Math. Z. **219**(2), 275–287 (1995)
340. Stroock, D.W.: An introduction to Markov processes. Graduate Texts in Mathematics 230. Berlin: Springer (2005)
341. Stroyan, K.D., Bayod, J.M.: Foundations of infinitesimal stochastic analysis, *Studies in Logic and the Foundations of Mathematics*, vol. 119. North-Holland Publishing Co., Amsterdam (1986)
342. Stroyan, K.D., Luxemburg, W.A.J.: Introduction to the theory of infinitesimals, *Pure and Applied Mathematics*, vol. 72. Academic Press [Harcourt Brace Jovanovich Publishers], New York (1976)
343. Sturm, K.T.: The geometric aspect of Dirichlet forms. In: New directions in Dirichlet forms, *AMS/IP Stud. Adv. Math.*, vol. 8, pp. 233–277. Amer. Math. Soc., Providence, RI (1998)
344. Taira, K.: Semigroups, boundary value problems and Markov processes. Springer Monographs in Mathematics. Springer-Verlag, Berlin (2004)
345. Takeda, M.: Two classes of extensions for generalized Schrödinger operators. Potential Anal. **5**(1), 1–13 (1996)
346. Takeda, M.: L^p-independence of the spectral radius of symmetric Markov semigroups. In: F. Gesztesy, H. Holden, J. Jost, S. Paycha, M. Röckner, S. Scarlatti (eds.) Stochastic processes, physics and geometry: new interplays, II (Leipzig, 1999), *CMS Conf. Proc.*, vol. 29, pp. 613–623. Amer. Math. Soc., Providence, RI (2000)
347. Takeda, M.: Conditional gaugeability and subcriticality of generalized Schrödinger operators. J. Funct. Anal. **191**(2), 343–376 (2002)
348. Takeda, M.: A formula on scattering length of positive smooth measures. Proc. Amer. Math. Soc. **138**(4), 1491–1494 (2010). DOI 10.1090/S0002-9939-09-10172-7. URL http://dx.doi.org/10.1090/S0002-9939-09-10172-7
349. Trutnau, G.: Stochastic calculus of generalized Dirichlet forms and applications. Ph.D. thesis, Bielefeld: Univ. Bielefeld, Fakultät für Mathematik, 79 p. (1999)
350. Trutnau, G.: Stochastic calculus of generalized Dirichlet forms and applications to stochastic differential equations in infinite dimensions. Osaka J. Math. **37**(2), 315–343 (2000)

351. Uemura, T.: On an extension of jump-type symmetric Dirichlet forms. Electron. Comm. Probab. **12**, 57–65 (electronic) (2007)
352. Vogt, H., Voigt, J.: Wentzell boundary conditions in the context of Dirichlet forms. Adv. Differential Equations **8**(7), 821–842 (2003)
353. Wang, F.Y.: Logarithmic Sobolev inequalities: conditions and counterexamples. J. Operator Theory **46**(1), 183–197 (2001)
354. Wang, F.Y., Wu, B.: Quasi-regular Dirichlet forms on Riemannian path and loop spaces. Forum Math. **20**(6), 1084–1096 (2008)
355. Wang, F.Y., Wu, B.: Quasi-regular Dirichlet forms on free Riemannian path spaces. Infin. Dimens. Anal. Quantum Probab. Relat. Top. **12**(2), 251–267 (2009)
356. Weidmann, J.: Linear operators in Hilbert spaces. Translated from the German by Joseph Szücs, *Graduate Texts in Mathematics*, vol. 68. Springer-Verlag, New York (1980)
357. Weidmann, J.: Lineare Operatoren in Hilberträumen. Teil 1. Mathematische Leitfäden. B. G. Teubner, Stuttgart (2000)
358. Weidmann, J.: Lineare Operatoren in Hilberträumen. Teil II. Mathematische Leitfäden. B. G. Teubner, Stuttgart (2003)
359. Wielens, N.: The essential selfadjointness of generalized Schrödinger operators. J. Funct. Anal. **61**(1), 98–115 (1985)
360. Williams, D.: Diffusions, Markov processes, and martingales. Vol. 1. John Wiley & Sons Ltd., Chichester (1979)
361. Williams, D.: Probability with martingales. Cambridge Mathematical Textbooks. Cambridge University Press, Cambridge (1991)
362. Wolff, M., Loeb, P.A. (eds.): Nonstandard analysis for the working mathematician, *Mathematics and its Applications*, vol. 510. Kluwer Academic Publishers, Dordrecht (2000)
363. Wu, L.: Uniqueness of Nelson's diffusions. Probab. Theory Related Fields **114**(4), 549–585 (1999)
364. Yosida, K.: Functional analysis, *Grundlehren der Mathematischen Wissenschaften*, vol. 123, sixth edn. Springer-Verlag, Berlin (1980)
365. Zhang, T.: Variational inequalities and optimization for Markov processes associated with semi-Dirichlet forms. SIAM J. Control Optim. **48**(3), 1743–1755 (2009)
366. Zhao, Z.: Subcriticality and gaugeability of the Schrödinger operator. Trans. Amer. Math. Soc. **334**(1), 75–96 (1992). DOI 10.2307/2153973. URL http://dx.doi.org/10.2307/2153973

Notation Index

Logical notation:

$\forall$–for all
$\exists$–exists
$\in$–in
$\subset$–subset or equal
$\emptyset$–empty set

Other notation:

a.e.–almost every
A–infinitesimal generator of $\mathcal{E}(\cdot,\cdot)$
$\hat{A}$–infinitesimal co-generator of $\mathcal{E}(\cdot,\cdot)$
$\overline{A}$–infinitesimal generator of $\overline{\mathcal{E}}(\cdot,\cdot)$
$B_\rho(x)$–the open ball of radius ρ centered at x
$\mathrm{Cap}_\alpha^{(\delta)}(\cdot)$–$\alpha$-capacity of hyperfinite Dirichlet form
$\complement B$–complement of a subset $B \subset \mathbb{R}^d$ or $B \subset {}^*\mathbb{R}^d$
$\ast_{\nu \in M}$ ν–convolution of all measures ν on $\mathbb{R}^d$ contained in the set M
(also used to denote convolutions of internal finitely-additive measures)
$D(E)$–domain of $E(\cdot,\cdot)$
$D(\overline{E})$–domain of $\overline{E}(\cdot,\cdot)$
$\mathcal{D}(\mathcal{E})$–domain of $\mathcal{E}(\cdot,\cdot)$
$\mathcal{D}(\overline{\mathcal{E}})$–domain of $\overline{\mathcal{E}}(\cdot,\cdot)$
$D(F)$–domain of $F(\cdot,\cdot)$
$e(A)$–energy of internal additive functional A
$e(A,B)$–mutual energy of internal additive functionals A and B
$E(\cdot,\cdot)$–standard or quasi-regular Dirichlet forms
$\mathcal{E}(\cdot,\cdot)$–hyperfinite quadratic form or hyperfinite Dirichlet form
$\hat{\mathcal{E}}(\cdot,\cdot)$–co-form of $\mathcal{E}(\cdot,\cdot)$
$\overline{\mathcal{E}}(\cdot,\cdot)$–symmetric part of $\mathcal{E}(\cdot,\cdot)$
$\mathring{\mathcal{E}}(\cdot,\cdot)$–anti-symmetric part of $\mathcal{E}(\cdot,\cdot)$
$F(\cdot,\cdot)$–standard coercive closed form or Dirichlet forms

G_α–resolvent of hyperfinite quadratic forms
$\hat{G}_\alpha$–co-resolvent of hyperfinite quadratic forms
$\Gamma_\alpha(\cdot)$–α-capacity of standard Dirichlet form
H–hyperfinite dimensional linear space with inner product $\langle\cdot,\cdot\rangle$ generating norm $||\cdot||$
$\mathbb{N}=\{1,2,\cdots\}$–natural number set without zero
$\mathbb{N}_0=\{0,1,2,\cdots\}$–natural number set
$\{Q^t\}$–semigroup of hyperfinite Markov chain
$\{\hat{Q}^t\}$–co-semigroup of hyperfinite Markov chain
$\{\overline{Q}^t\}$–semigroup of $\overline{\mathcal{E}}(\cdot,\cdot)$
$\mathbb{R}$–1-dimensional Euclidean space
$\mathbb{R}_+=[0,\infty)$–all nonnegative real numbers
$\mathbb{R}_{>0}=(0,\infty)$–all positive real numbers
$\{R_\alpha\}$–resolvent of standard Dirichlet forms
$\{\hat{R}_\alpha\}$–co-resolvent of standard Dirichlet forms
$S=\{s_0,s_1,\cdots,s_N\}$–state space
$S_0=\{s_1,s_2,\cdots,s_N\}$–state space without trap s_0
$T=\{k\Delta t|k\in{}^*\mathbb{N}_0\}$–nonstandard time-line
X and $\hat{X}$–dual hyperfinite Markov chains

Index

Editor in Chief: Franco Brezzi

Editorial Policy

1. The UMI Lecture Notes aim to report new developments in all areas of mathematics and their applications - quickly, informally and at a high level. Mathematical texts analysing new developments in modelling and numerical simulation are also welcome.

2. Manuscripts should be submitted (preferably in duplicate) to
Redazione Lecture Notes U.M.I.
Dipartimento di Matematica
Piazza Porta S. Donato 5
I – 40126 Bologna
and possibly to one of the editors of the Board informing, in this case, the Redazione about the submission. In general, manuscripts will be sent out to external referees for evaluation. If a decision cannot yet be reached on the basis of the first 2 reports, further referees may be contacted. The author will be informed of this. A final decision to publish can be made only on the basis of the complete manuscript, however a refereeing process leading to a preliminary decision can be based on a pre-final or incomplete manuscript. The strict minimum amount of material that will be considered should include a detailed outline describing the planned contents of each chapter, a bibliography and several sample chapters.

3. Manuscripts should in general be submitted in English. Final manuscripts should contain at least 100 pages of mathematical text and should always include

   - a table of contents;
   - an informative introduction, with adequate motivation and perhaps some historical remarks: it should be accessible to a reader not intimately familiar with the topic treated;
   - a subject index: as a rule this is genuinely helpful for the reader.

4. For evaluation purposes, manuscripts may be submitted in print or electronic form (print form is still preferred by most referees), in the latter case preferably as pdf- or zipped ps-files. Authors are asked, if their manuscript is accepted for publication, to use the LaTeX2e style files available from Springer's web-server at
ftp://ftp.springer.de/pub/tex/latex/svmonot1/ for monographs
and at
ftp://ftp.springer.de/pub/tex/latex/svmultt1/ for multi-authored volumes

5. Authors receive a total of 50 free copies of their volume, but no royalties. They are entitled to a discount of 33.3% on the price of Springer books purchased for their personal use, if ordering directly from Springer.

6. Commitment to publish is made by letter of intent rather than by signing a formal contract. Springer-Verlag secures the copyright for each volume. Authors are free to reuse material contained in their LNM volumes in later publications: A brief written (or e-mail) request for formal permission is sufficient.